Tension Structures

Tension Structures

Behavior and Analysis

John W. Leonard
Professor, Civil and Ocean Engineering
Oregon State University
Corvallis, Oregon

McGraw-Hill Book Company

New York   St. Louis   San Francisco   Auckland
Bogotá   Hamburg   London   Madrid   Mexico
Milan   Montreal   New Delhi   Panama
Paris   São Paulo   Singapore
Sydney   Tokyo   Toronto

Library of Congress Cataloging-in-Publication Data

Leonard, John W. (John William)
  Tension structures.

  Outgrowth of a series of lectures presented
during 1983 at the Politecnico di Milano and a
special course taught during 1985 in the Civil
Engineering Dept. of the Oregon State University.
  Includes index.
  1. Lightweight construction.   2. Cable structures.
I. Title.
TA633.L46 1988   624.1′77   87-3950
ISBN 0-07-037226-8

1234567890   DOC/DOC   89210987

ISBN 0-07-037226-8

*The editors for this book were Betty Sun and Dennis Gleason,
the designer was Naomi Auerbach, and the production
supervisor was Richard A. Ausburn. It was set in Century Schoolbook
by University Graphics, Inc.*

Printed and bound by R. R. Donnelley & Sons Company

*To the memory of my parents
Catherine and Frank,*

*and to the future of my children,
Kelly, Stacey, Austin, and Blake*

Contents

Part 3   Statics and Dynamics of Membrane Structures

Preface

Tension structures are ones in which the main load-carrying members transmit loads to the foundation or support system by tensile stresses with no compression or flexure allowed. They are load-adaptive in that members change geometry to accommodate changes in load rather than increase stress levels.

Tension structures could be comprised of membranes, cables, or both. They include air-supported structures, pneumatic shells, prestressed membranes, cable networks, tethers, suspension cables, guyed towers and tents, and ocean platforms and breakwaters. Land-based applications include temporary shelters and warehouses, tents, hanging roofs, and suspension bridges and systems. Sea-based applications include moored vessels and buoys, trawl lines and nets, towed arrays, floating breakwaters, inflatable breakwaters and storage tanks, and tension-leg and catenary-leg platforms.

Tension structures are well suited to support distributed loads. Equilibrium states developable under dead loads and such live loads as ocean currents and drift forces due to waves are studied. Because of their reduced stiffness characteristics, tension structures are susceptible to concentrated loads and dynamic effects. Various dynamic loads are considered, including wind and wave action. Frequency-domain and nonlinear time-domain methods of solution will be developed.

The static and dynamic behavior of continuous cable and membrane segments are studied. Matrix formulations of discretized systems of cables and membranes are developed for complicated structural systems. Numerical methods are emphasized for both cable and membrane systems, including finite element models, perturbation and incremental techniques, and numerical integration with Newton-Raphson iterations. The formulations and solution techniques are illustrated with numerous simple and more complicated example problems involving land-based and sea-based applications.

This book is the outgrowth of a series of special lectures presented during 1983 at the Politecnico di Milano and of a special course taught during 1985 to seniors and graduate students in the Civil Engineering Department of the Oregon State University. The perceptive comments of the students attending those lectures and courses led to significant improvements in this text. The encouragement and criticisms given by Professors Giulio Maier and Leone Corradi of the Politecnico di Milano are much appreciated. The entire manuscript was typed by Laurie Campbell. Her cheerful willingness to undertake seemingly continual revisions based on my illegible handwriting has earned my enduring gratitude.

Some of the results reported in this book were obtained during the course of research under contracts or grants since 1967 from the National Aeronautics and Space Administration, the National Science Foundation, The National Bureau of Standards, The U.S. Naval Civil Engineering Laboratory, and the National Oceanographic and Atmospheric Administration. Their financial support enabled this book. The contributions of past graduate students were essential to that research: C.-T. Li, Vijay Verma, David Ma, Ray Young, M. C. Huang, Hang Tuah, J. F. Lee, R. B. Chiou, and especially Alain Lo.

John W. Leonard

Characterization of Tension Structures

Tension structures are ones in which the main load-carrying members transmit applied loads to the foundations or other supporting structures by direct tensile stress without flexure or compression. Their cross-sectional dimensions and method of fabrication are such that their shear and flexural rigidities, as well as their buckling resistance, are negligible. There are two broad classes of tension structures: cable structures [10, 12],* comprised of uniaxially stressed members, and membrane structures [9, 13, 14], comprised of biaxially stressed members.

The general class of cable structures can be further divided into four subclasses [12]:

1. Single cables in which single cable segments, or several simply connected segments, are subjected to loads predominantly in a single plane of action, e.g., suspension cables, tether or mooring lines, guy lines for towers or tents

2. Cable trusses in which prestressed segments are multiply connected in a single plane and loaded in that same plane, e.g., cable-stayed bridges, double-layer cable-supported roofs

---

* Numbers in brackets denote entries in Appendix A: Bibliography.

3. Cable nets in which prestressed segments are multiply connected in a curved surface (synclastic or anticlastic) and loaded predominantly normal to that surface, e.g., hanging roofs, suspended nets

4. Cable networks in which cable segments are multiply connected to form a three-dimensional framework, e.g., suspension networks, trawl nets, multiple-leg systems

There are four subclasses of membrane structures [9]:

1. Air-supported structures in which an enclosing membrane is supported by a small differential air (or fluid) pressure, e.g., stadia roofs, inflated temporary shelters or storehouses

2. Inflated structures in which highly pressurized tubes or dual-walled mats are used as structural members in a space structure, e.g., inflated beams, columns, or arches; dual-walled shells; air cushion roofs

3. Prestressed membranes in which fabric or rubberlike sheets are stretched over rigid frameworks and columns to form enclosures or diaphragms, e.g., tents, masted roofs

4. Hybrid systems in which membrane panels span between primary load-carrying members such as prestressed cables and rigid members, e.g., reinforced fabric roofs, fluid storage tanks

Because of the inherently nonlinear nature of tension structures, conventional linear analysis, which assumes small elastic deformations and displacements, is often not applicable to cable and membrane structures. Over the past two decades there has been considerable development of analysis techniques and computer codes for tension structures. Simplified procedures for preliminary designs are evolving. Relatively inexpensive techniques and computer codes must be selected for the preliminary sizing and evaluation of candidate design concepts. After an initial design has been selected and detailed, refined analysis procedures must be used to evaluate the nonlinear dynamic response of the structure to both expected and extreme load conditions [55, 56, 60, 75, 76, 113–118].

In Part 1 of this book the static behavior of cable systems is considered. First, in Chap. 2, singly connected segments are studied under different loading conditions. Next, in Chap. 3, multiply connected systems are studied using methods of finite element analysis. Nonlinear solutions are emphasized.

In Part 2 the dynamic behavior of cable systems is considered. In Chap. 4 linear vibrations are studied: methods of analysis are reviewed and responses of finite element models for cables are considered. In Chap. 5 nonlinear dynamics of cable systems are studied. Nonlinear solution methods are reviewed and an isoparametric cable element is developed. Special attention is given to hydrodynamic loadings.

Membrane structures are considered in Part 3. Linearized and approximate formulations are the subject of Chap. 6 for both static and dynamic behavior. A partially nonlinear formulation is the subject of Chap. 7, in which the effects of prestress on the in-service behavior of membranes of revolution are studied. Fully nonlinear formulations for general membranes are considered in Chap. 8, where finite element models with superparametric and isoparametric elements are used to treat static and dynamic loads.

1.1 Applications of Tension Structures

Tension structures are well suited to support broadly distributed dead loads and live loads such as wind, ocean currents, and drift forces due to waves. It should not be surprising that lightweight tension structures resemble biological forms, since such forms also support loads by tension in pneumatically prestressed skins and fibers.

Some of the advantages of tension members for use as structural components are [9, 14]

1. They are lightweight and collapsible and therefore easy to transport and erect.
2. They can be prefabricated in a factory, have low installation costs, and are potentially relocatable.
3. For air-supported structures, the primary load-carrying mechanism is the habitable environment itself, i.e., a pressurized mixture of gases.
4. The environmental loads are efficiently carried by direct stress without bending.
5. They are load-adaptive in that the members change geometry to better accommodate changes in load patterns and magnitudes.

The history of early applications of cable and membrane structures as architectural forms and as engineering systems is described in Refs. 1 to 12. Land-based applications [5] include temporary shelters and warehouses, tents, hanging roofs, and suspension systems. Sea-based applications [15] include moored vessels and buoys, trawl lines and nets, towed arrays, floating hospitals and other logistical support facilities, floating or submerged breakwaters or storage tanks, and tension-leg or catenary-leg platforms.

Hybrid tension structures could be used as initial shelters in the erection of structures in hostile environments, e.g., in cold regions or under the ocean surface. The shelters could later be stiffened from the inside to form more permanent facilities. An inflatable shell, possibly made of vinyl-coated fabric or thin metallic film, could be submerged in packaged form with all air hoses and airlocks attached. Then with minimal subsurface labor, it could be pressurized to the desired shape. Depending on the material used in the construction of the shell and on the depth of its intended use, the inflated dome would provide either a semipermanent environment or temporary falsework and shelter within which concrete could be sprayed to construct a permanent atmospheric environment. If a dual-walled form were used, the pressurized air (or water) within the walls could be displaced by fiber-reinforced concrete pumped from the surface.

Recent thought has been given to the feasibility of using pressurized membranes to roof small bases or towns, the membrane being supported primarily by the circulating interior air.

1.2   Behavior of Tension Structures

Because of their reduced stiffness characteristics, tension structures are susceptible to large motions due to concentrated loads and dynamic effects. They respond in a nonlinear fashion to both prestressing forces and in-service forces, regardless of linearity of material or loads. Prestressing forces are those forces (edge loads, self-weight, or pressure) which act on a predominant configuration of static equilibrium for the structure. They stabilize the structure and provide stiffness against further deflection. The response of a tension structure to prestressing forces is always nonlinear in that the equilibrium configurations, as well as the state of stress, are dependent on those forces.

In-service forces are those variable live loads, static or dynamic, which the structure may be expected to encounter during its service life. They are superposed upon the prestressing forces. The response to in-service forces may be nonlinear or quasi-linear, depending on the directions and magnitudes of the in-service forces relative to the state of stress in, and configuration of, the prestressed structure. The response is not strictly linear and therefore superposition of results for different in-service loading conditions is not strictly valid and, if done, must be done carefully.

It is usually sufficient to consider only linear (possibly piecewise linear) material behavior for tension structures [15]. There are instances, however, where nonlinear material characteristics should be considered: hyperelastic and viscoelastic behavior of polymer cables [16] and membranes [17]; nonisotropic woven fabrics [19]; and thermal-elastic and elastoplastic behavior under extreme loads [20]. Another potential source of nonlinearities of response is the interaction of tension structures with hydrostatic and hydrodynamic loads. Not only are the magnitudes of drag force nonlinear, but they are also nonconservative in that directions of pressure loads are dependent on orientations of the cable axes and membrane surfaces, which may undergo considerable rotation during loading [21]. Hydrodynamic loads on cables and membranes are the subject of Secs. 5.3 and 8.4.

1.3 Phases of Behavior

In this section the mechanics of tension structures are considered with the objective of identifying similarities and dissimilarities of their response to that of conventional structural systems. The implications of strictly tensile behavior and of potentially large displacements on analysis and design techniques are discussed.

The physical behavior of a tension structure during the application of loads can be divided into three primary phases [26]. The first phase is deployment, in which the cable or membrane system unfolds from its compact configuration into a state of incipient straining. The second phase is prestressing, in which the cable or membrane system deforms into a predominant equilibrium configuration under the action of dead weight, pressure, or other fixed lifetime loads. The final, or in-service, phase is the stage in which the fully prestressed system is subjected to variable live or dynamic loads during its service life.

1.3.1 Deployment phase

In the deployment phase the cable or membrane unfolds with external forces counterbalanced by inertial forces only. During this dynamic process the structure is stress-free and the static and kinematic behavior is like that of a collection of rigid particles constrained by the topology of the fabricated cable or membrane segments.

The expansion from the compact configuration to the state of incipient straining is highly nonlinear, and no unique solution has been shown to exist. This need not be of concern in that stresses are negligible during this phase. However, one should take care to avoid "trapped" wrinkles or kinks which might lead to local tears or knots, high rates of deployment in relation to lack of stability under unexpected external loads, and transient overstresses due to high initial velocities and accelerations when the structure is fully deployed.

The behavior in the deployment phase is primarily a problem in mathematical topology. The field equations involved are those of differential geometry of isometric kinematics, and of boundary constraints [22]. Equations of statics and constitutive relation do not enter the picture until the state of incipient straining is reached and the prestressing phase begins.

1.3.2 Prestressing phase

The second phase in the erection of a tension structure is prestressing, wherein the structure undergoes displacement from its state of incipient straining, hereafter referred to as the initial state, into a static equilibrium state dictated by the fabricated geometry of the structure and the prestressing forces. Since the displacements during this phase are large, this is a nonlinear problem, but unique solutions for the stresses and displacements are possible [11, 26].

Geometry.   In the stress analysis of tension structures a reference geometry for the stress calculations must be defined. In most solution procedures one can easily specify the geometry for which all such calculations are made. However, this is not a trivial specification for tension structures. Stress calculations are desired for the prestressed state and that shape may be significantly different in dimension or configuration from the initial state.

Two alternatives are possible: (1) prescribe initial geometry, (2) prescribe prestressed geometry. The first alternative is more desirable from the fabrication point of view. This, however, leads to the more difficult problem of stress analysis in that it is necessary by incremental or iterative procedures to determine the nonlinear solution for the prestressed geometry and the stresses thereon. With the second alternative, a simpler procedure for stress analysis is possible. However, the initial state required to obtain the prescribed prestressed shape exactly may be impractical to fabricate [24, 25].

The descriptions of the differential geometry of space curves and of surfaces is an important aspect of the mechanics of tension structure. Because of dead weight, cable segments are continuously curved—in the absence of other distributed loads they form catenary curves. Counterstressed networks of cables are prestressed configurations which do not depend on external loads, e.g., weight, to maintain their shapes. The counterstressing is provided by prestretching combinations of cables curved in opposition to each other: the cables for which the principal loads are directed away from the center of curvature are called sagging cables; the counterstressing cables are called hogging cables and have an opposite curvature [5, 10, 12].

Three-dimensional curves are formed whenever distributed loads act out of the plane formed by the gravity vector and the chord connecting the endpoints of a cable segment. Since fluid loadings on cables are directed along the tangent, normal, and binormal vectors at each point of the cable [21] and since the magnitudes of their loads are dependent on the orientation of the base vectors relative to the direction of fluid velocity, rates of change of the base vector with arc length need to be formulated using the Frenet formulas [27].

For membrane structures, surface curvature is described by the so-called gaussian curvature G [5, 27], which is the inverse product of the two principal radii of curvature of the surface. If $G > 0$, the surface is called synclastic, elliptic, or positive-gaussian, and the centers of curvature in the two principal directions lie on the same side of the surface. If $G < 0$, the surface is called anticlastic, hyperbolic, saddlelike, or negative-gaussian, and the centers of curvature lie on opposite sides of the surface. If $G = 0$, the surface is called developable, parabolic, or zero-gaussian, and at least one radius of curvature is infinite.

Synclastic surfaces (e.g., spheres) require external loads, e.g., pressure or dead weight, to maintain the prestressed configuration. Anticlastic surfaces (e.g., hyperboloids) do not require external loads: the opposing curvature provides counterstressing such as for sagging-hogging cable combinations. Zero-gaussian surfaces (e.g., cylinders) are said to be developable because they are the only ones that can form curved surfaces from fabricated flat surfaces without straining. Thus, they are the most common surfaces adopted for air-supported structures. More complicated and architecturally interesting surfaces are obtainable by joining segments of flat curves [6]. This introduces geometric discontinuities in the prestressed configuration and localized seaming stresses along joints.

Statics. The prestressing phase is a static equilibrium problem in that the state of stress and shape due to a predominant static load is sought. If the reference geometry is specified on the initial configuration, the equilibrium equations are nonlinear since the loads and stresses act on the prestressed configuration which has unknown locations, orientations, and curvatures of line and area segments. Because of the assumptions of negligible bending rigidity, transverse loads are balanced by gradients in curvature mutiplied by the internal tensions. Thus iterative solutions are necessary to determine the stress state; assumed displacements lead to calculable stresses which lead, in turn, to new assumptions for displacements.

If the reference geometry is completely specified on the prestressed configuration, e.g., specified inflated shape or specified cable sag, the equilibrium equations will include only unknown stresses. In many cases, e.g., shells of revolution or singly connected cable segments, this will be a statically determinate problem and the internal stress resultants can be determined without recourse to kinematic and constitutive equations. However, in cases where closed cable loops are formed or where in-plane membrane shear is possible, the equations are statically indeterminate and displacements must be considered.

Kinematics. Because of the extreme flexibility of tension structures (no bending resistance, small cross sections), large displacements occur during the prestressing phase and nonlinear strain-displacement relations should be used. Strains will be small, but relative rotations are large and thus second-order terms of displacement gradients are significant, i.e., line segments may not change much in length, but they do translate and rotate appreciably due to transverse loads.

There are two definitions of strain possible: (1) *engineering strain,* the change in differential length divided by original length in the reference state, and (2) *true strain,* or metric strain, which is one-half the difference in squares of differential length divided by the square of the original length in the reference state. The true strain is a more accurate representation of the kinematics, but since small strains are usually assumed, the engineering strain is most commonly used because of its clearer physical significance. The definition of original lengths in the reference state is significant since if an energy principle is used to determine the equations of statics, the strain must be referred to the same state as the stress.

In cases where the prestressed configuration is the reference configuration, the kinematic and constitutive relations are used to determine the geometry of the initial state. This is an inverse problem [28], i.e., given a solution, what problem was solved? For example, if a sagged cable configuration is specified, the equations of statics are used in the form of a moment analogy [10] to determine segment tensions, and the kinematic and constitutive equations are used to determine the unstressed lengths of the segments.

Material behavior.   During the prestressing phase semirigid translations and rotations of differential segments predominate over strain effects. Thus, it is often sufficient for preliminary design purposes to assume inextensible behavior [20, 34–37] for most engineering materials unless the system is highly redundant. In such cases incremental procedure can be used and piecewise linear elastic behavior can be assumed. It is only in the last stages of prestressing and in the subsequent in-service phase that constitutive equations play an important role.

Constraints.   Boundary conditions for the prestressing phase involve prescription of surface tractions due to external prestressing forces, of edge forces, and of support motions. Dead-weight loads on cables and membranes are conservative forces. Pressure loads from internal gases or from hydrostatic loads are nonconservative in that they change directions and magnitudes as the system of cables and membranes undergoes finite deformation. Thus, if the initial shape is taken as the reference configuration, the external force terms in the equilibrium equation will be displacement-dependent. If the prestressed shape is used as the reference configuration, the load nonlinearity will instead occur in the equations used to determine the initial shape.

Tension structures have negligible bending and buckling resistance. The boundary restraints must be consistent with that behavior. Clamped edges cannot be realized; instead, the cables and membranes will undergo localized kinking. Also, if the boundaries of a membrane structure were constrained such that one of the principal stresses would have to be compressive, the membrane would instead wrinkle transverse to the direction of the predicted compressive stress. Care must be taken in the design of boundary supports or restraining members such that wrinkling is minimized during the finite deflections that occur in the prestressing plane.

The lack of compressive strength of membranes also has implications in the selection of a desired configuration for the prestressed membrane. There are limitations on possible surface geometries for air-supported structures in that there are certain shapes [5] which imply compressive stresses even under internal pressure and which, therefore, cannot be realized without extensive wrinkling.

1.3.3   In-service phase

The third phase in the behavior of a tension structure is the in-service phase, wherein various static live loads, e.g., snow loads, and various dynamic loads, e.g., wind or wave loads, which are expected to occur during the service life of the structure are superposed on the prestressed configuration. Depending on the relative magnitudes of those loads compared to the prestresses, one can consider the behavior in this phase linear or nonlinear [40]. The prestressing stiffens the structure, and the additional deflections due to in-service loads are considerably smaller than the prestressing displacements.

The geometry of the prestressed configuration can be used as the basis for stress calculations during the in-service phase in that usually only slight perturbations of that shape occur. If large excursions under added load are expected, then an incremental loading technique [38, 39] can be used in which occasional recalculations of reference geometry and prestress are made before additional increments of load are considered. This same procedure can be used to predict a new prestressed configuration if there is significant change in the predominant loads on the tension structure.

The equations of statics or equations of motion for the in-service phase are linearized equations in that a fixed geometry is considered and additional displacements are small. Some nonlinear effects are included by accounting for the effect of prestressing on the membrane or cable stiffness, i.e., "geometric" stiffness contributions are considered [41]. In dynamic analyses of tensile structures it has been found that the magnitude of prestress has an effect not only on the frequencies of vibration but also on the mode shapes [42, 43].

The behavior during the in-service phase is not in general statically determinate. The kinematic and constitutive equations must be incorporated into the equations of motion. Piecewise linear material behavior is often assumed. An important aspect in the analysis of in-service behavior is the calculation of reductions in tensile stresses due to added loads [44, 45]. Slackening of cables and wrinkling of membranes should be avoided, or if unavoidable, their extent and effect on global stability of the structure predicted. Flutter and dynamic wrinkling of cable-supported and air supported roofs under wind loads have been shown to be important considerations in the design of such structures [46, 47].

1.4   Materials of Construction

1.4.1   Cables [12, 34, 48, 49, 59, 60]

Cables have been made of steel, Kevlar (registered DuPont trademark; a synthetic aramid fiber), fiberglass, and polyester.

Structural strands and structural ropes are commonly utilized as cables. A strand consists of steel wires wound helically around a center wire in symmetrical layers. A rope consists of several strands wound helically around a core. A high-tensile breaking strength is a primary property of the wire rope. There are other important properties

1. Small cross section

2. Low weight

3. Long fatigue life

4. Resistance to corrosion and abrasion

5. High flexibility

6. Good stretch and rotational behavior

These properties depend on the rope manufacture and wire control.

Cables act principally as axial elements; however, because of the helical wires, a torque may be induced as the helical wires try to "unwind" during axial loading [48]. The effects of induced or externally applied torque may be significant [182, 183]; induced torque decreases the ultimate strength. A torque-balanced cable is one designed to yield zero or very small amounts of rotation under load. In addition to the stresses in the wires due to the axial force, the wound wires are subjected to bending stresses which are difficult to evaluate because of relative movements of the individual strands.

Cable materials typically have linear stress-strain relationships over only a portion of their usable strength. Beyond the elastic limit, the proportional relationships do not hold. Breaking-strength efficiency is the ratio of cable strength to the sum of the individual wire strengths and is greater for ropes and strand lay. The breaking-strength efficiency is reduced as the number of wires in the strand is increased. A rope made up of brittle wires will be less able to bear overstressing due to unequal distribution of strains and consequently will develop a lower breaking-strength efficiency than could be obtained with more ductile wire.

1.4.2 Membranes [13, 29, 50–54]

Various materials can be used in the fabrication of a tension structure, such as hyperelastic (rubberlike) materials, fabrics, composites, etc.

Hyperelastic materials. Hyperelastic materials [50] are highly nonlinear in nature and can sustain large strains. Because of their low stiffness and strength properties they have limited applications.

Fabrics [51]. Fabric should have high tensile strength and should be flexible and light in weight. It should have good dimensional stability, be highly resistant to the propagation of minor accidental damage, and be easily and cheaply seamed or jointed in a reliable manner. Finally, the fabric should be resistant to environmental degradation and should be fire-resistant. To date, fabric has relatively low resistance to cyclic folding and wrinkling.

Fabric is specified as having a particular width and weight per square yard. Thickness may be derived from the density of the material. Common weights range from 4 to 50 oz/yd^2 (1.3 to 16.6 Pa) with derived thicknesses of 0.001 to 0.060 in (0.03'to 1.52 mm).

Strengths of fabrics are described in terms of strip tensile, grab tensile, tongue tear, trapezoidal tear, and adhesion. Because most fabrics are orthotropic, properties are typically specified for each weave direction in fabrics.

Strip tensile:	20 to 1000 lb/in (35 to 1750 N/cm)
Grab tensile:	Typically 30 percent higher than strip tensile
Tear strength:	1 to 200 lb/in (2 to 350 N/cm)
Adhesion strength:	2 to 40 lb/in^2 (0.1 to 2 N/mm^2)

Strip tensile strength is a measure of the fabric's resistance to tensile failure. Tear strength is an indicator of the fabric's resistance to abrasion and tears. Adhesion strength relates to the resistance of the fabric to delamination.

Composites. Fiber-reinforced membranes have higher strengths. The coated fabrics have higher tensile strengths than the laminates because fabrics with a closer weave may be used. Coated fabrics also have better adhesion. Laminates have scrims with an open weave to permit the "strike-through" necessary for good film adhesion. This method of construction does result in relatively higher tear strengths as the individual fibers are freer to bunch to resist tearing.

Some properties of commonly used fabrics are

1. *Nylon.* Strong with good elastic recovery; affected by moisture to a greater extent than is polyester; affected by sunlight and oxidation; flammable.

2. *Polyester.* Not quite as strong as nylon, and a little more expensive; unaffected by moisture and resistant to environmental degradation; higher modulus of elasticity than nylon, and has a slightly better dimensional stability; flammable.

3. *Glass.* Strong with complete dimensional stability, and exceptional resistance to all forms of environmental degradation; poor abrasion resistance; translucent; flame-resistant; no sewn seams.

4. *Polypropylene.* Strong; outstanding resistance to chemical degradation; adhesive problems reported; coating materials must be cured at low temperature.

It is not possible at present to satisfy simultaneously all requirements in a single material. The best solution is a coated fabric in which each of the components is selected to fulfill a particular set of conditions, and the final material is much better than the sum of the components.

Properties of various coating materials are

1. *Vinyl.* Low cost with good mechanical and weathering properties; deterioration is caused by loss of plasticizer; can be made transparent at the cost of reduced service life; stiffens in cold environments; can be heat sealed.

2. *Hypalon.* Excellent resistance to acid, oxidation, ozone, heat, and sunlight; good mechanical properties; excellent abrasion resistance; low water absorption; good color retention; low cost; poor low-temperature resistance; requires cemented or seam joints.

3. *Neoprene.* Very similar to Hypalon in resistance and weatherability; better tear resistance and adhesion to base fabric; low cost; requires cemented or seam joints.

4. *Fluoreolastomer.* Good to excellent mechanical properties; outstanding resistance to all types of deterioration, particularly at high temperatures; transparent; high cost.

5. *Polyurethane.* Excellent mechanical properties and superior abrasion resistance; excellent adhesion to base fabric; reasonable weather

resistance; only fair acid and flame resistance; transparent; can be heat sealed; medium cost.

Teflon-coated fiberglass has been used in fabric roofs [57, 58] and is a durable, high-strength material. The fabric is translucent, flame-resistant, tearproof, and waterproof. The fabric is coated with Teflon fluorocarbon resin, and therefore the coated material is self-cleaning and a 20-year life expectancy is anticipated. The Teflon dispersion contains additives that improve flexibility, abrasion resistance, and solar transmission.

Statics of Cable Segments and Systems

Static Analysis of Cable Segments

Cable structures respond in a nonlinear fashion to both prestressing and in-service forces. Prestressing forces are those forces (end loads or self-weight) which exist in a predominant static equilibrium configuration of the structure. They stabilize the structure and provide stiffness against further load. In-service forces are those variable live loads, static or dynamic, which the structure may be expected to encounter during its service life. They are superposed upon the prestressing forces. The response of a cable structure to prestressing forces is always nonlinear in that the static equilibrium configurations, as well as the state of stress, are dependent on the prestressing forces. The response to in-service forces may be nonlinear or quasi-linear, depending on the directions and magnitudes of the in-service forces relative to the state of stress in, and configuration of, the prestressed structure. The response is not strictly linear, and therefore superposition of results for different in-service forces is not strictly valid.

We shall first consider single cable segments in order to better understand the nonlinear nature of cable response and to arrive at exact and approximate solutions which will be applicable to several types of cable structures. These applications include suspended or cable-stayed bridges and roofs, guyed towers, cable-logging systems, and tow or mooring lines for various ocean structures such as buoys, ships, and platforms.

The planar response of single segments is such that the loading, the unstressed geometry, and the stressed geometry all lie in a single plane.

We will first consider segments for which the self-weight of the segment can be neglected in comparison to the concentrated prestressing and in-service loads. Next, distributed loads, such as self-weight, acting in a single direction will be considered in combination with concentrated in-service loads. Last, three-directional uniformly distributed loads will be considered and exact solutions thereto used to illustrate several complexities in the prediction of nonlinear cable behavior.

2.1   Response of Segments with Self-Weight Neglected

Let us first consider cases in which there are no distributed loads, only concentrated intermediate and end forces. This means that we neglect, temporarily, the effects of self-weight of the cable, which we will show subsequently to be an important determinant of cable behavior. Nevertheless, we will make this assumption here for the reason that simplified solutions can be developed which still demonstrate the essential nonlinearities of cable behavior and which have practical applications, including tautly stressed cable stays for bridges and roofs, and segments of cable trusses. We will further assume that the material is linearly elastic, i.e., tension T is related to extension $(S - S_0)$ by

$$T = \frac{AE}{S_0}(S - S_0) \qquad (2.1.1)$$

where A = cross-sectional area
E = modulus of elasticity
S = deformed length
S_0 = unstressed length

2.1.1   Single concentrated load

The simplest example of a transversely loaded cable is that of a cable of unstressed length S_0 spanning a distance L, as shown in Fig. 2.1.1a. If we apply a transverse load P at midspan, the cable will stretch until an equilibrium position as shown in Fig. 2.1.1b is obtained, with the midpoint having deflected a distance z from the chord connecting points 1 and 2.

Elementary statics relates P to the vertical component V of tension T and to the horizontal component H. The sums of moments about points 1 and 3 give

$$V = \frac{P}{2} \qquad (2.1.2)$$

$$H = \frac{P}{2}\left(\frac{L}{2z}\right) \qquad (2.1.3)$$

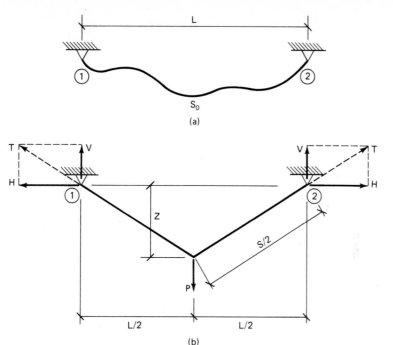

Figure 2.1.1   Cable with central concentrated load.

Thus, the tension is

$$T = \frac{P}{2}\sqrt{1 + \left(\frac{L}{2z}\right)^2}$$

(2.1.4)

However, the problem is not statically determinate in that the deflection z is unknown. An auxiliary relation is needed.

The stretched length S is related to the tension and the deflection by two equations. First, from Eq. (2.1.1)

$$S = S_0\left(1 + \frac{T}{AE}\right)$$

(2.1.5)

Second, from geometric considerations

$$S = 2\sqrt{\left(\frac{L}{2}\right)^2 + (z)^2} = 2z\sqrt{1 + \left(\frac{L}{2z}\right)^2}$$

(2.1.6)

Thus, combining Eqs. (2.1.4) to (2.1.6), we obtain

$$2z \sqrt{1 + \left(\frac{L}{2z}\right)^2} = S_0 \left[1 + \frac{P}{2AE} \sqrt{1 + \left(\frac{L}{2z}\right)^2} \right] \quad (2.1.7)$$

which is a single nonlinear equation for z. It is not easy to obtain an explicit equation for z as a function of P without recourse to numerical methods.

Let us instead consider the inverse problem of determining P as a function of z. An inverse problem is one in which you are given the solution and are asked what problem was solved [26, 28]. Such an approach is often useful for nonlinear problems of cable and membrane behavior. In fact, it closely resembles the architectural design problem of specifying a desired equilibrium configuration for the tension structure and determining the prestressing forces to obtain that configuration [10, 23, 29]. In this instance we ask the question, Given a deflection z, what force P is necessary to maintain that deflection? To answer the question, we rewrite Eq. (2.1.7) as

$$P = 2AE \left(\frac{2z}{L}\right) \left[\frac{L}{S_0} - \frac{1}{\sqrt{1 + (2z/L)^2}} \right] \quad (2.1.8)$$

Then, with a specified z, Eq. (2.1.8) determines P and Eq. (2.1.4) determines T.

We must go beyond this question and determine what the additional deflection Δz would be if an additional load ΔP were applied. That is, considering P to be the prestressing force, what is the response Δz to an in-service load ΔP? To calculate that response we need to know the prestressed stiffness of the structure. For a small additional load ΔP, by a Taylor expansion [61] to the first order in Δz

$$(P + \Delta P) = P + \frac{\partial P}{\partial z}\bigg|_z (\Delta z) \quad (2.1.9)$$

where $K = \partial P/\partial z$, evaluated at z, is the stiffness coefficient in the prestressed configuration dictated by P and z. Differentiating Eq. (2.1.8),

$$K = \frac{\partial P}{\partial z} = \frac{4AE}{L} \left\{ \frac{L}{S_0} - \frac{1}{[1 + (2z/L)^2]^{3/2}} \right\} \quad (2.1.10)$$

Thus, by Eq. (2.1.9), the in-service response for small ΔP is

$$\Delta z = \frac{\Delta P}{4AE/L\{L/S_0 - 1/[1 + (2z/L)^2]^{3/2}\}} \quad (2.1.11)$$

Prestressing stiffens the structure, as can be seen from Eqs. (2.1.10) and (2.1.11), where $\partial P/\partial z$ increases as z increases. Plotted in Fig. 2.1.2 are the

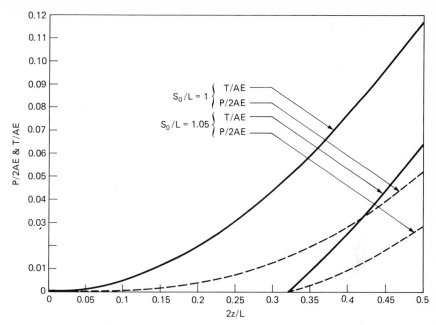

Figure 2.1.2   Load-deflection curve for central concentrated load.

nondimensionalized tension and load as functions of $(2z/L)$ for two special cases of $S_0/L = 1.0$ and 1.05. Note that for $S_0/L = 1$ the initial stiffness $(z = 0)$ is zero, i.e., without prestress the deflection is initially unbounded. Further, as z (and thus P) increases, the importance of the nonlinear term in Eq. (2.1.10) decreases and the response z, given by Eq. (2.1.11), tends to be linear. Because of the dependency of $K = \partial P/\partial z$ on z, one should be cautious of superimposing loads.

This simple example is useful in developing solutions for the two cases shown in Fig. 2.1.3 of end loads applied to a single inclined cable stay. For

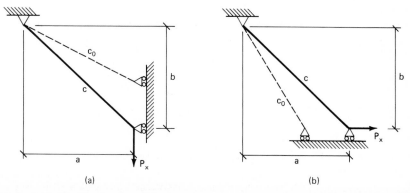

Figure 2.1.3   End loads on a cable stay.

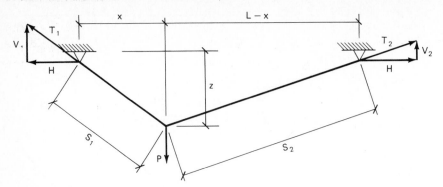

Figure 2.1.4 Concentrated load at variable location.

a vertical load P_y as in Fig. 2.1.3a, replace in Eqs. (2.1.4) to (2.1.11) P with $P_y/2$, S_0 with $2c_0$, L with $2a$, and z with b. For a horizontal load P_x as in Fig. 2.1.3b, replace P with $P_x/2$, S_0 with $2c_0$, L with $2b$, and z with a.

Figure 2.1.4 shows a cable with a transverse load at an arbitrary location x from the left support. This models the problem of a movable load on a horizontal span. The same approach can be used for this problem as was used for the midspan load. By elementary statics

$$V_1 = P\frac{L - x}{L} \tag{2.1.12a}$$

$$V_2 = P\frac{x}{L} \tag{2.1.12b}$$

$$H = P\frac{x(L - x)}{Lz} \tag{2.1.12c}$$

and thus the tensions in the two segments are

$$T_1 = \frac{Pab}{\gamma c_1} \tag{2.1.13a}$$

$$T_2 = \frac{Pab}{\gamma c_2} \tag{2.1.13b}$$

where $a = x/L$, $b = 1 - a$, $\gamma = z/L$, and

$$c_1 = \frac{1}{\sqrt{1 + (\gamma/a)^2}} \tag{2.1.14a}$$

$$c_2 = \frac{1}{\sqrt{1 + (\gamma/b)^2}} \tag{2.1.14b}$$

Again, we relate the deformed lengths S_1 and S_2 to the unstressed lengths S_{01} and S_{02} by geometry and by Eq. (2.1.1):

$$S_1 = S_{01}\left(1 + \frac{T_1}{AE}\right) = \sqrt{x^2 + z^2} = L\frac{a}{c_1} \qquad (2.1.15a)$$

$$S_2 = S_{02}\left(1 + \frac{T_2}{AE}\right) = \sqrt{(L-x)^2 + z^2} = L\frac{b}{c_2} \qquad (2.1.15b)$$

Thus,

$$\frac{S_{01}}{L} = \frac{1}{c_1/a + Pb/\gamma AE} \qquad (2.1.16a)$$

$$\frac{S_{02}}{L} = \frac{1}{c_2/b + Pa/\gamma AE} \qquad (2.1.16b)$$

Since $S_0 = S_{01} + S_{02}$ is the unstressed length, the equation relating P and $\gamma\, (= z/L)$ is found from Eqs. (2.1.16) to be

$$\frac{S_0}{L} = \frac{1}{c_1/a + Pb/\gamma AE} + \frac{1}{c_2/b + Pa/\gamma AE} \qquad (2.1.17)$$

which, remembering c_1 and c_2 are nonlinear functions of γ, is a nonlinear equation of γ. The inverse solution for P as a function of γ can be found by solving the quadratic form of Eq. (2.1.17):

$$\frac{P}{AE} = \frac{\gamma}{2ab}\left[\frac{L}{S_0} - c_1 - c_2 \right.$$
$$\left. + \sqrt{\left(\frac{L}{S_0} + c_1 - c_2\right)^2 + 4a\left(\frac{L}{S_0}\right)(c_2 - c_1)}\right] \qquad (2.1.18)$$

Therefore, for a specified γ determine P from Eq. (2.1.18) and T_1 and T_2 from Eqs. (2.1.13a and b). The stiffness coefficient $K = \partial P/\partial z = 1/L\ (\partial P/\partial \gamma)$ can be determined by taking the derivative of Eq. (2.1.18) with respect to γ.

For the special case of $S_0 = L$ solutions for P and T_1 are plotted in nondimensional form in Fig. 2.1.5 for a range of $0 \leqslant \gamma \leqslant 0.1$. Also shown in Fig. 2.1.5 are the deflected shapes of the same cable as a load (magnitude $= 0.006AE$) moves across the span. It can be seen that for $x \geqslant L/2$, $T_1 \geqslant T_2$, and the coordinate system can be reflected in order to obtain solutions for $x \geqslant L/2$.

A useful observation that holds true for all transversely loaded cable segments can be made from Figs. 2.1.4 and 2.1.5 and from the equations

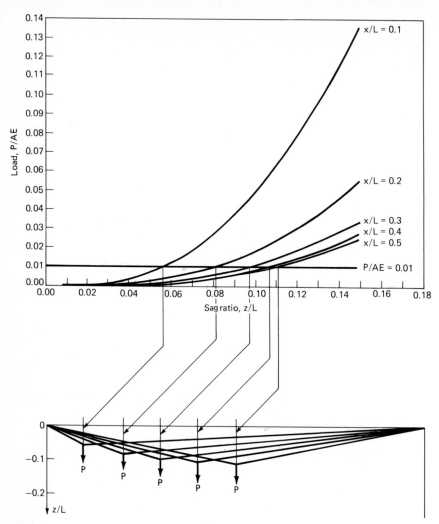

Figure 2.1.5   Response curves for variable-location concentrated load.

of statics, Eqs. (2.1.12). The horizontal component of tension is constant throughout. The deflected shape assumed by the cable is the same as the shape of the moment diagram of an equivalent simply supported beam loaded in the same manner as the cable. Equation (2.1.12c), rewritten as

$$Hz = \frac{Px(L - x)}{L} = M_x \qquad (2.1.19)$$

is the equation for moment at the load in the equivalent beam. However, since neither H nor z is known, the scale of the moment diagram is

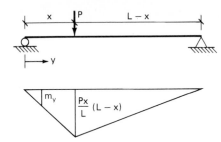

Figure 2.1.6 Moment diagram for simply supported span.

unknown. If for a cable problem we can specify either H (constant throughout) or z at any single point, then the profile of the cable can be determined from the equivalent moment diagram for a simply supported beam loaded in the same fashion. For this example, the moment diagram is shown in Fig. 2.1.6 with positive moment plotted on the tension side of the beam. Thus, at any location y along the cable span,

$$Hz = m_y = \begin{cases} \dfrac{Px(L-x)}{L} \dfrac{(y)}{x} & 0 \leq y \leq x \\[4mm] \dfrac{Px(L-x)}{L} \left(\dfrac{L-y}{L-x}\right) x & x \leq y \leq L \end{cases} \qquad (2.1.20)$$

yields the desired solution once H or z at any single point is determined.

2.1.2 Multiple concentrated loads

The analogy of the deflected profile of a cable segment to the moment diagram for a simply supported beam is a useful tool in the determination of the prestressed configurations of cables subjected to multiple loads. A semi-inverse method can be used in which either the horizontal component H of tension is specified or the deflected profile at a key point is specified. Then the rest of the cable profile, the segmental tensions, and the initial lengths can be determined, which are necessary to maintain that profile for that set of applied loads. Unfortunately, when in-service loads are applied, or when no locations of points on the profile are known, the moment analogy method fails and we must resort to iterative methods.

The moment analogy method can be developed from consideration of the generalized problem, shown in Fig. 2.1.7, of a set of transverse loads P_i applied at points x_i along the horizontal projection of the span x_{n+1}. The chord connecting the endpoints 0, $n+1$ makes an angle θ with the x axis ($\tan \theta = z_{n+1}/x_{n+1}$). By statics we observe that the horizontal com-

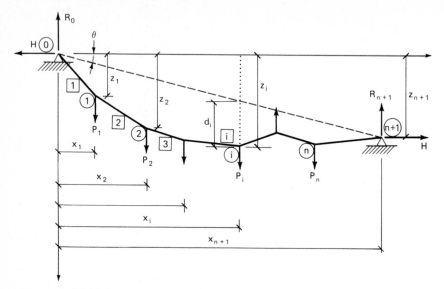

Figure 2.1.7 Multiple concentrated loads.

ponent of tension in each segment is constant, and the vertical reactions are

$$R_0 = \frac{1}{x_{n+1}} \left[\sum_{i=1}^{n} (x_{n+1} - x_i)P_i + Hz_{n+1} \right] \qquad (2.1.21a)$$

$$R_{n+1} = \frac{1}{x_{n+1}} \left(\sum_{i=1}^{n} x_i P_i - Hz_{n+1} \right) \qquad (2.1.21b)$$

If the cable were instead a simply supported beam directed along the chord, the moment at any point i would be (set $H = 0$ in reaction equations)

$$m_i = \frac{1}{x_{n+1}} \left[\sum_{j=1}^{i} x_j(x_{n+1} - x_i)P_j + \sum_{j=i+1}^{n} x_i(x_{n+1} - x_j)P_j \right] \qquad (2.1.22)$$

The moments M_i in the real cable must be zero at every point and thus

$$M_i = \frac{1}{x_{n+1}} \left[\sum_{j=1}^{i} x_j(x_{n+1} - x_i)P_j + \sum_{j=i+1}^{n} x_i(x_{n+1} - x_j)P_j \right.$$

$$\left. + Hz_{n+1}x_i \right] - Hz_i$$

$$0 = m_i + H(x_i \tan \theta - z_i)$$

$$0 = m_i - Hd_i \qquad (2.1.23)$$

where d_i = distance of cable from chord at point i. Therefore, at every point

$$d_i = \frac{m_i}{H} \qquad (2.1.24)$$

If we know H, all d_i can be calculated by Eq. (2.1.24). If we know the distance d_I at a particular point $i = I$, we can calculate

$$H = \frac{m_I}{d_I} \qquad (2.1.25a)$$

$$d_i = \left(\frac{m_i}{m_I}\right) d_I \qquad (2.1.25b)$$

The semi-inverse moment analogy method will be illustrated with two examples: one with supports on the same level and one with supports on two levels. Figure 2.1.8 shows an approximation of a suspension bridge in which the distributed load q is assumed to be transmitted equally by nine suspenders to the suspension cable (assumed weightless). In a subsequent section we will reconsider this problem including distributed load effects directly. Also shown in Fig. 2.1.8 is the moment diagram for a simply supported beam subjected to nine equally spaced concentrated loads of magnitude $P = qa$. The equivalent shear V_i in each segment and the equivalent moments m_i at each point are listed in Table 2.1.1. Because of symmetry we need only list those values in the left half of the span.

By Eq. (2.1.24) the deflected profile z_i at the ith suspender is given by

$$z_i = \frac{m_i}{H} \qquad (2.1.26)$$

If we specify the sag ratio f, then the deflected profile at suspender 5 is $z_5 = fL = 10fa$. Thus

$$H = \frac{m_5}{z_5} = \frac{25}{20}\left(\frac{P}{f}\right) \qquad (2.1.27)$$

and the deflected profile at the other points is given by Eq. (2.1.26) and is as listed in Table 2.1.1. The tension in each segment is

$$T_i = \sqrt{H^2 + V_i^2} \qquad (2.1.28)$$

where V_i is the vertical component of tension in each segment and, in this case, equals the shear in the equivalent beam. Knowing H from Eq. (2.1.27) and V_i from Table 2.1.1, we can calculate the tensions as listed in Table 2.1.1.

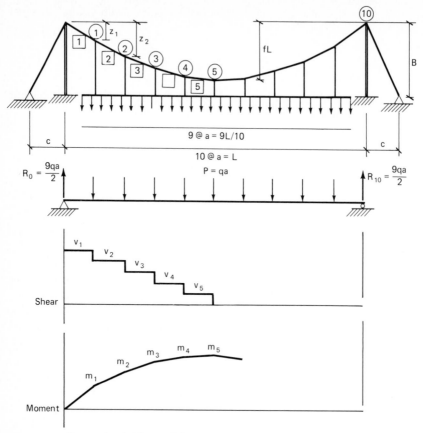

Figure 2.1.8   Suspension bridge model.

The unstressed length of each segment can be calculated from Eq. (2.1.1) and the tension in the backstay from $T_B = \sqrt{B^2 + C^2}\,(H/C)$. The maximum tension occurs in the segment with maximum slope, which usually occurs adjacent to a support. Plotted in Fig. 2.1.9 is the maximum tension as a function of sag ratio f. It can be seen that small sag ratios induce a dramatic increase in tension. Also, note that the profile of the deflected suspender points lie on a quadratic curve $z = Lf\,(25 - x)/a$. This is true, as will be shown subsequently, for cables subjected to loads distributed uniformly over a horizontal projection.

The moment analogy method can be applied to cable segments with supports at different levels such as the example shown in Fig. 2.1.10. In this case we wish to determine the location of points in an unstressed cable at which to attach equal and opposite loads such that the central one-third span remains level. We will first calculate H and the tensions in each span. Then we can calculate the lengths of each unstressed segment.

TABLE 2.1.1 **Suspension Cable Results**

Segment	Point	V_i	m_i	z_i	T_i
1		$9\left(\dfrac{P}{2}\right)$			$\dfrac{P}{2}\sqrt{\left(\dfrac{2.5}{f}\right)^2 + 81}$
	1		$9\left(\dfrac{PL}{20}\right)$	$9\left(\dfrac{Lf}{25}\right)$	
2		$7\left(\dfrac{P}{2}\right)$			$\dfrac{P}{2}\sqrt{\left(\dfrac{2.5}{f}\right)^2 + 49}$
	2		$8\left(\dfrac{PL}{10}\right)$	$16\left(\dfrac{Lf}{25}\right)^2$	
3		$5\left(\dfrac{P}{2}\right)$			$\dfrac{P}{2}\sqrt{\left(\dfrac{2.5}{f}\right)^2 + 25}$
	3		$21\left(\dfrac{PL}{20}\right)$	$21\left(\dfrac{Lf}{25}\right)$	
4		$3\left(\dfrac{P}{2}\right)$			$\dfrac{P}{2}\sqrt{\left(\dfrac{2.5}{f}\right)^2 + 9}$
	4		$12\left(\dfrac{PL}{10}\right)$	$24\left(\dfrac{Lf}{25}\right)$	
5		$\dfrac{P}{2}$			$\dfrac{P}{2}\sqrt{\left(\dfrac{2.5}{f}\right)^2 + 1}$
	5		$25\left(\dfrac{PL}{20}\right)$	$25\left(\dfrac{Lf}{25}\right)$	

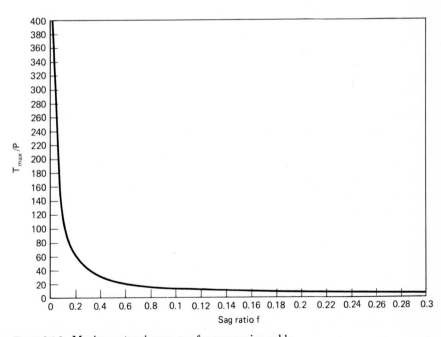

Figure 2.1.9 Maximum tension vs. sag for suspension cable.

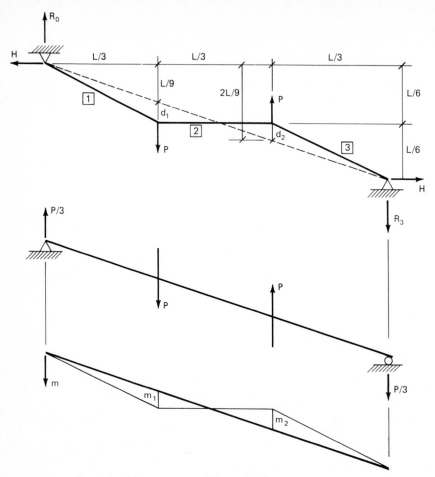

Figure 2.1.10 Example with supports at different levels.

Figure 2.1.10 shows the moment diagram for the equivalent simply supported bean, with

$$m_1 = -m_2 = \frac{P}{3}\left(\frac{L}{3}\right)$$

Thus, by Eq. (2.1.24)

$$d_1 = -d_2 = \frac{PL}{9H}$$

In the central segment H is determined from

$$d_1 = \frac{L}{6} - \frac{L}{9} = \frac{m_1}{H}$$

$$H = 2P$$

and the vertical reaction components are

$$R_0 = R_3 = P$$

The tensions are

$$T_1 = T_3 = \sqrt{H^2 + R_0^2} = P \sqrt{5}$$

$$T_2 = H = 2P$$

and the unstressed lengths should be

$$S_{01} = S_{03} = \frac{S_1}{1 + T_1/AE} = \frac{\sqrt{(L/3)^2 + (L/6)^2}}{1 + P \sqrt{5}/AE} = \frac{L/6 \sqrt{5}}{1 + P \sqrt{5}/AE}$$

$$S_{02} = \frac{S_2}{1 + T_2/AE} = \frac{L/3}{1 + 2P/AE}$$

In cases where neither H nor any of z_i are known, the moment analogy fails. Further, even if we have used that method to determine the pre-stressed configuration for a single set of prevailing loads, in-service loadings will usually change the geometry and tensions in such a fashion that the method cannot be used for the various sets of in-service loads. An iterative method is needed.

Consider again Fig. 2.1.7. Isolated in Fig. 2.1.11 are free-body diagrams of two adjacent segments and of the point at which they are connected. In that figure the possibility of a horizontal load Q_i, as well as a vertical load P_i, has been included. Horizontal loads imply that the horizontal components of tension are no longer constant.

By static equilibrium of the point i and of segment i

$$T_i = \sqrt{H_i^2 + V_i^2} \qquad (2.1.29a)$$

$$V_{i+1} = V_i - P_i \qquad (2.1.29b)$$

$$H_{i+1} = H_i - Q_i \qquad (2.1.29c)$$

Thus, in Fig. 2.1.7, if we were lucky enough to know H_1 and V_1 at the point 0, Eqs. (2.1.29) enable us to propagate solutions for V_i, H_i, and T_i through all the segments, each in succession $i = 1$ to n. Knowing the unstressed lengths ΔS_{0i}, the coordinates of point 0 (x_0, z_0) and the tension

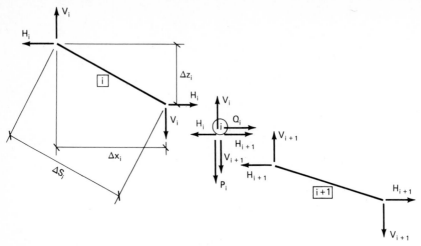

Figure 2.1.11   Free-body diagrams of load points and segments.

T_i in all the segments, we could calculate the coordinates of all the connection points from

$$\Delta x_i = \frac{H_i}{T_i}(\Delta S_i) \qquad (2.1.30a)$$

$$\Delta z_i = \frac{V_i}{T_i}(\Delta S_i) \qquad (2.1.30b)$$

$$x_{i+1} = x_i + \Delta x_i \qquad (2.1.30c)$$

$$z_{i+1} = z_i + \Delta z_i \qquad (2.1.30d)$$

where the stretched segment length ΔS_i is

$$\Delta S_i = \left(1 + \frac{T_i}{AE}\right)\Delta S_{0i} \qquad (2.1.30e)$$

On the other hand, if the unstressed segment lengths ΔS_{0i} were unknown, but Δx_i were known, Eqs. (2.1.30) would be recast as

$$\Delta z_i = \frac{V_i}{H}\Delta x_i \qquad (2.1.31a)$$

$$z_{i+1} = z_i + \Delta z_i \qquad (2.1.31b)$$

$$\Delta S_i = \frac{T_i}{H_i}\Delta x_i \qquad (2.1.31c)$$

$$\Delta S_{0i} = \frac{\Delta S_i}{1 + T_i/AE} \qquad (2.1.31d)$$

and the cumulative unstressed length to the ith point would be

$$S_{0i} = \sum_{j=1}^{i} (\Delta S_{0j}) \qquad (2.1.31e)$$

Unfortunately, we usually are not lucky enough to know H_1 and V_1 at the point 0. But the above equations suggest an iterative solution method which consists of the following steps:

1. Assume H_1 and V_1.
2. Use Eqs. (2.1.29) to calculate V_i, H_i, T_i.
3. Use Eqs. (2.1.30) or (2.1.31) to calculate x_i, z_i, S_i.
4. Compare the computed x_{n+1}, z_{n+1}, S_{n+1} to the exact values at the endpoint $n + 1$.
5. If they are different, assume new H_1 and V_1 and repeat from step 1.

This is a "shooting" method [61], i.e., a trial solution is propagated from the initial point to the terminal point with assumed initial values. The quality of the trial solution is evaluated at the terminal point by comparison to the true boundary conditions on z_{n+1} and either x_{n+1} or S_{n+1}. What is lacking in this iterative method is a scheme for improving successive trial solutions H_1 and V_1. Such a scheme will be provided by the Newton-Raphson method [61, 71, 72], which is described under "Multiple Concentrated Loads." For now, a trial-and-error scheme combined with judgment will be described.

In the trial-and-error scheme, several trial solutions are attempted until the terminal $i = n$ trial solutions bracket the exact answers. Subsequent trial solutions should be based on interpolations of previous trials to reduce the bracketing errors. It is important to note that the equations are not linear and therefore the interpolation will not be linear.

Consider as a simple example the case of a single concentrated load placed at midspan, as previously considered in Fig. 2.1.1. For an unstressed length $S_0 = L$, Eqs. (2.1.29) give

$$H_1 = H_2$$

$$V_1 = -V_2 = \frac{P}{2}$$

$$T_1 = T_2 = \sqrt{H_1^2 + \frac{P^2}{4}}$$

Equations (2.1.30) give

$$\Delta S_1 = \Delta S_2 = \left(1 + \frac{1}{AE} \sqrt{H_1^2 + \frac{P^2}{4}}\right) \frac{L}{2}$$

$$\Delta x_1 = \Delta x_2 = \frac{H_1 L}{\sqrt{H_1^2 + P^2/4}} \left(1 + \frac{1}{AE} \sqrt{H_1^2 + \frac{P^2}{4}}\right)$$

$$\Delta z_1 = -\Delta z_2 = \frac{P}{2\sqrt{H_1^2 + P^2/4}} \left(1 + \frac{1}{AE} \sqrt{H_1^2 + \frac{P^2}{4}}\right)$$

$$x_{n+1} = \frac{H_1 L}{\sqrt{H_1^2 + P^2/4}} \left(1 + \frac{1}{AE} \sqrt{H_1^2 + \frac{P^2}{4}}\right)$$

$$z_{n+1} = 0$$

Thus, since z_{n+1} is identically zero, the iteration on successive assumptions of H_1 proceeds until $x_{n+1} = L$. In terms of nondimensional variables $h = H_1/AE$, $p = P/2AE$, the critical equation is

$$h \left(1 + \frac{1}{\sqrt{h^2 + p^2}}\right) - 1 \overset{?}{=} 0 \qquad (2.1.32)$$

Two particular values of p were considered in Table 2.1.2. There, successive iterates were attempted in order to satisfy Eq. (2.1.32). Selection of successive iterates was based on sign changes in Eq. (2.1.32).

TABLE 2.1.2   Iterative Solution for Concentrated Load

	$p = 0.01$		$p = 0.001$
h	$h \left(1 + \dfrac{1}{\sqrt{h^2 + p^2}}\right) - 1$	h	$h \left(1 + \dfrac{1}{\sqrt{h^2 + p^2}} - 1\right)$
0.01	−0.38	0.001	−0.292
0.02	−0.09	0.002	−0.103
0.03	−0.02	0.003	−0.05
0.04	+0.01	0.004	−0.026
0.036	−0.0005	0.01	+0.005
0.037	+0.0024	0.009	+0.003
0.0361	−0.0002	0.008	0.000
0.0362	+0.0001		

2.2   Response of Cable Segments to Distributed Loads

In Sec. 2.1 it was assumed that all loads were concentrated at discrete points along the cable segment and, therefore, the segment consisted of straight lines between load points. All effects of distributed loads were neglected, including self-weight of the cable segment. Because the deflected geometry of a cable segment is sensitive to load patterns as well as magnitudes, it is important to consider distributed loads. In fact, it is much too conservative to simply replace distributed loads by lumped effects at discrete points, as will be seen subsequently.

In this section the response of cable segments to various distributed loads assumed to be acting in a single coordinate direction will be treated. First, we will consider that only distributed loads are present and serve to prestress the segment. Second, we will consider the combined effects of distributed and concentrated loads. It is important to remember, because of the nonlinear geometric characteristics of cable behavior, that responses to separate loadings cannot be simply superimposed.

2.2.1   Distributed transverse loads

Two types of transverse distributed loads are common: loads distributed uniformly along the arc length of the segment, e.g., self-weight, and loads distributed uniformly along the horizontal span of the segment, e.g., permanent suspended loads. The first type leads to the classic catenary solution and, in the limit as the sag ratio is reduced, tends to the simpler solution to the second type.

Load uniform over horizontal span.   For the case of a load distributed uniformly over the horizontal span of a cable segment, we will adopt the horizontal coordinate x as the independent variable. The dependent variables will be the two components of tension H and V, the vertical deflection z, and the deformed and undeformed arc lengths S and S_0. A definition sketch is shown in Fig. 2.2.1 along with a free-body diagram of a differential length of the cable segment.

Equilibrium of forces on the differential length gives

$$\frac{dH}{dx} = 0 \rightarrow H = H_0 \tag{2.2.1a}$$

$$\frac{dV}{dx} = -p \rightarrow V = V_0 - px \tag{2.2.1b}$$

and, therefore,

$$T = \sqrt{H^2 + V^2} = \sqrt{H_0^2 + (V_0 - px)^2} \tag{2.2.1c}$$

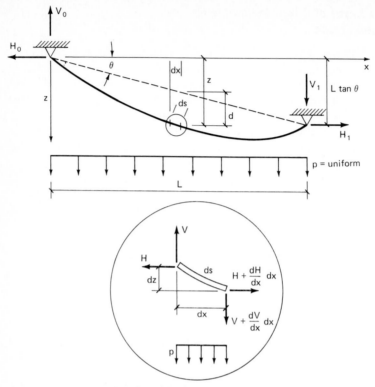

Figure 2.2.1 Uniform distributed load over horizontal span.

where H_0 and V_0 are integration constants with physical significance of components of tension T_0 at $x = 0$.

$$T_0 = \sqrt{H_0^2 + V_0^2} \qquad (2.2.1d)$$

Note that $V = v + H_0 \tan \theta$, where v is the vertical shear in the equivalent simply supported beam spanning along the chord. Also, $V_1 = V_0 - pL$ guarantees overall equilibrium. The horizontal component of tension is constant throughout the span. Both H_0 and V_0 are as yet undetermined.

The geometry of the differential length satisfies the following constraints:

$$\frac{dz}{dx} = \frac{V}{H} = \frac{V_0 - px}{H_0} \rightarrow z = \frac{x}{H_0}\left(V_0 - \frac{px}{2}\right) \qquad (2.2.2a)$$

$$\frac{ds}{dx} = \frac{T}{H} \rightarrow s = \frac{H_0}{2p}\left[\frac{V_0 T_0 - VT}{H_0^2} + \log\left(\frac{T - V}{T_0 - V_0}\right)\right] \qquad (2.2.2b)$$

where the boundary conditions $z = 0$ and $s = 0$ at $x = 0$ have been used to evaluate the constants of integration. The boundary condition $z = L \tan \theta$ at $x = L$ can be used to evaluate the constant V_0 in Eq. (2.2.2a):

$$V_0 = H_0 \tan \theta + \frac{pL}{2} \qquad (2.2.3a)$$

and thus

$$V = H_0 \tan \theta + \frac{pL}{2}\left(1 - 2\frac{x}{L}\right) \qquad (2.2.3b)$$

$$T = H_0 \sqrt{1 + \left[\tan \theta + \frac{pL}{2H_0}\left(1 - \frac{2x}{L}\right)\right]^2} \qquad (2.2.3c)$$

$$z = x \tan \theta + d \qquad (2.2.3d)$$

where d is the vertical distance from the chord to the cable and is a parabolic equation:

$$d = \frac{px}{2H_0}(L - x) \qquad (2.2.3e)$$

Note that $d = m_x/H_0$ where m_x is the bending moment in the equivalent simply supported beam. Equations (2.2.3) can be used in Eq. (2.2.2b) to evaluate the deformed lengths.

The constant H_0 remains to be determined. If we adopt the inverse approach, H_0 can be defined such that a specified sag $d_m = fL$ of the cable below the chord is obtained at midspan. In that case, Eq. (2.2.3d) gives

$$\frac{L \tan \theta}{2} + fL = \frac{L}{2}\tan \theta + \frac{pL^2}{8H_0}$$

which implies, in terms of the sag ratio f, that

$$H_0 = \frac{pL}{8f} \qquad (2.2.4a)$$

$$V = \frac{pL}{2}\left(\frac{\tan \theta}{4f} + 1 - 2\frac{x}{L}\right) \qquad (2.2.4b)$$

$$T = \frac{pL}{8f} \sqrt{1 + \left[\tan \theta + 4f\left(1 - \frac{2x}{L}\right)\right]^2} \qquad (2.2.4c)$$

$$z = x\left[\tan \theta + 4f\left(1 - \frac{x}{L}\right)\right] \qquad (2.2.4d)$$

Note that V and T are maximum at $x \geqslant 0$ if $\tan \theta \leqslant 0$.

If, on the other hand, the sag ratio is not known a priori, e.g., when an in-service load is applied, we have to determine H_0 from a nonlinear transcendental equation generated from the boundary condition on total length of the cable segment between supports. The unstressed length s_0 of cable to the point x on the span is determined by applying Eq. (2.1.1) to a differential segment in combination with Eq. (2.2.2b). Thus

$$\frac{ds}{dx} = \left(1 + \frac{T}{AE}\right)\frac{ds_0}{dx} = \frac{T}{H} \qquad (2.2.5a)$$

For all practical purposes T/AE is small, so let us assume $T/AE \ll 1$, in which case

$$\frac{ds_0}{dx} = \left(1 - \frac{T}{AE}\right)\frac{T}{H} \qquad (2.2.5b)$$

is the governing differential equation for s_0.

Rather than solving Eq. (2.2.5b) directly for H_0, we will instead determine the sag ratio f and then use Eq. (2.2.4a) to specify H_0. It is convenient to define nondimensional variables:

$$v = \tan\theta + 4f\left(1 - 2\frac{x}{L}\right)$$

$$\rho = \frac{pL}{2AE}$$

Then, the solution to Eq. (2.2.5b) can be shown to be

$$c)\left(\frac{s_0}{L}\right) = \frac{1}{16f}\left\{\left[(v_0\sqrt{1+v_0^2} - v\sqrt{1+v^2}) + \log\left(\frac{v_0 + \sqrt{1+v_0^2}}{v + \sqrt{1+v^2}}\right)\right]\right.$$
$$\left. + \frac{\rho}{2f}\left[(v - v_0) + \frac{1}{3}(v^3 - v_0^3)\right]\right\} \qquad (2.2.5c)$$

where the boundary condition $s_0 = 0$ at $x = 0$ has been used to eliminate the arbitrary constant of integration and where $v_0 = \tan\theta + 4f$. Evaluating Eq. (2.2.5c) at $x = L$ and letting $v_1 = \tan\theta - 4f$, we find the total unstressed length S_0 to be

$$\left(\frac{S_0}{L}\right) = \frac{1}{16f}\left\{\left[(v_0\sqrt{1+v_0^2} - v_1\sqrt{1+v_1^2}) + \log\left(\frac{v_0 + \sqrt{1+v_0^2}}{v_1 + \sqrt{1+v_1^2}}\right)\right]\right.$$
$$\left. + \frac{\rho}{2f}\left[(v_1 - v_0) + \frac{1}{3} + (v_1^3 - v_0^3)\right]\right\} \qquad (2.2.6a)$$

which is the pertinent nonlinear transcendental equation to determine the sag ratio f. Note that v_0 and v_1 are nonlinear functions of f.

Equation (2.2.6a) could be solved directly for f using various numerical procedures [61], e.g., trial-and-error or Newton-Raphson, for determining roots for nonlinear algebraic equations. Such procedures are considered in "Multiple Concentrated Loads." Let us now examine the special case of $\tan \theta = 0$, i.e., supports on the same level. Then Eq. (2.2.6a) becomes

$$\left(\frac{S_0}{L}\right) = \frac{1}{16f}\left\{\left[4f\sqrt{1+(4f)^2} + \log\left(\frac{\sqrt{1+(4f)^2}+4f}{\sqrt{1+(4f)^2}-4f}\right)\right] - 4\rho\left[1+\frac{(4f)^2}{3}\right]\right\} \quad (2.2.6b)$$

For small sag ratios, $f \ll 1$, binomial and Taylor expansions can be used in Eq. (2.2.6b) to obtain

$$(4f)^3 - 2\rho(4f)^2 - 6\left(\frac{S_0}{L}-1\right)(4f) - 6\rho = 0 \quad (2.2.6c)$$

which is a cubic equation for $(4f)$. Because of the assumption $f \ll 1$, it only has one real root which can be determined by Cardan's method:

Given: $x^3 + ax^2 + bx + c = 0$

Real root: $x = y - \dfrac{a}{3}$

Where: $y = A + B$

$$A = \left(\frac{\sqrt{Q}-q}{2}\right)^{1/3}$$

$$B = \left(\frac{\sqrt{Q}+q}{2}\right)^{1/3}$$

$$q = 2\left(\frac{a}{3}\right)^3 - \frac{ab}{3} + c$$

$$P = -\frac{a^2}{3} + b$$

$$Q = \left(\frac{P}{3}\right)^3 + \left(\frac{q}{2}\right)^2 \geq 0$$

Thus, if we are given an elastic cable segment with unstressed length S_0, Eqs. (2.2.6c) can be solved for the sag ratio f. Then, Eq. (2.2.4) can be used to determine the rest of the pertinent variables.

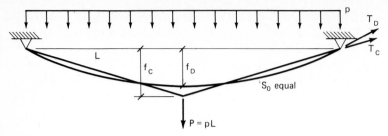

Figure 2.2.2   Comparison of distributed and concentrated loads.

If the cable is inextensible, i.e., $\rho = 0$, Eq. (2.2.6c) has the solution

$$f = \sqrt{\frac{3}{8}\left(\frac{S_0}{L} - 1\right)} \tag{2.2.7a}$$

If additional terms in the binomial expansions of Eq. (2.2.6b) were used, a more accurate expression for the inextensible sag ratio would be obtained:

$$f = \left(\frac{5}{24}\left\{1 - \left[1 - \frac{18}{5}\left(\frac{S_0}{L} - 1\right)\right]^{1/2}\right\}\right)^{1/2} \tag{2.2.7b}$$

In a similar fashion, but with more tedious manipulations, we could find for an inextensible cable with an inclined chord

$$f = \sqrt{\frac{3}{8}\left(\frac{S_0}{L} - 1 - \frac{1}{2}\tan^2\theta\right)} \tag{2.2.7c}$$

If f were specified, as discussed previously, Eq. (2.2.7c) yields an equation for the unstressed length required:

$$S_0 = L\left(1 + \frac{8}{3}f^2 + \frac{1}{2}\tan^2\theta\right) \tag{2.2.7d}$$

It is instructive to compare the parabolic segment solution to that of a single midspan concentrated load with magnitude equal to the total distributed load $P = pL$. See Fig. 2.2.2 for a definition sketch. For an inextensible cable with a central concentrated load the sag ratio f_C is

$$f_C = \frac{1}{2}\sqrt{\left(\frac{S_0}{L}\right)^2 - 1}$$

The maximum tension is

$$T_C = \frac{pL}{2} \sqrt{1 + \frac{1}{4f_C^2}} = \frac{pL}{2} \sqrt{\frac{(S_0/L)^2}{(S_0/L)^2 - 1}}$$

For a uniform distributed load

$$f_D = \sqrt{\frac{3}{8}\left(\frac{S_0}{L} - 1\right)}$$

and the maximum tension is

$$T_D = \frac{pL}{2} \sqrt{1 + \frac{1}{16f_D^2}} = \frac{pL}{2} \sqrt{1 + \frac{1}{6(S_0/L - 1)}}$$

The ratio

$$\frac{T_D}{T_C} = \left\{1 - \frac{1}{6}\left[\frac{5 - S_0/L}{(S_0/L)^2}\right]\right\}^{1/2}$$

indicates that the tension due to a distributed load is less than that due to an equivalent concentrated load, especially for smaller sag ratios, e.g., $T_D = 0.46T_C$ for $f_D = 0.1$. The ratio of sags

$$\frac{f_C}{f_D} = \sqrt{\frac{4}{3}\left(1 + \frac{S_0}{L}\right)}$$

indicates, as expected, that the sag of the concentrated load is the larger of the two.

Once the equilibrium state has been established for the parabolic arc, i.e., f and H are specified, an additional in-service change in unstressed length may occur, e.g., due to temperature change, or an additional load, similarly distributed, may be applied. We therefore desire to calculate a stiffness coefficient for the horizontal component of tension. If we take the partial derivatives of Eq. (2.2.4a) with respect to p and f,

$$\frac{\partial H_0}{\partial p} = +\frac{H_0}{p} \tag{2.2.8a}$$

$$\frac{\partial H_0}{\partial f} = -\frac{H_0}{f} \tag{2.2.8b}$$

Thus, if an additional sag $L\,\Delta f$, occurs, e.g., due to temperature increase, with no increase in load, Taylor expansion of Eq. (2.2.4a) can be used to determine the new horizontal force:

$$\overline{H}_0 = H_0 + \Delta H = H_0 + \frac{\partial H}{\partial f} \Delta f = H_0 \left(1 - \frac{\Delta f}{f} \right)$$

We see that the force is decreased.

For the case of a small additional load Δp, the sag will change by an amount Δf. Since the unstressed length remains the same, we can use Eqs. (2.2.6) to calculate an approximate Δf corresponding to Δp. Let $\epsilon = 4f$ in Eq. (2.2.6c), and since Eq. (2.2.6c) is a function of both sag ratio (ϵ) and load ($\rho = pL/2AE$), we can write its Taylor expansion as

$$g\,(\varepsilon,\, \rho) = \varepsilon^3 - 2\rho\varepsilon^2 - 6 \left(\frac{S_0}{L} - 1 \right) \varepsilon - 6\rho = 0$$

$$\overline{g}\,(\overline{\varepsilon},\, \overline{\rho}) = g(\varepsilon,\, \rho) + \frac{\partial g}{\partial \varepsilon} \Delta\varepsilon + \frac{\partial g}{\partial \rho} \Delta\rho = 0$$

Therefore

$$4\,\Delta f = \Delta\varepsilon = -\frac{\partial g/\partial \rho}{\partial g/\partial \varepsilon} \Delta\rho = \left[\frac{6 + 2\varepsilon^2}{3\varepsilon^2 - 4\rho\varepsilon - 6(S_0/L - 1)} \right] \frac{L\,\Delta p}{2AE} \quad (2.2.9)$$

The horizontal force is likewise a function of both load and sag ratio. Its Taylor expansion is

$$\overline{H}_0 = H_0 + \Delta H_0 = H_0 + \frac{\partial H}{\partial p} \Delta p + \frac{\partial H}{\partial f} \Delta f \quad (2.2.10)$$

Substituting Eqs. (2.2.8) and (2.2.9) into Eq. (2.2.10), we obtain

$$\overline{H}_0 = H_0 \left(1 + \frac{\Delta p}{p} \left\{ 1 - \frac{H_0}{AE} \left[\frac{6 + 2\varepsilon^2}{3\varepsilon^2 - 4\rho\varepsilon - 6(S_0/L - 1)} \right] \right\} \right) \quad (2.2.11)$$

which is an approximate value of the horizontal force due to a slight increase in transverse uniform load. The corresponding increase in sag ratio is given by Eq. (2.2.9).

Catenary segment. If the load is uniformly distributed along the arc length of the cable segment, rather than over the horizontal span, a different solution is obtained. A natural choice of independent variable is arc length. In a subsequent section that choice will be used for bidirectional loadings. In this section, in order to compare results with those for a parabolic segment, we will again use the horizontal coordinate x as the independent variable. Let q denote the intensity of uniform load per unit length of the segment. Then equilibrium of forces on a differential length, as in Fig. 2.2.1, gives

$$\frac{dH}{dx} = 0 \rightarrow H = H_0 \quad \text{(constant throughout)} \quad (2.2.12a)$$

$$\frac{dV}{dx} = -q\,\frac{ds}{dx} \quad (2.2.12b)$$

Also, since the tension must be directed along the tangent to the arc

$$V = H_0\,\frac{dz}{dx} \quad (2.2.12c)$$

Substitute the derivative of Eq. (2.2.12c) into Eq. (2.2.12b) and use the fact that $ds^2 = dx^2 + dz^2$ to obtain

$$\frac{d^2z}{dx^2} + \frac{q}{H_0}\sqrt{1 + \left(\frac{dz}{dx}\right)^2} = 0 \quad (2.2.12d)$$

which is the classic catenary equation for the deflected profile z of the arc. Once we obtain a solution to Eq. (2.2.12d), Eq. (2.2.12c) can be used to determine V and the tension will be given by

$$T = H_0\sqrt{1 + \left(\frac{dz}{dx}\right)^2} \quad (2.2.12e)$$

Integrating Eq. (2.2.12d) twice and applying the boundary conditions $z = 0$ at $x - 0$ and $z = L\tan\theta$ at $x = L$, we obtain

$$z = \frac{H_0}{q}\left\{\cosh(\gamma + \beta) - \cosh\left[\gamma + \beta\left(1 - 2\frac{x}{L}\right)\right]\right\} \quad (2.2.13)$$

where

$$\beta = \frac{qL}{2H_0} \quad (2.2.14a)$$

$$\gamma = \sinh^{-1}\left(\tan\theta\,\frac{\beta}{\sinh\beta}\right) \quad (2.2.14b)$$

The sag ratio at midspan is

$$f = \frac{1}{2\beta}[\cosh(\gamma + \beta) - \cosh\gamma] - \frac{1}{2}\tan\theta \quad (2.2.14c)$$

Substituting Eqs. (2.2.13) into Eqs. (2.2.12c and e), we obtain the vertical force and tension as

$$V = H_0 \sinh \left[\gamma + \beta \left(1 - 2\frac{x}{L} \right) \right] \tag{2.2.15a}$$

$$T = H_0 \cosh \left[\gamma + \beta \left(1 - 2\frac{x}{L} \right) \right] \tag{2.2.15b}$$

The stretched length to a point x on the horizontal span is obtained from $ds = dx \, [1 + (dz/dx)^2]^{1/2}$ (and $s = 0$ at $x = 0$) as

$$\frac{s}{L} = \frac{V_0 - V}{qL} \tag{2.2.16a}$$

where V_0 is the vertical force at $x = 0$. The total stretched length is

$$\frac{S}{L} = \frac{V_0 - V_1}{qL} \tag{2.2.16b}$$

where V_1 is the vertical force at $x = L$.

The unstretched total length is determined from the differential equation

$$\frac{ds_0}{dx} = \frac{T/H_0}{1 + T/AE} \approx \left(\frac{T}{H_0} \right) \left[1 - \frac{H_0}{AE} \left(\frac{T}{H_0} \right) \right] \tag{2.2.17a}$$

with boundary conditions $s_0 = 0$ at x and $s_0 = S_0$ at $x = L$. Integrating Eq. (2.2.17a), we obtain

$$\frac{S_0}{L} = \frac{V_0 - V_1}{qL} - \frac{qL}{2AE} \left[\frac{H_0}{qL} + \frac{V_0 T_0 - V_1 T_1}{(qL)^2} \right] \tag{2.2.17b}$$

The behavior of a catenary segment is described by Eqs. (2.2.13) to (2.2.17) in terms of an as yet unknown horizontal force H_0 (or $\beta = qL/2H_0$). If the sag ratio f were specified, the nonlinear transcendental equation, Eq. (2.2.14c), could be solved numerically for β and hence H_0. If the total unstretched length S_0 were specified, Eq. (2.2.17b) could be solved numerically for H_0.

Let us write the equations for the special case of $\theta = 0$, i.e., level supports. In that case

$$z = \frac{H_0}{q} \left\{ \cosh \beta - \cosh \left[\beta \left(1 - 2\frac{x}{L} \right) \right] \right\} \tag{2.2.18a}$$

$$f = \frac{1}{2\beta} (\cosh \beta - 1) \tag{2.2.18b}$$

$$V = H_0 \sinh \left[\beta \left(1 - \frac{2x}{L} \right) \right] \tag{2.2.18c}$$

$$T = H_0 \cosh \left[\beta \left(1 - \frac{2x}{L} \right) \right] \tag{2.2.18d}$$

$$\frac{S_0}{L} = \frac{\sinh \beta}{\beta} - \frac{qL}{4AE} \left(1 + \cosh \beta \, \frac{\sinh \beta}{\beta} \right) \tag{2.2.18e}$$

For small sag ratios the solution of the catenary segment is closely approximated by that of a parabolic segment [10, 62, 63]. This can be determined by examination of Eqs. (2.2.13) to (2.2.15) for the case of small β and θ. Using the Taylor expansions of $\sinh \varepsilon$ and $\cosh \varepsilon$

$$\sinh \epsilon \approx \epsilon + \frac{\varepsilon^3}{6}$$

$$\cosh \epsilon \approx 1 + \frac{\varepsilon^2}{2}$$

and recognizing that $\gamma \simeq \tan \theta$, we see that Eq. (2.2.13) becomes

$$z \approx \gamma x + \beta x \left(1 - \frac{x}{L} \right) = x \tan \theta + d$$

which is the same as Eq. (2.2.3d), the solution for a parabolic segment. Likewise, for a catenary with a small sag ratio

$$V = H_0 \left[\gamma + \beta \left(1 - \frac{2x}{L} \right) \right] = H_0 \tan \theta + \frac{qL}{2} \left(1 - 2 \frac{x}{L} \right)$$

$$T = H_0 \left\{ 1 + \frac{1}{2} \left[\gamma + \beta \left(1 - 2 \frac{x}{L} \right) \right]^2 \right\}$$

$$= H_0 \sqrt{ 1 + \left[\tan \theta + \frac{qL}{2H_0} \left(1 - \frac{2x}{L} \right) \right]^2 }$$

$$f = \left(\frac{\gamma}{2} - \frac{1}{2} \tan \theta \right) + \frac{\beta}{4} = \frac{qL}{8H_0}$$

which are the same as Eqs. (2.2.3b and c) and (2.2.4a), respectively, for a parabolic segment. Figure 2.2.3 shows the difference between a parabolic and catenary segment for various sag ratios [10]. The difference is negligible. Therefore, in all practical cases it is sufficient to use the simpler equations for a parabolic segment as presented in Sec. 2.2.1 when the sag ratio is small. For large sag ratios, as may occur in ocean applications, the catenary equations should be used. However, in those instances, the formulation presented in Sec. 2.3.1 should be used in that the alternative

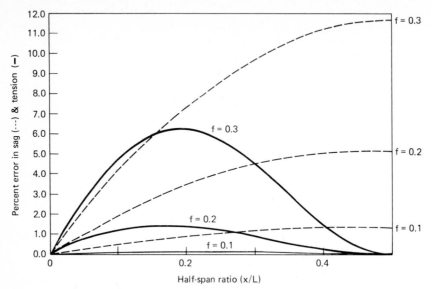

Figure 2.2.3 Difference between sags and tensions for parabolic and catenary arcs with same spans and H_0.

choice of arc length as independent variable and the inclusion of alternate boundary conditions allow for a more general treatment of catenary problems.

Let us consider an example problem in which the span length changes slightly. Shown in Fig. 2.2.4 is a planar model of a guyed tower. An actual tower would have several guys spaced circumferentially. The guys provide lateral support at the top of the tower. Each guy is prestressed until a horizontal component of tension H_0 is obtained, e.g., by tightening the cable until a premarked section of cable is at the support.

First, in the prestressing phase (Fig. 2.2.4a) we find the maximum tension in each guy to be

$$T_{\max} = H_0 \cosh (\gamma + \beta) \approx H_0 \sqrt{1 + \left(\frac{D}{L} + \frac{qL}{2H_0} \right)^2}$$

The compressive force in the column is

$$P_0 = W + 2H_0 \sinh (\gamma + \beta) \approx 2H_0 \tan \theta + qL + W$$

and the initial sags are

$$f_0 = \frac{qL}{8H_0}$$

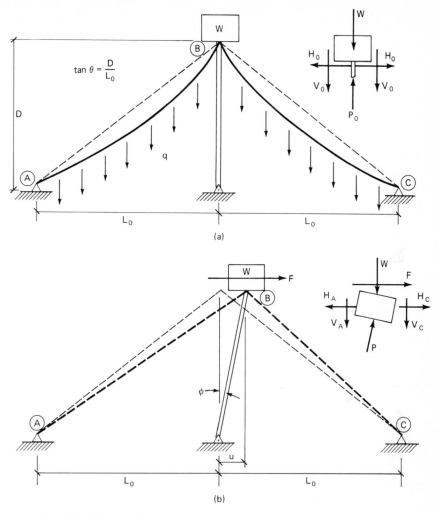

Figure 2.2.4 Model of guyed tower example.

As an example of an in-service load, apply a horizontal load F at the top of the column. This will induce a motion u at the top of the column. For small u neither the height D nor the angle θ will change appreciably. However, the horizontal forces, and hence the sag ratios, must change to maintain equilibrium. Since the new horizontal forces, H_A and H_C, are nonlinear functions of $\Delta L = \pm u$ (the span lengths of each cable have changed), let us find a linearized stiffness coefficient for the system. The horizontal force is given by Eq. (2.2.19) for taut cables:

$$H = \frac{qL}{8f} \qquad H_0 = \frac{qL_0}{8f_0}$$

Hence

$$H = H_0 + \Delta H = H_0 + \frac{\partial H_0}{\partial L_0} \Delta L + \frac{\partial H_0}{\partial f_0} \Delta f$$

$$= H_0 \left(1 + \frac{\Delta L}{L_0} - \frac{\Delta f}{f_0} \right) \qquad (2.2.20a)$$

We need another relationship between ΔL and Δf. Since the unstressed lengths remain constant, we can use the equation for the unstressed length of a parabolic segment as a function of L and f to obtain that relationship:

$$g_0(L_0, f_0) = S_0 - L_0 \left[\left(1 + \frac{1}{2} \tan^2 \theta + \frac{8}{3} f_0^2 \right) \right.$$

$$\left. - \left(\frac{qL_0}{2AE} \right) \left(\frac{1}{4f_0} \right) \left(1 + \tan^2 \theta + \frac{16}{3} f_0^2 \right) \right] = 0 \quad (2.2.20b)$$

By Taylor expansion

$$g(L, f) = g_0(L_0, f_0) + \frac{\partial g_0}{\partial f_0} \Delta f + \frac{\partial g_0}{\partial L_0} \Delta L = 0$$

Hence

$$\Delta f = \frac{f_0(S_0/L - \tan^2 \theta) - qL_0/8AE(1 + \tan^2 \theta + {}^{16\!/\!3} f_0^2)}{(S_0/L_0 - 1) - \frac{1}{2} \tan^2 \theta - 8f_0^2 + qL_0/2AE({}^{8\!/\!3} f_0)} \frac{\Delta l}{L_0} \qquad (2.2.21)$$

Now substituting Eq. (2.2.21) into Eq. (2.2.20a), we obtain

$$H = H_0 \left(1 + K_0 \frac{\Delta L}{L_0} \right) \qquad (2.2.22a)$$

where

$$K_0 = 1 + \frac{\tan^2 \theta - S_0/L + (qL_0/8AE)(1/f_0)(1 + \tan^2 \theta + {}^{16\!/\!3} f_0^2)}{(S_0/L_0 - 1) - \frac{1}{2} \tan^2 \theta - 8f_0^2 + qL_0/2AE ({}^{8\!/\!3} f_0)}$$

$$(2.2.22b)$$

Consider equilibrium of the free body of the top of the tower. Although the vertical components of tension in each guy change, the column force remains unchanged if only small motions are allowed. This can be seen by taking Taylor expansions of V_A and V_C and noting that $\Delta L = \pm u$ in the two guys. Horizontal equilibrium requires

$$F = H_A - H_C - \frac{P_0}{\tan \theta} \frac{u}{L_0}$$

where it has been assumed that $\sin \phi = u/D$. The forces H_A and H_C are given by Eq. (2.2.22) as

$$H_A = H_0 \left(1 + K_0 \frac{u}{L_0} \right) \qquad H_C = H_0 \left(1 - K_0 \frac{u}{L_0} \right)$$

and therefore,

$$F = \left(2K_0 H_0 - \frac{P_0}{\tan \theta} \right) \frac{u}{L_0}$$

yields the desired stiffness relationship between F and u.

2.2.2 Combine concentrated and distributed loads

A description of cable segment behavior under concentrated loads can now be developed that is more accurate than that provided in Sec. 2.1 where distributed load effects were neglected. First, we will consider a single concentrated load and also present results for the case of an additional in-service load uniformly distributed over a portion of the span. Then a numerical procedure will be developed to treat multiple concentrated loads.

Single concentrated load. Figure 2.2.5 shows an in-service concentrated load of magnitude P superimposed upon the prestressed state due to a uniform load with intensity q. Each of the sections can be treated as a parabolic arc. The moment analogy can be used to calculate the displacement of the cable below each of the chord sections connecting points 0, 1, and 2. The simple beam moment due to the distributed load is

$$m_0 = \frac{qx(L - x)}{2} \qquad 0 < x < L$$

The simple beam moments due to the concentrated load are

$$m_1 = P(1 - \alpha)x \qquad 0 < x < \alpha L$$

$$m_2 = P\alpha(L - x) \qquad \alpha L < x < L$$

First, determine the solution to the prestressing phase. For a specified sag ratio f_0, Eq. (2.2.4) gives

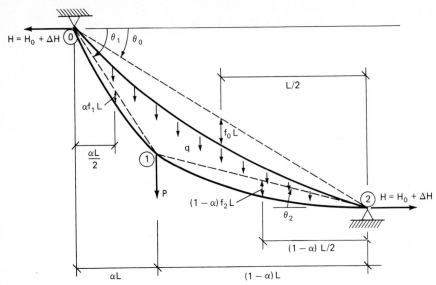

Figure 2.2.5 Catenary with concentrated load.

$$H_0 = \frac{qL}{8f_0} \tag{2.2.23a}$$

$$z_0 = x \tan \theta_0 + \frac{m_0}{H_0} \tag{2.2.23b}$$

$$T_0 = H_0 \left\{ 1 + \left[\tan \theta_0 + 4f_0 \left(1 - \frac{2x}{L} \right) \right]^2 \right\}^{1/2} \tag{2.2.23c}$$

and the unstressed length is, by Eq. (2.2.20b),

$$S_0 = L \left[\left(1 + \frac{1}{2} \tan^2 \theta_0 + \frac{8}{3} f_1^2 \right) \right.$$

$$\left. - \frac{H_0}{AE} \left(1 + \tan^2 \theta_0 + \frac{16}{3} f_0^2 \right) \right] \tag{2.2.23d}$$

During the in-service phase after the load P is added, the solution in section 1 $(0 < x < \alpha L)$ is

$$z_1 = x \tan \theta_0 + \frac{m_0 + m_1}{H} \tag{2.2.24a}$$

with sag ratio

$$f_1 = \frac{[(\overline{m}_1 + \overline{m}_0)/H + \alpha L/2 (\tan \theta_0 - \tan \theta_1)]}{\alpha L}$$

in which

$$\tan \theta_1 = \tan \theta_0 + \frac{(\overline{m}_1 + \overline{m}_0)/\alpha L}{H}$$

$$= \tan \theta_0 + (1 - \alpha) \frac{Q}{2H} \qquad (2.2.24b)$$

where we have defined $Q = qL + P$, $\overline{m}_1$ and $\overline{m}_0$ are m_1 and m_0 evaluated at $x = \alpha L/2$, and $\overline{\overline{m}}_1$ and $\overline{\overline{m}}_0$ are m_1 and m_0 evaluated at $x = \alpha L$. Thus

$$f_1 = \alpha f_0 \left(\frac{H_0}{H} \right) \qquad (2.2.24c)$$

and the unstressed length of section 1 is given by

$$S_1 = \alpha L \left[\left(1 + \frac{1}{2} \tan^2 \theta_1 + \frac{8}{3} f_1^2 \right) \right.$$

$$\left. - \frac{H}{AE} \left(1 + \tan^2 \theta_1 + \frac{16}{3} f_1^2 \right) \right] \qquad (2.2.24d)$$

Similarly, for section 2 ($\alpha L < x < L$)

$$z_2 = x \tan \theta_0 + \frac{m_0 + m_2}{H} \qquad (2.2.25a)$$

$$\tan \theta_2 = \tan \theta_0 - (\alpha) \frac{Q}{2H} \qquad (2.2.25b)$$

$$f_2 = (1 - \alpha) f_0 \frac{H_0}{H} \qquad (2.2.25c)$$

$$S_2 = (1 - \alpha) L \left[\left(1 + \frac{1}{2} \tan^2 \theta_2 + \frac{8}{3} f_2^2 \right) \right.$$

$$\left. - \frac{H}{AE} \left(1 + \tan^2 \theta_2 + \frac{16}{3} f_2^2 \right) \right] \qquad (2.2.25d)$$

The solution is not complete until we specify H in Eqs. (2.2.24) and (2.2.25). To do that we impose the condition $S_0 = S_1 + S_2$. Combining Eqs. (2.2.23d), (2.2.24d), and (2.2.25d), we obtain

$$\frac{8}{3} f_0^2 - \frac{H_0}{AE} \left(1 + \tan^2 \theta_0 + \frac{16}{3} f_0^2 \right)$$

$$= \left(\frac{8}{3} f_0^2 \right) \left(\frac{H_0}{H} \right)^2 \left[(1 - 3\alpha) \left(1 - \frac{Q^2}{q^2 L^2} \right) \right]$$

$$- \frac{H_0}{AE} \frac{H}{H_0} \left\{ 1 + \tan^2 \theta_0 + \frac{16}{3} f_0^2 \left(\frac{H_0}{H} \right)^2 \left[(1 - 3\alpha)(1 - \alpha) \left(1 - \frac{Q^2}{q^2 L^2} \right) \right] \right\}$$

$$(2.2.26)$$

which, for an extensible cable, constitutes a cubic equation for H. For the inextensible cable, the solution to Eq. (2.2.26) is

$$H = H_0 \left\{ 1 - 3\alpha (1 - \alpha) \left[1 - \left(\frac{Q}{qL} \right)^2 \right] \right\}^{1/2} \qquad (2.2.27)$$

which is the essential factor for the complete determination of the behavior of a cable segment with a concentrated load at an arbitrary point on the span.

Partial uniform load superposed.   Figure 2.2.6 shows an in-service uniform load of intensity q_1 superimposed over a central portion of the span. This would model the intermediate stage in the erection of a suspension bridge. This problem can be treated as three parabolic arcs, and symmetry properties can be used to great advantage. Again, the moment analogy can be used to specify the deflection of the cable:

$$z_1 = \frac{q_0 L^2}{2H} \left(\frac{x}{L} \right) \left[\left(1 - \frac{x}{L} \right) + \beta (1 - 2\alpha) \right] \qquad x < \alpha L$$

$$z_2 = \frac{q_0 L^2}{2H} \left[\left(\frac{x}{L} \right) (1 + \beta) \left(1 - \frac{x}{L} \right) - \beta \alpha^2 \right] \qquad \alpha L < x < \frac{L}{2}$$

where $\beta = q_1/q_0$. The horizontal component H can be determined in terms of the prestressed H_0 or f_0 by the condition that the unstressed lengths of the three segments sum to equal the total unstressed length, i.e., $S_0 = S_1 + S_2 + S_3$. This gives

$$H = H_0[(1 + \beta)^2 - 4\alpha^2 \beta^2(3 - 4\alpha) - 4\alpha^2 \beta(3 - 2\alpha)]^{1/2}$$

for an inextensible cable. The central deflection w_0 is

$$\left(\frac{w_0}{L} \right) = \left(\frac{z_2}{L} - \frac{z_0}{L} \right) \Big|_{x=L/2} = \frac{z_2}{L} \Big|_{x=L/2} - f_0$$

$$= f \left\{ - \frac{(1 + \beta) - 4\alpha^2 \beta}{[(1 + \beta)^2 - 4\alpha^2 \beta^2(3 - 4\alpha) - 4\alpha^2 \beta(3 - 2\alpha)]^{1/2}} \right\}$$

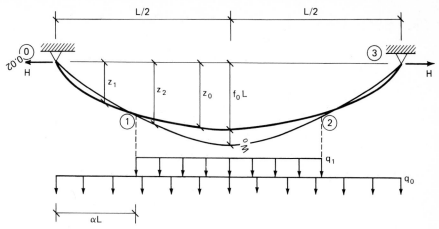

Figure 2.2.6 Two symmetric uniform loads.

Multiple concentrated loads. If, in addition to a distributed prestressing load, several vertical and horizontal concentrated loads, and possibly several different intensities of distributed in-service loads, are acting at various locations on the cable segment, a numerical treatment of the problem can be developed in an analogous fashion to that developed in Sec. 2.1.2 for multiple concentrated loads with distributed load effects neglected. In this instance, in addition to distributed load effects which may vary from section to section and in addition to horizontal loads, we will also include the possibility of the cross-sectional area and modulus of elasticity varying from section to section.

Let us divide the cable segment into several elements as shown in Fig. 2.2.7. Within each element the uniform load, area, and modulus of elas-

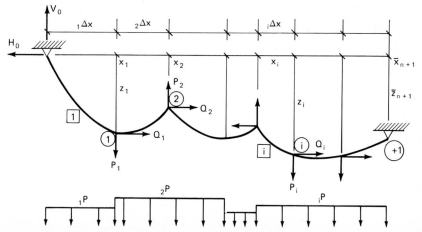

Figure 2.2.7 Multiple concentrated and distributed loads.

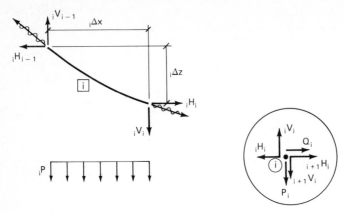

Figure 2.2.8 Free-body diagrams of element and node.

ticity will be assumed invariable. A new element will begin at each discontinuity of load, area, or modulus of elasticity. Although the overall sag ratio for the cable segment may be large, we will restrict our attention to those cases in which the sag ratios of the individual elements are small. Thus, the parabolic approximation to a catenary is adopted as described under "Single Concentrated Load." A similar treatment to that described here can be developed for catenary segments in which large sag ratios are admitted. We will also limit our attention to cases in which the horizontal spans of the elements are specified a priori, rather than cases in which the unstressed lengths of the elements are specified. Again, an alternate treatment for those cases could be developed in an analogous manner.

Consider the definition sketch shown in Fig. 2.2.8. A leading subscript denotes an element number and a following subscript denotes a nodal number where adjoining elements are connected. Given the lengths $_i\Delta x$ we seek values of components of tension $_iH_{i-1}$ and $_iV_{i-1}$ at the beginning of an element; components of tension $_iH_i$ and $_iV_i$ at the end of an element; and unstressed length $_i\Delta S_0$ and relative displacement $_i\Delta z$ for an element. With those values, the behavior within an element is describable using the equations under "Single Concentrated Load."

From the statics of the typical element shown in Fig. 2.2.8, Eqs. (2.2.1) give

$$_iH_i = {}_iH_{i-1} \tag{2.2.28a}$$

$$_iV_i = {}_iV_{i-1} - {}_ip\ {}_i\Delta X \tag{2.2.28b}$$

$$_iT_i = \sqrt{{}_iH_i^2 + {}_iV_i^2} \tag{2.2.28c}$$

$$iT_{i-1} = \sqrt{{}_iH_{i-1}^2 + {}_iV_{i-1}^2} \tag{2.2.28d}$$

The deflected profile $_i\Delta z$ relative to the beginning of the element and the unstressed length of the element are given by Eqs. (2.2.2a) and (2.2.6a) as

$$_i\Delta z = \frac{_i\Delta x}{_iH_i}\left(_iV_i + {_ip}\frac{_i\Delta x}{2}\right) \tag{2.2.29a}$$

$$_i\Delta s_0 = \left\{\frac{_iH_i}{2_ip}\left[\frac{_iV_{i-1}\,_iT_{i-1} - _iV_{ii}T_i}{_iH_i^2} + \log\left(\frac{_iV_{i-1} + T_{i-1}}{_iV_i + _iT_i}\right)\right]\right.$$

$$\left. - \frac{_iH_i\,_i\Delta x}{_iA_iE}\left[1 + \frac{_iV_{i-1}^2 + _iV_{i-1i}V_i + _iV_i^2}{3_iH_i^2}\right]\right\} \tag{2.2.9b}$$

From consideration of statics of the node i shown in Fig. 2.2.8,

$$_{i+1}H_i = {_iH_i} - Q_i \tag{2.2.30a}$$

$$_{i+1}V_i = {_iV_i} - P_i \tag{2.2.30b}$$

Also, the cumulative profile z_i and unstressed length s_{0i} to node i are

$$z_i = \sum_{j=1}^{i-1} {_j\Delta z} \tag{2.2.31a}$$

$$s_{0i} = \sum_{j=1}^{i-1} {_j\Delta s_0} \tag{2.2.31b}$$

Thus, if we knew $H_0 = {_1H_0}$ and $V_0 = {_1V_0}$, Eqs. (2.2.28) and (2.2.30) would give us components of tension in every element by recursive calculations. Furthermore, Eqs. (2.2.29) and (2.2.31) would give us the locations of, and the unstressed lengths to, each node once the components of tensions were calculated. Unfortunately, $_1H_0$ and $_1V_0$ are unknown. If we assumed incorrect values, the recursive calculations of z_i and s_{0i} from $i = 0$ to $n + 1$ would lead us to incorrect values of z_{n+1} and $S_{0,n+1}$ at the terminus of the segment. We can measure the errors in closure at that point, $(\bar{z}_{n+1} - z_{n+1})$ and $(\bar{S}_0 - S_{0,n+1})$.

An iterative procedure is indicated by Eqs. (2.2.28) to (2.2.31):

1. Assume $_1H_0$ and $_1V_0$ (prestressing values).
2. Calculate $_iH_i$, $_iV_i$; $_iH_{i+1}$, $_iV_{i+1}$ by Eqs. (2.2.28) and (2.2.30).
3. Calculate z_i, s_{0i} by Eqs. (2.2.29) and (2.2.31).
4. Calculate the closure errors $(\bar{z}_{n+1} - z_{n+1})$ and $(\bar{S}_0 - S_{0,n+1})$.
5. If the closure errors are sufficiently small, terminate the iteration; if not, assume new values of $_1H_0$, $_1V_0$ and repeat from step 2.

A rational procedure is needed in step 5 to obtain improved estimates of H_0 and V_0. The trial-and-error approach described previously could be used, but an improved method is available. The Newton-Raphson [61, 73, 72] method is a convergence acceleration method in which successive iterates are based on gradients of the closure errors.

Given a set of N nonlinear equations

$$\{g_k(y_i)\} = \{0\} \qquad k = 1, N; i = 1, N \qquad (2.2.32)$$

and an initial estimate $\{y_i^*\}$ to $\{y_i\}$, Taylor expansion of Eq. (2.2.32) gives

$$\{g_k\} = \{g_k^*\} + \frac{\partial\{g_k^*\}}{\partial\{y_i^*\}}(\{y_i\} - \{y_i^*\}) \qquad (2.2.33)$$

where $\{g_k^*\}$ and $[J_{ki}^*] = \partial\{g_k^*\}/\partial\{y_i^*\}$ are evaluated at $\{y_k^*\}$. Thus, setting $\{g_k\} = 0$ in Eq. (2.2.33), we obtain an improved estimate $\{y_i^*\} + \{\Delta y_i\}$ from

$$\{\Delta y_i\} = -[J_{ki}^*]^{-1}\{g_k^*\} \qquad (2.2.34)$$

where $[J_{ki}^*]^{-1}$ is the inverse of $[J_{ki}^*]$. Successive approximations are obtained by calculating corrections $\{\Delta y_i\}$ by Eq. (2.2.34) and setting

$$\{y_i^*\} = \{y_i^*\} + \{\Delta y_i\} \qquad (2.2.35)$$

This method can be used for any number of nonlinear equations for which the gradients $[J_{ki}^*]$ can be evaluated analytically. In fact, for problems in earlier sections in which roots to nonlinear transcendental equations were sought, the Newton-Raphson method can be fruitfully employed.

In the present instance we have two closure error equations

$$\{g\} = \left\{ \begin{array}{c} z_{n+1} - \bar{z}_{n+1} \\ S_{n+1} + \bar{S}_0 \end{array} \right\}$$

which are nonlinear functions of H_0 and V_0. The matrix $[J]$ is given by

$$[J] = \begin{bmatrix} J_{11} & J_{12} \\ J_{21} & J_{22} \end{bmatrix} = \begin{bmatrix} \dfrac{\partial z_{n+1}}{\partial H_0} & \dfrac{\partial z_{n+1}}{\partial V_0} \\ \dfrac{\partial S_{0,n+1}}{\partial H_0} & \dfrac{\partial S_{0,n+1}}{\partial V_0} \end{bmatrix}$$

Noting that $\partial_i H_k/\partial H_0 = 1$, then $\partial_i V_k/\partial V_0 = 1$, and taking the indicated partial derivatives of Eq. (2.2.29), we obtain

$$J_{11} = -\sum_{j=1}^{n}\left(\frac{_j\Delta z}{_j H_j}\right)$$

$$J_{12} = + \sum_{j=1}^{n} \left(\frac{{}_j\Delta x}{{}_jH_j} \right)$$

$$J_{21} = - \sum_{j=1}^{n} \left(\frac{{}_j\Delta S}{{}_jH_j} + \frac{2{}_j\Delta x}{{}_jA_jE} - \frac{1}{{}_jp} \left\{ \log \left(\frac{{}_jV_{j-1} + {}_jT_{j-1}}{{}_jV_j + {}_jT_j} \right) \right. \right.$$

$$\left. \left. + {}_jH_j^2 \left[\frac{{}_jT_j \,({}_jV_j + {}_jT_j) - {}_jT_{j-1}({}_jV_{j-1} + {}_jT_{j-1})}{2{}_jT_{jj}T_{j-1}({}_jV_j + {}_jT_j)({}_jV_{j-1} - {}_jT_{j-1})} \right] \right\} \right)$$

$$J_{22} = + \sum_{j=1}^{n} \left\{ \frac{1}{{}_jp} \left[\frac{({}_jT_{j-1} - {}_jT_j)({}_jT_{jj}T_{j-1} - {}_jH_j^2)}{{}_jH_{jj}T_{jj}T_{j-1}} \right] \right.$$

$$\left. - \frac{{}_j\Delta x}{{}_jA_jE} \, \frac{{}_jV_j + {}_jV_{j-1}}{{}_jH_j} \right\}$$

Thus, improved estimates of H_0 and V_0 for step 5 in the iterative process are given by

$$\begin{Bmatrix} H_0 \\ V_0 \end{Bmatrix} = \begin{Bmatrix} H_0 \\ V_0 \end{Bmatrix} + \begin{Bmatrix} \Delta H \\ \Delta V \end{Bmatrix}$$

with

$$\begin{Bmatrix} \Delta H \\ \Delta V \end{Bmatrix} = + \begin{bmatrix} J_{11} & J_{12} \\ J_{21} & J_{22} \end{bmatrix}^{-1} \begin{Bmatrix} \bar{z} - z_{n+1} \\ \bar{S} - S_{0,n+1} \end{Bmatrix}$$

The iteration may be terminated when closure errors $\begin{Bmatrix} \bar{z} - z_{n+1} \\ \bar{S}_0 - S_{0,n+1} \end{Bmatrix}$ are sufficiently small.

2.3   Three-Dimensional Behavior of Cable Segments

Because of the predominance of dead weight as a loading on flexible cable segments, the behavior of most cables tends to be planar and to have singly directed loads. However, in some land-based applications, e.g., cross winds on transmission lines, and in many sea-based applications, e.g., current and towing drag on mooring lines, the three-dimensional behavior of cable segments should be considered. Three-dimensional effects of concentrated loads with distributed loads neglected can be treated by a direct extension of the numerical method presented in Sec. 2.1.2 and will not be treated here.

In this section we will consider a curvilinear cable segment arbitrarily oriented in a three-dimensional cartesian space and subjected to nonuni-

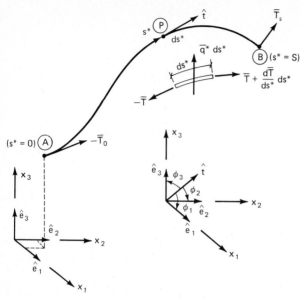

Figure 2.3.1   Three-dimensional curved cable segment.

form distributed loads in all three directions. An analytical solution to the special case of uniform loads will be generated and illustrated by several two-dimensional bidirectional loading cases, including the classic catenary solution with arc length as an independent coordinate. A numerical procedure will then be presented for instances in which the loads are non-uniformly distributed along the length of the segment. The possibility of nonlinear interaction of the load with the cable response will be included.

Figure 2.3.1 shows a cable segment spanning two points A and B located in a three-dimensional cartesian space x_i, $i = 1, 2, 3$, with unit base vectors $\hat{e}_i$. We will adopt unstretched arc length s to the material point P as the independent variable. The stretched length s^* to the material point P is related to the unstretched length by the strain-displacement relation

$$\frac{(ds^*)^2 - (ds)^2}{2(ds)^2} = \varepsilon \tag{2.3.1}$$

where ε is the nonlinear strain ($\varepsilon = 0$ for inextensible materials). Hence

$$\frac{ds^*}{ds} = \sqrt{1 + 2\varepsilon} = 1 + \varepsilon + \frac{1}{2}\varepsilon^2 = 1 + \gamma \tag{2.3.2a}$$

where γ is engineering strain $[= (ds^* - ds)/ds]$, which is related to the magnitude of tension in a hookean material by $\gamma = T/AE$. We will assume small strains. So

$$\frac{ds^*}{ds} = 1 + \frac{T}{EA} + \cdots \qquad (2.3.2b)$$

will govern.

Equilibrium of the stretched differential length in Fig. 2.3.1 requires

$$\frac{d\overline{T}}{ds^*} + \overline{q}^* = 0 \qquad (2.3.3)$$

where $\overline{T}$ is the tension vector, given in terms of its cartesian components T_i by

$$\overline{T} = T_i \hat{e}_i \qquad (2.3.4)$$

Note that the summation convention over the range 1, 2, 3 has been adopted for repeated subscripts. The vector $\overline{q}^*$ is the distributed load per unit length of stretched cable. It is more convenient to define a distributed load vector $\overline{q}$ per unit length of unstressed cable, so that

$$\overline{q} \, ds = \overline{q}^* \, ds^* \qquad (2.3.5a)$$

$$\overline{q} = q_i \hat{e}_i \qquad (2.3.5b)$$

Therefore, changing variables in Eq. (2.3.3) and casting into component form, we have

$$\frac{dT_i}{ds} + q_i = 0 \qquad (2.3.6)$$

The magnitudes of $\overline{T}$ and $\overline{q}$ are defined as

$$T = \{T_i T_i\}^{1/2} \qquad (2.3.7a)$$

$$Q = \{q_i q_i\}^{1/2} \qquad (2.3.7b)$$

If q_i are not functions of x_i (as yet unknown), Eq. (2.3.6) can be integrated directly:

$$T_i = T_{i0} + \int q_i \, ds \qquad (2.3.8)$$

where T_{i0} are constants of integration with physical significance of components of tension at $s = 0$:

$$T_0 = \{T_{i0} T_{i0}\}^{1/2} \qquad (2.3.9)$$

To determine x_i and locate the deflected material point P, we write the directed differential length

$$d\bar{s}^* = ds^* \hat{t} = dx_i \hat{e}_i \qquad (2.3.10a)$$

where $\hat{t}$, is the unit tangent vector at point P:

$$\hat{t} = \theta_i \hat{e}_i \qquad (2.3.10b)$$

where the θ_i are direction cosines of $\hat{t}$. Therefore, by Eqs. (2.3.10) and (2.3.2b)

$$\frac{dx_i}{ds} = \frac{dx_i}{ds^*}\frac{ds^*}{ds} = \theta_i \frac{ds^*}{ds} = \theta_i \left(1 + \frac{T}{EA}\right) \qquad (2.3.11)$$

Because the tension vector is directed along the unit tangent vector, we can relate θ_i to T_i by

$$\overline{T} = T_i \hat{e}_i = T\hat{t} = T\theta_i \hat{e}_i$$

Therefore,

$$\theta_i = \frac{T_i}{T} \qquad (2.3.12)$$

Now, Eq. (2.3.11) can be rewritten as

$$\frac{dx_i}{ds} = \frac{T_i}{T}\left(1 + \frac{T}{EA}\right) \qquad (2.3.13)$$

Equations (2.3.2b), (2.3.6), and (2.3.13) constitute seven first-order non-linear differential equations for the stretched length, three tension components, and three coordinates at point P. Once the T_i are determined, the orientation at point P is specified by Eq. (2.3.12). The equations are nonlinear because of their dependence on the tension magnitude $T = \{T_i T_i\}^{1/2}$ and because the distributed loads q_i could be functions of x_i, θ_i, or both.

2.3.1 Uniform loads

Let us assume that the q_i are uniformly distributed along the cable length. This is a reasonable assumption for self-weight and translational inertial forces, and is a crude approximation of drag forces in a fluid medium with uniform flow, e.g., wind drag. With q_i uniform, Eqs. (2.3.6) and (2.3.13) can be integrated to obtain

$$T_i = T_{0i} - q_i s \qquad (2.3.14a)$$

$$x_i = x_{i0} + \frac{s}{AE}\left(T_{0i} - \frac{q_i s}{2}\right) + \frac{1}{Q^3}\left[T_{0m}(Q^2\,\delta_{im}\right.$$

$$\left. - q_i q_m\right)\log\left(\frac{QT - q_j T_j}{QT_0 - q_j T_{0j}}\right) - q_i Q(T - T_0)\right] \quad (2.3.14b)$$

where δ_{im} is the Kronecker delta; $\delta_{im} = 1$ if $i = m$, and is otherwise 0. The factor Q is the magnitude of $\bar{q}$:

$$Q = \{q_i q_i\}^{1/2}$$

The constants of integration T_{0i} and x_{0i} have the physical significance of components of tension and the coordinates of point A where $s = 0$.

$$T_0 = \{T_{0i} T_{0i}\}^{1/2}$$

The stretched length is obtained by substituting Eq. (2.3.14a) into Eq. (2.3.2b); integrating and applying the boundary condition $s^* = 0$ at $s = 0$:

$$s^* = s\left(1 + \frac{T}{2AE}\right) + \frac{1}{2AEQ^3}\left[QT_{0i}q_i(T_0 - T)\right.$$

$$\left. + (Q^2 T_0^2 - T_{0i}q_i T_{0j}q_j)\log\left(\frac{QT - T_{0k}q_k + Q^2 s}{QT_0 - T_{0m}q_m}\right)\right] \quad (2.3.14c)$$

The constants T_{0i} and x_{0i} remain to be determined. Typically the origin is placed at $s = 0$ so that $x_{0i} = 0$. Then conditions at the other end are used to determine T_{0i}. If forces are specified at both ends, they must be such that overall equilibrium is maintained, i.e., $T_{Bi} = T_{0i} - q_i S$ where S is the total length A to B, and the shape x_i is completely specified to within a rigid body motion x_{i0}, i.e., set $x_{i0} = 0$ in Eq. (2.3.14b).

Catenary equation.   If we set $q_1 = q_2 = 0$ and $q_3 = q$ in Eqs. (2.3.14), we obtain ($Q = q$). Figure 2.3.2 showed a catenary segment.

$$T_1 = T_{01} \quad\quad\quad\quad\quad\quad\quad\quad\quad\quad\quad\quad\quad (2.3.15a)$$

$$T_2 = T_{02} \quad\quad\quad\quad\quad\quad\quad\quad\quad\quad\quad\quad\quad (2.3.15b)$$

$$T_3 = T_{03} - qs \quad\quad\quad\quad\quad\quad\quad\quad\quad\quad (2.3.15c)$$

$$x_1 = T_{01}\left[\frac{s}{AE} + \frac{1}{q}\log\left(\frac{T - T_3}{T_0 - T_{03}}\right)\right] \quad\quad (2.3.15d)$$

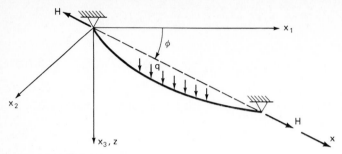

Figure 2.3.2   Catenary segment.

$$x_2 = T_{02} \left[\frac{s}{AE} + \frac{1}{q} \log \left(\frac{T - T_3}{T_0 - T_{03}} \right) \right] \tag{2.3.15e}$$

$$x_3 = \frac{s}{AE} \left(T_{03} - \frac{qs}{2} \right) - \frac{T - T_0}{q} \tag{2.3.15f}$$

These equations correspond to catenary behavior in a plane oriented at an angle $\phi = \tan^{-1}(T_{02}/T_{01})$. Let $z = x_3$, and define

$$H = \sqrt{T_{01}^2 + T_{02}^2} = \sqrt{T_1^2 + T_2^2}$$

$$V = T_{03} - qs$$

$$x = \sqrt{x_1^2 + x_2^2}$$

Then, Eqs. (2.3.15) become

$$T = \sqrt{H^2 + V^2} \tag{2.3.16a}$$

$$V = V_0 - qs \tag{2.3.16b}$$

$$x = \frac{Hs}{AE} + \frac{H}{q} \log \left(\frac{T - V}{T_0 - V_0} \right) \tag{2.3.16c}$$

$$z = \frac{s}{AE} \left(V_0 - \frac{qs}{2} \right) - \frac{T - T_0}{q} \tag{2.3.16d}$$

which are the equations of a catenary using arc length as the independent variable.

Catenary problems with length as variable.   In many ocean applications there are instances when a portion of the catenary is intended to lie along the seabed in the prestressed condition. Under in-service conditions the

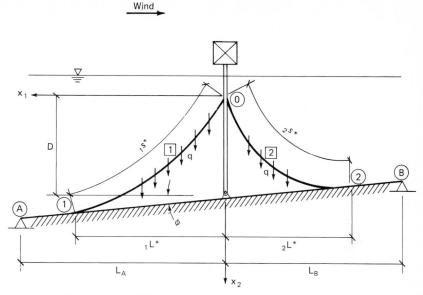

Figure 2.3.3 Guyed tower with slack guys.

catenary segment lifts off the bottom and provides additional restoring force to the system. For example, consider the model of a guyed tower shown in Fig. 2.3.3. The catenary guys sag to points 1 and 2 and then run along the bottom to pile caps at A and B. When the platform is subjected to lateral forces, e.g., wind, section 1 increases in length while section 2 decreases in length. Restoring forces are provided by the increase in horizontal tension component in section 1 over that in section 2. Let us allow for a sloping bottom and place the origin at point 0. We will consider only section 1 here. Results for section 2 can be developed in a similar fashion.

Equations (2.3.14) can be used with

$$Q = q \qquad q_1 = q_3 = 0 \qquad q_2 = q \qquad x_{i0} = 0$$

where q is the buoyant weight per unit length of unstretched cable. Then, at point s along the cable

$$T_1 = T_{01} \tag{2.3.17a}$$

$$T_2 = T_{02} - qs \tag{2.3.17b}$$

$$x_1 = \frac{T_{01}s}{AE} + \frac{T_{01}}{q} \log\left(\frac{T - T_{02} + qs}{T_0 - T_{02}}\right) \tag{2.3.17c}$$

$$x_2 = \left(T_{02} - \frac{qs}{2}\right)\frac{s}{AE} - \frac{T - T_0}{q} \tag{2.3.17d}$$

$$s^* = s\left(1 + \frac{T}{2AE}\right) + \frac{1}{2AE}\left[\frac{T_{02}}{q}(T_0 - T) + \frac{T_{01}^2}{q}\log\left(\frac{T - T_{02} + qs}{T_0 - T_{02}}\right)\right]$$

$$(2.3.17e)$$

The constants T_{01} and T_{02} can be evaluated from conditions at the point where contact with the bottom occurs. At that point the cable is tangent to the bottom and we form the ratio

$$\frac{T_2}{T_1} = \tan \Phi \rightarrow \frac{T_{02}}{T_{01}} = \tan \Phi + \frac{qS}{T_{01}}$$

where S = unknown unstretched length from 0 to 1. Also, the coordinates of the contact point are, using Eqs. (2.3.17c and d) evaluated at $s = S$,

$$_1L^* = \frac{T_{01}}{q}\left\{\frac{qS}{AE}\right.$$

$$+ \log\left[\frac{\sqrt{1 + \tan^2 \Phi} - \tan \Phi}{\sqrt{1 + \left(\frac{qS}{T_{01}} + \tan \Phi\right)^2} - \left(\frac{qS}{T_{01}} + \tan \Phi\right)}\right]\right\}$$

$$(2.3.18a)$$

$$D + {}_1L^* \tan \Phi = \frac{T_{01}}{q}\left\{\left(\tan \Phi + \frac{qS}{2T_{01}}\right)\frac{qS}{AE}\right.$$

$$+ \left[\sqrt{1 + \left(\frac{qS}{T_{01}} + \tan \Phi\right)^2} - \sqrt{1 + \tan^2 \Phi}\right]\right\} \qquad (2.3.18b)$$

and the stretched length from 0 to 1 is

$$_1S^* = \frac{T_{01}}{q}\left(\left(\frac{qS}{T_{01}} - \frac{qS}{AE}\sqrt{1 + \tan^2 \Phi}\right)\right.$$

$$+ \frac{T_{01}}{2AE}\left\{\left(\tan \Phi + \frac{qS}{T_{01}}\right)\left[\sqrt{1 + \left(\frac{qS}{T_{01}} + \tan \Phi\right)^2}\right.\right.$$

$$- \sqrt{1 + \tan^2 \Phi} \Bigg]$$

$$+ \log \left[\frac{\sqrt{1 + \tan^2 \Phi} - \tan \Phi}{\sqrt{1 + \left(\dfrac{qS}{T_{01}} + \tan \Phi \right)^2} - \left(\dfrac{qS}{T_{01}} + \tan \Phi \right)} \right] \Bigg] \Bigg\} \Bigg)$$

$$(2.3.18c)$$

From Eqs. (2.3.18a and c) we can find the total stretched length of cable from 0 to the pile cap A as

$$S_A^* = S^* + \frac{L_A - {}_1L^*}{\cos \Phi} \left(1 - \frac{q}{AE} \frac{T_{01}}{q} \sqrt{1 + \tan^2 \Phi} \right) \quad (2.3.18d)$$

where we have assumed that the tension is constant from 1 to A and equals

$$T_{01} \sqrt{1 + \tan^2 \Phi}$$

In Eqs. (2.3.18a, b, and c), there are three unknown quantities, S, ${}_1L^*$, and T_{01}, if q, L_A, S_A, and D are specified. Thus a numerical procedure or an inverse method would have to be used since those equations are non-linear. The term ${}_1L^*$ could be eliminated from Eqs. (2.3.18b and d) by the use of Eq. (2.3.18a), but that would still leave two nonlinear equations for T_{01} and S.

Let us examine the special case of $\Phi = 0$, i.e., level bottom, and $1/EA = 0$, i.e., *inextensible* chain. Then Eqs. (2.3.18a through d) are considerably simplified to

$$D = \left(\frac{T_{01}}{q} \right) \left[\sqrt{1 + \left(\frac{qS}{T_{01}} \right)^2} - 1 \right] \quad (2.3.19a)$$

$$_1L^* = \left(\frac{T_{01}}{q} \right) \log \left[\frac{1}{\sqrt{1 + (qs/T_{01})^2} - qS/T_{01}} \right] \quad (2.3.19b)$$

$$S = (S_A - L_A) + \left(\frac{T_{01}}{q} \right) \log \left[\frac{1}{\sqrt{1 + (qS/T_{01})^2} - qS/T_{01}} \right] \quad (2.3.19c)$$

Define nondimensional variables

$$y = \frac{S}{D} \qquad \kappa = \left(\frac{S_A - L_A}{D} \right) \qquad z = \frac{T_{01}}{qD}$$

Then Eqs. (2.3.19a and c) become

$$1 = \sqrt{z^2 + y^2} - z \qquad (2.3.20a)$$

$$y = \kappa + z \log \frac{z}{\sqrt{z^2 + y^2} - y} \qquad (2.3.20b)$$

Solving Eq. (2.3.20a) for z in terms of y

$$z = \tfrac{1}{2}(y^2 - 1)$$

and substituting into Eq. (2.3.20b), we have

$$y = \kappa + (y^2 - 1)\frac{1}{2}\log\left(\frac{y+1}{y-1}\right)$$

Since $1/y = D/S$ is less than 1, we can take a Taylor expansion of the logarithm and obtain

$$\left(\frac{1}{y}\right)^3 + 5\left(\frac{1}{y}\right) - \frac{15}{2}\kappa = 0$$

which is a cubic equation for $(1/y)$. It has only one real root, given by

$$\left(\frac{1}{y}\right) = \left[-\frac{15}{4}\kappa + \frac{5}{3}\sqrt{\left(\frac{9\kappa}{4}\right)^2 + \frac{5}{3}} \right]^{1/3}$$

$$+ \left[-\frac{15}{4} \sqrt{\left(\frac{9\kappa}{4}\right)^2 + \frac{5}{3}} \right]^{1/3}$$

and thus

$$\frac{S}{D} = y$$

$$\frac{T_{01}}{qD} = \frac{1}{2}\sqrt{y^2 - 1}$$

is the solution for the arc length to the point of bottom contact and for the horizontal component of tension at the top of the tower.

A similar problem is that of calculating the excursion u of a buoy, as shown in Fig. 2.3.4, due to a force ΔF, e.g., wind, wave, drift, or current, applied in addition to a prestressing force $F_0 = T_{01}$.

During the excursion an additional length of cable $(S' - S)$ lifts off the

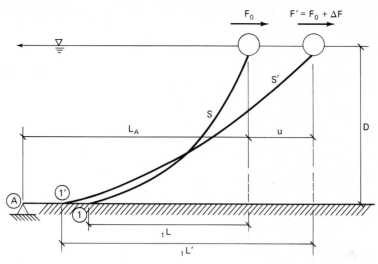

Figure 2.3.4 Buoy excursion.

sea floor. The prestressing phase due to force F_0 is given by (for inextensible cable)

$$\frac{S}{D} = \sqrt{\frac{2F_0}{qD} + 1}$$

$$\frac{{}_1L}{D} = \left(\frac{F_0}{qD}\right) \log \left(\frac{F_0/qD}{1 + F_0/qD - S/D}\right)$$

With an increased force $F' = F_0 + \Delta F$ the length of cable lifted off the bottom is

$$\frac{S'}{D} = \sqrt{\frac{2(F_0 + \Delta F)}{qD} + 1}$$

and horizontal distance is

$$\frac{{}_1L'}{D} = \frac{F_0 + \Delta F}{qD} \log \left[\frac{(F_0 + \Delta F)/qD}{1 + (F_0 + \Delta F)/qD - S'/D}\right]$$

Because the total length from the buoy to the pile cap remains the same, we find the excursion from

$$S + (L_A - {}_1L) = S' + (L_A + u - {}_1L')$$

Thus,

$$u = ({}_1L' - {}_1L) - (S' - S)$$

Bidirectional loads. Let us examine a problem in which uniform loads in two directions could be acting on a cable. If we make the crude approximation that the drag force on a cable is uniform along its length and acts only in opposition to a uniform horizontal velocity of the fluid, we can generate a solution to the problem of a vessel towing a body, as shown in Fig. 2.3.5, by use of Eqs. (2.3.14). The buoyant weight of the body is W. The drag force on the body is F_0, that on the cable is $q_1 = C_D V^2$ where V is the velocity of the vessel. We will attach the origin at the vessel and calculate the location (D, L) of the body and the cable tension at the vessel.

In this case, adopting Eqs. (2.3.14) and setting

$$q_1 = C_D V^2$$

$$q_2 = q \quad \text{(buoyant weight of cable)}$$

$$Q = \sqrt{q^2 + (C_D V^2)^2}$$

we evaluate the coefficients T_{01} and T_{02} by equilibrium at the body, i.e., at $s = S$, the total length of cable. Therefore,

$$T_{01} = F_D + C_D V^2 S$$

$$T_{02} = W + qS$$

and the tensions at the vessel, T_A, and at the body, T_B, are

$$T_A = \sqrt{T_{01}^2 + T_{02}^2}$$

$$T_B = \sqrt{F_D^2 + W^2}$$

The coordinates L and D of the body are

$$\begin{Bmatrix} L \\ D \end{Bmatrix} = \begin{Bmatrix} F_D + \dfrac{C_D V_S^2}{2} \\ W + \dfrac{qS}{2} \end{Bmatrix} \dfrac{S}{AE} + \dfrac{T_A - T_B}{Q^2} \begin{Bmatrix} C_D V^2 \\ q \end{Bmatrix}$$

$$+ \dfrac{F_D q - W C_D V^2}{Q^3} \begin{Bmatrix} q \\ -C_D V^2 \end{Bmatrix} \log \left(\dfrac{Q T_B - C_D V^2 F_D - qW}{Q T_A - Q^2 S - C_D V^2 F_D - qW} \right)$$

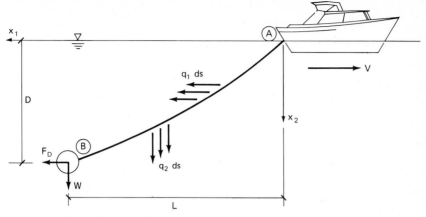

Figure 2.3.5 Towed body problem.

2.3.2 Varying loads

There are instances, especially when fluid loads are encountered, where the governing three-dimensional equations of the static behavior of cable segments [Eqs. (2.3.2b), (2.3.6), and (2.3.13)] include loads q_i which vary along the arc length as functions of s, θ_i ($= T_i/T$), or x_i. Some examples are shown in Fig. 2.3.6. In cases of fluid pressures the loads are nonlinear functions of T_i, i.e., the direction cosines $\theta_i = T_i/T$. Also, the magnitudes of q_i may be nonlinear as in the case of fluid drag on a cable (Fig. 2.3.6c). In such problems not only are the pressure and skin friction dependent on the squares of the normal and tangential relative velocities, but also the drag coefficients C_N and C_T are nonlinear functions of the normal Reynold's number and, hence, of the velocity and orientation θ_i [96].

It is possible to idealize an isolated section of a membrane long and straight in one direction as a cable segment transverse to that direction. In instances of the retention of water by membranes, either purposefully by inflatable dams [5] (Fig. 2.3.6a) or inadvertently by ponding of roof membranes [106] (Fig. 2.3.6b), the load is a function of depth of water and of orientation. Also, as the volume of water retained changes, the portion of structure subjected to load changes.

In cases where the loads vary in magnitude, location, and direction along the cable, a numerical solution to Eqs. (2.3.2b), (2.3.6), and (2.3.13) is advised. Those equations are first-order differential equations, and the boundary-value problem they pose can be treated as an iterative initial-value problem. That is, solutions can be propagated from one end of the segment to the other by numerical integration techniques (such as the Runge-Kutta method or Hamming's method) [61] *if* initial values of all the variables at the starting end are assigned numerical values. Typically,

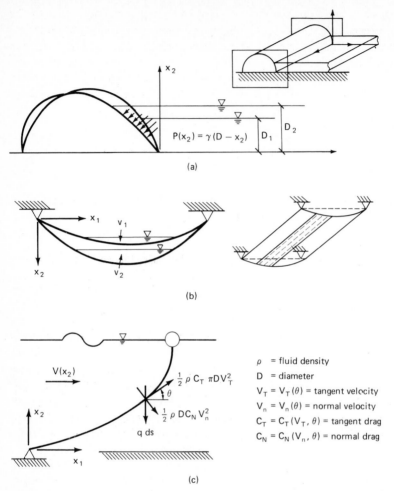

Figure 2.3.6 Nonlinear varying load examples. (*a*) Inflatable dam; (*b*) ponding of a membrane; (*c*) drag on a mooring cable.

not all of x_{i0} and T_{0i} are known. So, fictitious values would be assigned in linearly independent combinations. If the equations had been linear equations, then the partial solutions could be recombined at the terminating end of the segment so as to satisfy boundary conditions there. The factors of that recommendation would then lead to the proper selection of initial values of T_{0i} and x_{0i}. However, the equations are not linear.

But an iterative solution is possible. One approach is to use trial-and-error selection of initial values of T_{0i} and x_{0i}. Successive trial values can be based on minimization of closure errors in the boundary conditions at the terminating end [69, 70]. That approach is similar to that described previously under "Multiple Concentrated Loads."

Another approach is to quasi-linearize the differential equations by the Newton-Raphson method so as to always directly satisfy the boundary conditions at the terminating end by recombining linear partial solutions at that end as described above. The closure errors are embedded in the comparisons of the nonlinear and linearized solutions at every point along the cable segment. Successive iterates are then generated by the Newton-Raphson method as applied to first-order differential equations.

Given a set of nonlinear differential equations

$$\left\{\frac{dy_i}{dx}\right\} = \{g_i(x, y_j)\} \qquad (2.3.21)$$

where $\{g_i(x, y_j)\}$ are nonlinear functions of $\{y_j\}$, assume a trial solution $\{y_j^*\}$ and take a Taylor expansion of Eq. (2.3.21) to obtain

$$\left\{\frac{dy_i}{dx}\right\} = \{g_i^*(x, y_j^*)\} + \left[\frac{\partial g_i^*}{\partial y_j^*}\right](\{y_j\} - \{y_j^*\})$$

This gives a linearized set of equations for $\{y_i\}$:

$$\left\{\frac{dy_i}{dx}\right\} = \left[\frac{\partial g_i^*}{\partial y_j^*}\right]\{y_j\} + \left(\{g_i^*\} - \left[\frac{\partial g_i^*}{\partial y_j^*}\right]\{y_j^*\}\right) \qquad (2.3.22)$$

in terms of x and $\{y_j^*\}$. Solve Eq. (2.3.22) for $\{y_i\}$ and compare it to $\{y_i^*\}$. If they are not sufficiently close, replace $\{y_i^*\}$ with $\{y_i\}$ in Eq. (2.3.22) and repeat the process.

The application of this method to a problem of towing a cable in a fluid medium is described in Refs. 71 and 72.

Finite Element
Static Analysis of
Cable Networks

In Chap. 2, solution methods were presented for single segments of cables or for a series of singly connected segments. By singly connected we mean that the segments are connected in series and that at connection points only two adjoining segments meet. In this chapter we will study multiply connected systems. In a multiply connected system more than two segments may meet at a node and closed loops may be formed. Such systems are highly redundant, and the deformations of the individual segments play an important role in determining the stresses in the segments.

If the segments all lie in a single curved surface and are loaded principally transverse to that surface, then analytical and numerical methods based on a membrane analogy can be used. That is, the net is replaced by an equivalent prestressed membrane and solution methods available for that type of problem can be adopted. They will be considered in Chap. 6 when we study the statics of membrane structures.

More generally, multiply connected segments will form a three-dimensional network. Methods of redundant structural analysis are needed to predict both the prestressed configuration and the in-service response. The prestressed configuration and stress state can be determined either by an inverse method or by nonlinear analysis. Sometimes a combination of both is needed. Often, if in the in-service phase the added loads are small, the response can be modeled by linearized equations. But the essential nonlinearity of the system remains, and care must be exercised in superposing responses and extreme in-service loads must also be handled with nonlinear techniques.

Of the two principal methods of redundant structural analysis [108], flexibility and stiffness methods, stiffness methods have proved to be more useful in the analysis of highly redundant cable systems. In stiffness methods the fundamental unknowns of the equations formulated are the displacements of the nodal points connecting segments. In that class of methods, the lumped parameter method (analogous to the finite difference method) was the first to be applied to cable systems [12]. In the lumped parameter method all loads (concentrated and distributed) are "lumped" at the nodal points, and attention is focused on the static equilibrium of the nodes. The effects of the segments are replaced by forces in equivalent straight and weightless springs, which may be nonlinear, and a system of simultaneous algebraic equations is generated from the nodal equilibrium conditions.

The finite element method of stiffness analysis [94] has gradually supplanted the lumped parameter method in the analysis of cable systems [39, 43, 82–86]. The advantages of the finite element method are (1) the numerical modeling has a clearer physical interpretation; (2) the approximations for a particular problem are more flexible (refined approximating functions [90] can be used for higher-order effects such as segment curvature); (3) the refinement of the elements in regions of particular interest is more readily implemented; (4) comparable accuracy can be obtained with less computational effort and mesh refinement; (5) efficient matrix manipulation and solution routines can be used; and (6) variable grid sizes and layouts can be easily generated and changed. Also, the finite element method is less problem-dependent, i.e., with a set of standard cable elements as subroutines in a computer program it is possible to handle with a single program various configurations and loadings of networks and to include other structural elements, such as beams, in the model.

In Sec. 3.1 we will summarize the general aspects of the finite element method as applied to structural systems. Second, we will generate the requisite continuum formulation of the behavior of a differential segment of a cable element. Then by appropriate finite element interpolation we will derive the system of matrix equations for an arbitrary assemblage of cable elements. Solution techniques will be considered for linearized in-service behavior. Finally, nonlinear solution techniques will be presented. We will focus attention here on one simple finite element formulation of static problems. In Chap. 4 we will consider more refined finite element approximations for dynamic problems.

3.1   Finite Element Methodology

The finite element method is a numerical analysis technique by which an elastic or inelastic continuum can be approximated by a discrete number of finite-sized elements which retain the material properties of the continuum [94]. The elements are interconnected at a discrete number of points, called nodes, on the boundaries of the elements. The displacement and force functions can be approximated locally within each element by continuous functions which are defined in the element in terms of the displacements and forces at the nodes of that element. If these nodal values are regarded as unknown parameters, then the continuum problem with an infinite number of degrees of freedom is replaced by an approximate model with a finite number of degrees of freedom. In static problems a set of algebraic equations is derived relating the various nodal displacements and the equivalent external nodal loads. Once a solution for the nodal displacements is determined from the equations, the internal behavior of each element is characterized by the local approximations within the element.

A finite element solution can be roughly divided into four processes, each having its own source of potential error in the modeling of the true continuum behavior:

1. *Idealization of the continuum.*   The body is subdivided into an adequate number of finite elements. The number, shape, and size of the elements should be determined in such a way that the original body geometry is simulated as closely as possible. The error inherent in this process can be called idealization error and is a function of the number and type of elements representing the body, the number of nodes defining each element, and the approximation of each element's geometry, e.g., degree of curvature of the element.

2. *Discretization of displacements.*   Simple functions are selected to approximate the actual behavior at an arbitrary point within an element. Various functions can be adopted, e.g., polynomials and trigonometric functions. Since polynomials offer ease in mathematical manipulation, they have been most commonly selected. There are three desirable characteristics for a suitable displacement function. First, it should be compatible between adjacent elements. Second, potential rigid body motion of the element should be reproducible. Third, as the element size is shrunk, a state of uniform strain within the element should be predicted. The error in this process can be called discretization error and is a function of the number of nodes spanning the continuum and of the order of the approximating, or interpolating, functions within each element.

3. *Formulation of element stiffness.*   The stiffness matrix relates the nodal displacements to the applied forces transformed to the nodes. This

relationship is developed from a discrete form of the principle of virtual work which governs in the continuum. The sum of approximate energies in each element is assumed equal to the total energy of the real continuum. If this discrete sum of energies is minimized with respect to the nodal displacements, then overall equilibrium of the model is guaranteed and the nodal displacements calculated are the best answers available for the given approximating functions. The error inherent in this process can be called formulation error and is a function of the selected equations relating stress, strain, and displacement at a point in an element and of the accuracy and completeness of the energy formulation for each element.

4. *Solution of matrix equations.*   After assembly of the stiffness and force matrices of each element into matrices for the entire continuum, the final step is to apply discrete forms of the boundary conditions and solve the matrix equilibrium equations (or equations of motion) for the nodal displacements as a function of load (or time). The error inherent in this process can be called truncation error and is a function of digital accuracy of the computer hardware available, of the selected arithmetic precision, of the number and types of numerical operations performed, and of the size of increments in load (or time) selected or of the tolerances allowed in an iteration.

Various refinements have been developed in the finite element methodology to reduce various combinations of the errors inherent in the finite element model. We will consider only the simplest models for cable structures consistent with their nonlinear behavior, i.e., we will focus on formulation errors. In a subsequent chapter on the dynamics of cable structures we will address the other types of errors with more refined models.

Although we will be concerned with nonlinear behavior, it is instructive to review the modeling process for a linear continuum, i.e., small displacements and elastic material. The vector of displacements $\{u\}$ at a point $\{x\}$ within an element can be expressed in terms of its nodal values $\{d\}$ and a selected matrix of approximating functions $[N(x)]$ by

$$\{u\} = [N(x)]\{d\} \tag{3.1.1}$$

Likewise, the distributed loads $\{f\}$ at a point $\{x\}$ can be expressed in terms of nodal values $\{p\}$ by

$$\{f\} = [\overline{N}(x)]\{p\} \tag{3.1.2}$$

where $[\overline{N}(x)]$ is a matrix of approximating functions, often taken equal to $[N(x)]$ to develop a consistent formulation.

The linear strain-displacement relations from elasticity theory then give the strains $\{\varepsilon\}$ at a point as

$$\{\varepsilon\} = [\partial B_x] \{u\} = [\partial B_x] [N(x)] \{d\} \qquad (3.1.3a)$$

where $[\partial B_x]$ denotes a differential operator matrix with respect to $\{x\}$ acting on $[N(x)]$ to give

$$\{\varepsilon\} = [A(x)] \{d\} \qquad (3.1.3b)$$

The linear stress-strain relationship for the stresses $\{\sigma\}$ is

$$\{\sigma\} = [E]\{\varepsilon\} + \{\sigma_0\} = [E][A]\{d\} + \{\sigma_0\} \qquad (3.1.4)$$

where $[E]$ is a symmetric matrix of material properties and $\{\sigma_0\}$ is a vector of prestresses, assumed a priori.

The principle of virtual work states that the work done by the internal and external forces during an arbitrary small displacement $\{\delta u\}$ consistent with the kinematic boundary conditions must be zero to give an equilibrium state, i.e.,

$$\int_V ([\delta\varepsilon]^T \{\sigma\} - [\delta u]^T \{f\}) \, dv - [\delta D]^T \{P\} = 0 \qquad (3.1.5)$$

where a superscript T denotes transposition, the integral is taken over the total volume V of the continuum, $\{P\}$ are concentrated loads at discrete points (nodes) with corresponding virtual displacements $\{\delta D\}$, and $\{\delta\varepsilon\}$ is the virtual small strain:

$$\{\delta\varepsilon\} = [\partial B_x] \{\delta u\} \qquad (3.1.6)$$

If we equate the total volume integral for the continuum to the sum of volume integrals V_e of the approximated behavior within each element, we can substitute Eqs. (3.1.1) to (3.1.4) into Eq. (3.1.5) to obtain

$$\sum_e ([\delta d]^T \int_{V_e} [A]^T [E] [A] \, dv_e \{d\} + [\delta d]^T \int_{V_e} [A]^T \{\sigma_0\} \, dv_e$$

$$- [\delta d]^T \int_{V_e} [N]^T [\overline{N}] \, dv_e \{p\}) - [\delta D]^T \{P\} = 0 \qquad (3.1.7)$$

The nodal displacements $\{d\}$ and $\{\delta d\}$ of the separate elements are not independent. By compatibility between elements, i.e., connectivity,

$$\{d\} = [\Gamma_e] \{D\} \qquad (3.1.8a)$$

$$\{\delta d\} = [\Gamma_e] \{\delta D\} \qquad (3.1.8b)$$

where $[\Gamma_e]$ is a congruent transformation matrix which relates the individual element $\{d\}$'s to the system of nodal displacements $\{D\}$ for the assembled finite element model. Thus, Eq. (3.1.7) becomes

$$[\delta D\}^T ([K] \{D\} - \{Q\}) = 0 \qquad (3.1.9a)$$

where the system stiffness matrix is

$$[K] = \sum_e [\Gamma_e]^T [K_e] [\Gamma_e] \qquad (3.1.9b)$$

the system load vector is

$$\{Q\} = \{P\} + \sum_e [\Gamma_e]^T \int_{V_e} [N]^T [\overline{N}] \, dv_e \{p\}$$

$$- \sum_e [\Gamma_e]^T \int_{V_e} [A]^T \{\sigma_0\} \, dv_e \qquad (3.1.9c)$$

and the stiffness matrix of an individual element is

$$[K_e] = \int_{V_e} [A]^T [E] [A] \, dv_e \qquad (3.1.9d)$$

Since $\{\delta D\}$ is an arbitrary small displacement ($\{\delta D\} \neq 0$), Eq. (3.1.9a) can only be satisfied by

$$[K] \{D\} = \{Q\} \qquad (3.1.10)$$

which gives the required system of algebraic equations of equilibrium for the system to be solved for $\{D\}$. The assembly process represented in Eq. (3.1.7) by $\Sigma_e [\Gamma_e]^T [K_e] [\Gamma_e]$ is easily achieved in that $[\Gamma_e]$ is usually a collection of 0s and 1s depicting the connectivity of the elements. Thus, the summation can be replaced in a digital computer by a simple "book-keeping" scheme, or overlay process, in which submatrices of each element stiffness matrix are added to different locations in an accumulating overall stiffness matrix $[K]$.

In Sec. 3.2 we will develop the equivalent matrix equations for a cable structure including nonlinear terms in $\{\varepsilon\}$ and $\{\sigma\}$. The principle of virtual work, as given by Eq. (3.1.5), can still be used since the arbitrary virtual displacements $\{\delta u\}$ can be specified small enough that $\{\delta\varepsilon\}$ is still linear in $\{\delta u\}$ while the nonlinearities in $\{\varepsilon\}$ and $\{\sigma\}$ are retained.

3.2   Cable Elements

To arrive at stiffness equations for a cable network we must first consider the behavior of a cable continuum, i.e., formulate the relationships between stresses, strains, and displacements of a differential segment. Then we will introduce approximating functions into the formulations along with the principle of virtual work to form matrix representations.

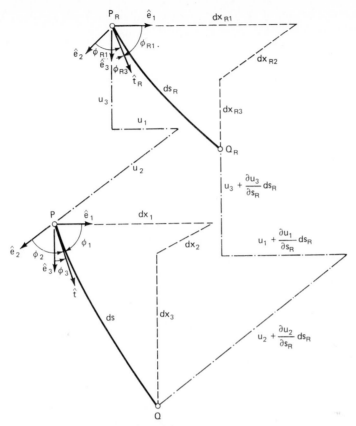

Figure 3.2.1 Motion of differential segment.

3.2.1 Continuum formulation

We will develop a description [43] of the geometry and stress state in a deformed configuration with respect to some reference state, e.g., the pre-stressed configuration of the cable network. Consider first the geometry of the deformation as shown in Fig. 3.2.1 where a differential segment of length ds_R in the reference state is shown. We will assume that the segment has undergone displacements $\bar{u}_R$ from an unstressed state and is now subject to further displacements $\bar{u}$. The direction cosines in the reference state are denoted by θ_{Ri}

$$\theta_{Ri} = \frac{dx_{Ri}}{ds_R} \tag{3.2.1a}$$

and those in the additionally deformed state by θ_i

$$\theta_i = \frac{dx_i}{ds} \tag{3.2.1b}$$

The displacements at end Q of the segment are taken as Taylor expansions of the displacements at end P. Thus, the components of the additionally deformed differential length are (see Fig. 3.2.1)

$$dx_i = dx_{Ri} + \frac{\partial u_i}{\partial s_R} ds_R \tag{3.2.2}$$

and the additionally stretched length is

$$
\begin{aligned}
ds &= \sqrt{dx_i\, dx_i} \\
&= \left(dx_{Ri}\, dx_{Ri} + 2dx_{Ri} \frac{\partial u_i}{\partial s_R} ds_R + \frac{\partial u_i}{\partial s_R} \frac{\partial u_i}{\partial s_R} ds_R^2 \right)^{1/2} \tag{3.2.3} \\
&= ds_R \left(1 + 2\theta_{Ri} \frac{\partial u_i}{\partial s_R} + \frac{\partial u_i}{\partial s_R} \frac{\partial u_i}{\partial s_R} \right)^{1/2}
\end{aligned}
$$

We will assume a nonlinear strain-displacement relationship:

$$
\begin{aligned}
\varepsilon &= \frac{ds^2 - ds_0^2}{2ds_0^2} = \frac{ds^2 - ds_R^2}{2ds_R^2} \left(\frac{ds_R}{ds_0} \right)^2 + \frac{ds_R^2 - ds_0^2}{2ds_0^2} \\
&= \frac{ds^2 - ds_R^2}{2ds_R^2} \lambda_R^2 + \varepsilon_R \tag{3.2.4a} \\
&= \gamma \lambda_R^2 + \varepsilon_R \tag{3.2.4b}
\end{aligned}
$$

where ds_0 = unstressed length
  λ_R = elongation ratio of ds_R to ds_0
  ε_R = strain in reference state

We have defined a relative strain γ. Substituting Eq. (3.2.3) into Eq. (3.2.4a), we obtain

$$\gamma = \theta_{Ri} \frac{\partial u_i}{\partial s_R} + \frac{1}{2} \frac{\partial u_i}{\partial s_R} \frac{\partial u_i}{\partial s_R} \tag{3.2.5}$$

Substituting Eqs. (3.2.2), (3.2.3), and (3.2.5) into Eq. (3.2.1b), we obtain an equation for the deformed direction cosines:

$$\theta_i = \frac{\theta_{Ri} + \partial u_i/\partial s_R}{\sqrt{1 + 2\gamma}} \tag{3.2.6}$$

Constitutive relation. We will adopt the conventional approach to the definition of a stress-strain relation for materials in which the "engineering" stress (force per unit unstressed area) and "engineering" strain (extension

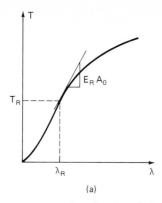

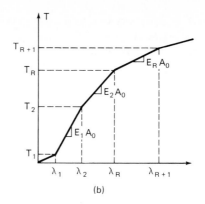

(a)   (b)

Figure 3.2.2 Constitutive relations.

ratio) are used. Since many materials used in cable fabrication, including steel as well as nylon, exhibit nonlinear tension-extension relations, we will assume a piecewise linear relationship beyond the reference state (see Fig. 3.2.2a). The tension T in the additionally deformed state can be related to the tension T_R in the reference state and to the extension ratios λ and λ_R of the two states by a piecewise relationship of the form

$$T = T_R + E_R A_0 \left(\lambda - \lambda_R \right) \tag{3.2.7}$$

where A_0 is the unstressed cross-sectional area, E_R is the instantaneous value of the modulus of elasticity at the reference state (conventionally based on unstressed area A_0), and λ is given by

$$\lambda = \frac{ds}{ds_0} = \frac{ds}{ds_R}\frac{ds_R}{ds_0} = \sqrt{1 + 2\gamma}\,\lambda_R \tag{3.2.8}$$

Therefore, the T-γ relationship is given by

$$T = T_R + E_R A_0 \lambda_R \left(\sqrt{1 + 2\gamma} - 1 \right) \tag{3.2.9}$$

It should be noted that we have assumed T and T_R to have the same modulus E_R, and if λ is much larger than λ_R for a nonlinear material, significant error could be thereby introduced. One way around that difficulty is to proceed in an incremental fashion with periodic updates of the reference state (hence T_R and E_R) as in Fig. 3.2.2b.

Virtual work expressions. Since both the reference and additionally deformed states are in equilibrium, we can write virtual work equations for both states. Using those equations we will be able to derive an incremental version of virtual work relating the additional displacements u_i to the additional in-service loads. Both a nonlinear form and a piecewise linear form of the incremental virtual work will be presented.

For cables there is a single stress component in each state, $\sigma_R = T_R/A_R$ and $\sigma = T/A$; and a single strain component in each state, ε_R and $\varepsilon =$ Eq. (3.2.4b). Equation (3.1.5) becomes, for virtual displacements δu_i from the two states,

$$\int_{S_R} \left(\delta\varepsilon_R \frac{T_R}{A_R} - \delta u_i\, f_{Ri} \right) A_R\, ds_R - \delta D_i\, P_{Ri} = 0 \qquad (3.2.10a)$$

$$\int_{S} \left(\delta\varepsilon \frac{T}{A} - \delta u_i\, f_i \right) A\, ds - \delta D_i\, P_i = 0 \qquad (3.2.10b)$$

where f_{Ri} and f_i are the components of distributed load per unit volume of the reference and additionally deformed states, and $\delta\varepsilon_R$, $\delta\varepsilon$ are the small virtual strains in the two states due to δu_i, i.e.,

$$\delta\varepsilon_R = \theta_{Ri} \frac{\partial(\delta u_i)}{\partial s_R} \qquad (3.2.11a)$$

$$\delta\varepsilon = \theta_i \frac{\partial(\delta u_i)}{\partial s} \qquad (3.2.11b)$$

Thus, Eqs. (3.2.10) become

$$\int_{S_R} \left[\frac{\partial(\delta u_i)}{\partial s_R} \theta_{Ri} T_R - \delta u_i\, q_{Ri}^* \right] ds_R - \delta D_i\, P_{Ri} = 0 \qquad (3.2.12a)$$

$$\int_{S} \left[\frac{\partial(\delta u_i)}{\partial s} \theta_i T - \delta u_i\, q_i^* \right] ds - \delta D_i\, P_i = 0 \qquad (3.2.12b)$$

where $q_{Ri}^* = f_{Ri} A_R$ and $q_i^* = f_i A$ are distributed loads per unit length in s_R and s, respectively.

To check the validity of Eq. (3.2.12) we should be able to recover the equilibrium equations of a differential segment [Eq. (2.3.3) of Sec. 2.3] by writing the Euler equation of the variational statement embodied in Eqs. (3.2.12). For example, integrating by parts the first term in Eq. (3.2.12a) we obtain

$$\int_{s_R=0}^{S_R^*} \left[\frac{\partial(\delta u_i)}{\partial s_R} T_R \theta_{Ri} - \delta u_i\, q_{Ri}^* \right] ds_R$$

$$= \left[\delta u_i\, (T_R \theta_{Ri}) \right] \Big|_0^{S_R^*} - \int_{s_R=0}^{S_R^*} \delta u_i \left[\frac{\partial(T_R \theta_{Ri})}{\partial s_R} + q_{Ri}^* \right] ds_R = 0 \qquad (3.2.13)$$

Boundary conditions on u_i and $T_{Ri} = T_R \theta_{Ri}$ are recovered from evaluation of the first term on the right-hand side of Eq. (3.2.13) at the endpoints s_R

$= 0$, S_R^*. Since $\delta u_i \neq 0$, and adopting the notation of Sec. 2.3 ($T_i = T_R \theta_{Ri}$), we obtain the equilibrium equations by setting the integrand of the right-hand side of Eq. (3.2.13) equal to zero, i.e.,

$$\frac{\partial T_i}{\partial s_R} + q_{Ri}^* = 0$$

which is the component form of Eq. (2.3.3) in Sec. 2.3. In the finite element method we do not attempt to obtain the Euler equations explicitly, but instead minimize the variational statement by the Rayleigh-Ritz procedure [107] using approximating functions for u_i and δu_i [and hence $\partial(\delta u_i)/\partial s_R$] as discussed in Sec. 3.1.

The integrand, load, and virtual strain in Eq. (3.2.12b) for the additionally deformed state must be restated in terms of the known geometry and stress of the reference state. By Eqs. (3.2.3) and (3.2.5)

$$ds = ds_R \sqrt{1 + 2\gamma}$$

Also, the loads q_{Ri}^* and q_i^* are most conveniently defined in terms of distributed loads q_{Ri} and q_i per unit length of the unstressed configuration. Let

$$ds_R \, q_{Ri}^* = q_{Ri} \, ds_0$$

$$ds \, q_i^* = q_i \, ds_0$$

and therefore,

$$q_{Ri} = \frac{q_{Ri}}{\lambda_R}$$

$$q_i^* = \frac{q_i}{\lambda_R \sqrt{1 + 2\lambda}}$$

Thus, substituting Eq. (3.2.9) for T and Eq. (3.2.6) for θ_i, we obtain the virtual work statements for the two states as

$$\int_{S_R} \left(\frac{\partial \delta u_i}{\partial s_R} T_R \theta_{Ri} - \delta u_i \frac{q_{Ri}}{\lambda_R} \right) ds_R - \delta D_i \, P_{Ri} = 0 \qquad (3.2.14a)$$

$$\int_{S_R} \left\{ \frac{\partial \delta u_i}{\partial s_R} [T_R + E_R A_0 \lambda_R [\sqrt{1 + 2\gamma} - 1]] \frac{\theta_{Ri} + \partial u_i/\partial s_R}{\sqrt{1 + 2\gamma}} \right.$$

$$\left. - \delta u_i \frac{q_i}{\lambda_R} \right\} ds_R - \delta D_i \, P_i = 0 \qquad (3.2.14b)$$

in which γ is the nonlinear relative strain given by Eq. (3.2.5).

In the next section we will propose an approximate finite element solution procedure in which increments of load are applied:

$$\Delta q_i = q_i - q_{Ri}$$

$$\Delta P_i = P_i - P_{Ri}$$

To enable that procedure we will require an incremental version of Eq. (3.2.14b) for the additionally deformed state. This is obtained by subtracting Eq. (3.2.14a) from Eq. (3.2.14b), i.e.,

$$\int_{S_R} \left(\frac{\partial \delta u_i}{\partial s_R} \left\{ \frac{E_R A_0 \lambda_R - T_R}{\sqrt{1 + 2\lambda}} \left[(\sqrt{1 + 2\lambda} - 1) \, \theta_{Ri} - \frac{\partial u_i}{\partial s_R} \right] \right. \right.$$

$$\left. \left. E_R A_0 \lambda_R \frac{\partial u_i}{\partial s_R} \right\} - \delta u_i \frac{\Delta q_i}{\lambda_R} \right) ds_R - \delta D_i \, \Delta P_i = 0 \quad (3.2.15)$$

Equation (3.2.15) can be used for determining the additional large displacement response u_i due to additional in-service loads Δq_i and ΔP_i.

In the sequel, when small in-service loads are considered, we will make the assumption that the additional displacements u_i and strain γ are small. Thus, approximately

$$\gamma \simeq \theta_{Ri} \frac{\partial u_i}{\partial s_R}$$

We also note that $\sqrt{1 + 2\gamma} - 1 \simeq \gamma$ and $1/\sqrt{1 + 2\gamma} \simeq 1 - \gamma$. In that case the piecewise linear versions of the incremental virtual work, Eq. (3.2.15), can be obtained by retaining only first-order terms as

$$\int_{S_R} \left\{ \frac{\partial \delta u_i}{\partial s_R} \left[T_R \delta_{ij} + (E_R A_0 \lambda_R - T_R) \, \theta_{Ri} \theta_{Rj} \right] \frac{\partial u_j}{\partial s_R} \right.$$

$$\left. - \delta u_i \frac{\Delta q_i}{\lambda_R} \right\} ds_R - \delta D_i \, \Delta P_i = 0 \quad (3.2.16)$$

For a prestressed reference state with known geometry θ_{Ri} and tension T_R, Eq. (3.2.16) is a linear equation for the small additional displacements u_i due to Δq_i and ΔP_i.

A large displacement in-service problem can be treated in a piecewise linear fashion using Eq. (3.2.16). In that instance, after solution of Eq. (3.2.16) for u_i due to a small portion of the added load, a new reference state $S_{\bar{R}}$ can be determined from Eqs. (3.2.5), (3.2.6), (3.2.7), and (3.2.8) as

$$\overline{\gamma} = \theta_{Ri} \frac{\partial u_i}{\partial s_R} + \frac{1}{2} \left(\frac{\partial u_i}{\partial s_R} \right) \left(\frac{\partial u_i}{\partial s_R} \right) \tag{3.2.17a}$$

$$\lambda_{\overline{R}} = \sqrt{1 + 2\overline{\gamma}} \, \lambda_R \tag{3.2.17b}$$

$$\theta_{\overline{R}i} = \left(\theta_{Ri} + \frac{\partial u_i}{\partial s_R} \right) \frac{\lambda_R}{\lambda_{\overline{R}}} \tag{3.2.17c}$$

$$T_{\overline{R}} = T_R + E_R A_0 \left(\lambda_{\overline{R}} - \lambda_R \right) \tag{3.2.17d}$$

and a new $E_{\overline{R}}$ selected from the T vs. λ relationship. Additional loads Δq_i and ΔP_j can then be applied and Eq. (3.2.16) used, with R replaced by $\overline{R}$, to calculate an additional displacement u_i. We must be cautious, however, in that small errors will accumulate in each increment of load. The alternative is to treat the nonlinear equations, Eq. (3.2.15), in an iterative fashion.

3.2.2 Finite element approximation (linearized)

In this section we will derive a simple linearized finite element model of the small-displacement in-service response to added loads. Thus, we will adopt Eq. (3.2.16) as our governing virtual work expression. The simplest idealization of a cable network is as a system of interconnected straight-line elements, each element having uniform material, stress, and geometric properties over its length (see Fig. 3.2.3). Thus, we will assume that

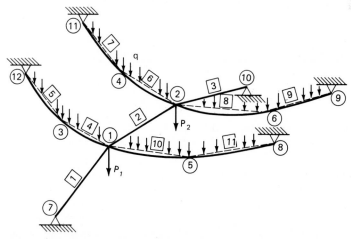

Figure 3.2.3 Straight element idealization of system.

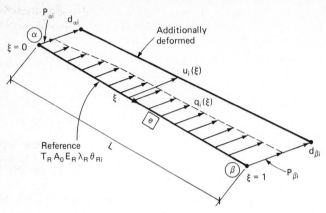

Figure 3.2.4 Straight element definition.

each element in the reference state has a constant modulus E_R, pretension T_R, cross-sectional area A_0, extension ratio λ_R, and direction cosines θ_{Ri}. Of course, these values can differ from element to element. In Chap. 5 when we consider an isoparametric finite element formulation, those properties will no longer need to be assumed constant in an element. The most severe restriction inherent in our straight-element assumption is on the direction cosines which in a real network, e.g., Fig. 3.2.3, are continuous functions of arc length in the element. We will have discontinuities in slopes at nodes, e.g., nodes 3 to 6 in Fig. 3.2.3. Thus, we will require several straight elements to model strongly curved segments of networks.

A typical element e is shown in Fig. 3.2.4 spanning between two nodes α and β with coordinates $x_{\alpha i}$ and $x_{\beta i}$, respectively. Henceforth in this chapter, repeated Greek subscripts are exempted from the indicial and summation conventions. The direction cosines θ_{Ri} of the reference state for the element can be determined from the nodal coordinates as

$$\theta_{Ri} = \frac{x_{\beta i} - x_{\alpha i}}{L}$$

where L is the length of the element:

$$L = \sqrt{(x_{\beta i} - x_{\alpha i})(x_{\beta i} - x_{\alpha i})}$$

In Fig. 3.2.4 we have also defined a nondimensional arc length coordinate $0 \leqslant \xi = s_R/L \leqslant 1$.

Consistent with our straight-element assumption we will assume that the additional displacements u_i vary as first-order polynomials over the arc length of an element. Thus, we will have a first-order isoparametric formulation. Define approximating functions $\eta_\alpha(\xi)$ and $\eta_\beta(\xi)$ such that

$$u_i = \eta_\alpha \, d_{i\alpha} + \eta_\beta \, d_{i\beta} \qquad (3.2.18)$$

where $d_{i\alpha}$, $d_{i\beta}$ are nodal values of u_i at $\xi = 0, 1$, respectively. We require η_α and η_β to be first-order polynomials such that

$$\eta_\alpha \, (\xi = 0) = 1 \qquad \eta_\alpha \, (\xi = 1) = 0$$

$$\eta_\beta \, (\xi = 0) = 0 \qquad \eta_\beta \, (\xi = 1) = 1$$

Thus,

$$\eta_\alpha = 1 - \xi \qquad (3.2.19a)$$

$$\eta_\beta = \xi \qquad (3.2.19b)$$

will be our assumed displacement functions. Some useful expressions are

$$\frac{\partial \eta_\alpha}{\partial \xi} = - \frac{\partial \eta_\beta}{\partial \xi} = - 1 \qquad (3.2.20a)$$

$$\int_0^1 \eta_\alpha^2 \, d\xi = \int_0^1 \eta_\beta^2 \, d\xi = 2 \int_0^1 \eta_\alpha \eta_\beta \, d\xi = \frac{1}{3} \qquad (3.2.20b)$$

In matrix notation we have

$$\{u\} = \begin{Bmatrix} u_1 \\ u_2 \\ u_3 \end{Bmatrix} = [N_\alpha \; N_\beta] \begin{Bmatrix} d_\alpha \\ d_\beta \end{Bmatrix} \qquad (3.2.21a)$$

where we have defined submatrices

$$\{d_\alpha\} = \begin{Bmatrix} d_{\alpha 1} \\ d_{\alpha 2} \\ d_{\alpha 3} \end{Bmatrix} \qquad \{d_\beta\} = \begin{Bmatrix} d_{\beta 1} \\ d_{\beta 2} \\ d_{\beta 3} \end{Bmatrix} \qquad (3.2.21b)$$

$$[N_\alpha] = \eta_\alpha \, [I] \qquad [N_\beta] = \eta_\beta \, [I] \qquad (3.2.21c)$$

in which $[I]$ is a 3×3 identity matrix. Similarly, the virtual displacements are

$$\{\delta u\} = [N_\alpha \; N_\beta] \begin{Bmatrix} \delta d_\alpha \\ \delta d_\beta \end{Bmatrix} \qquad (3.2.22)$$

With the assumption of small additional displacements $\{u\}$ we can approximate the strain (Eq. 3.2.5) by

$$\gamma = \theta_{Ri} \frac{\partial u_i}{\partial s_R} = \frac{\theta_{Ri}}{L} \frac{\partial u_i}{\partial \xi} = \frac{1}{L} [\theta_R]^T \frac{\partial}{\partial \xi} \{u\} \qquad (3.2.23a)$$

where $[\theta_R]^T = [\theta_{R1}, \theta_{R2}, \theta_{R3}]$ is the matrix of direction cosines in the reference state. Therefore, by Eqs. (3.2.20) and (3.2.21)

$$\gamma = \frac{1}{L} [\theta_R]^T [-I \; I] \begin{Bmatrix} d_\alpha \\ d_\beta \end{Bmatrix} = \frac{1}{L} [\theta_R]^T (\{d_\beta\} - \{d_\alpha\}) \quad (3.2.23b)$$

which is seen to be constant along the length of an element. For a rigid body motion $\{d_\alpha\} = \{d_\beta\}$ we have zero strain. Thus the three criteria for an approximating function discussed in Sec. 3.1 are satisfied by Eqs. (3.2.19).

We will assume small increments of load Δq_i to be approximated in the same fashion as for u_i, i.e.,

$$\{\Delta q\} = \begin{Bmatrix} \Delta q_1 \\ \Delta q_2 \\ \Delta q_3 \end{Bmatrix} = [N_\alpha \; N_\beta] \begin{Bmatrix} p_\alpha \\ p_\beta \end{Bmatrix} \quad (3.2.24)$$

where $\{p_\alpha\} = \begin{Bmatrix} p_{\alpha 1} \\ p_{\alpha 2} \\ p_{\alpha 3} \end{Bmatrix}$

$$\{p_\beta\} = \begin{Bmatrix} p_{\beta 1} \\ p_{\beta 2} \\ p_{\beta 3} \end{Bmatrix}$$

and are nodal intensities of $\{\Delta q_i\}$ at $\xi = 0, 1$, respectively.

The linearized virtual work statement, Eq. (3.2.16), can be written in matrix form as

$$\sum_e \frac{1}{L} \int_0^1 \frac{\partial}{\partial \xi} [\delta u]^T [B] \frac{\partial}{\partial \xi} \{u\} \, d\xi$$

$$- L \int_0^1 [\delta u]^T \left(\frac{1}{\lambda_R}\right) \{\Delta q\} \, d\xi - [\delta D]^T \{P\} = 0 \quad (3.2.25)$$

where the summation is taken over all the elements, $\{\delta D\}$ is the vector of nodal virtual displacements, $\{P\}$ are concentrated nodal loads, and

$$[B] = (T_R) [I] + (E_R \, A_0 \, \lambda_R - T_R) \{\theta_R\} [\theta_R]^T \quad (3.2.26)$$

Note that $\partial \{u\}/\partial \xi$ and $[B]$ are not functions of ξ.

Substituting Eqs. (3.2.19) to (3.2.24) into Eq. (3.2.25), we obtain

$$\sum_e [\delta d_\alpha^T \; \delta d_\beta^T] \left(\frac{1}{L} \begin{bmatrix} -I \\ I \end{bmatrix}\right) [B] [-I \; I] \begin{Bmatrix} d_\alpha \\ d_\beta \end{Bmatrix}$$

$$-\frac{L}{\lambda_R} \int_0^1 \begin{bmatrix} N_\alpha^T \\ N_\beta^T \end{bmatrix} [N_\alpha \ N_\beta] \ d\xi \ \begin{Bmatrix} p_\alpha \\ p_\beta \end{Bmatrix} \Bigg) - [\delta D]^T \{P\} = 0 \quad (3.2.27a)$$

The integral in Eq. (3.2.27a) can be evaluated with the help of Eq. (3.2.20b). Thus, the virtual work is

$$\sum_e [\delta d_\alpha^T \ \delta d_\beta^T] \ ([_eK] \{d\} - \{_eQ\}) - [\delta D]^T \{P\} = 0 \quad (3.2.27b)$$

in which the element stiffness matrix is

$$[_eK] = \frac{1}{L} \begin{bmatrix} B & -B \\ -B & B \end{bmatrix} \quad (3.2.28)$$

and contributions of distributed element loads to the external concentrated nodal loads are

$$\{_eQ\} = \left(\frac{L}{6\lambda_R}\right) \begin{bmatrix} 2p_\alpha + p_\beta \\ p_\alpha + 2p_\beta \end{bmatrix} \quad (3.2.29)$$

The element stiffness matrices $[_eK]$ and element contributions $\{_eQ\}$ to the external concentrated loads can be assembled into the system stiffness matrix $[K]$ and $[Q]$ by use of the connectivity matrix $[\Gamma]$, as in Sec. 3.1, Eqs. (3.1.8) to (3.1.10).

Subsequent to the assembly and then solution for $\{D\}$, the added nodal displacements, we can determine a new reference state $\bar{R}$ by use of the matrix forms of Eqs. (3.2.17):

Strain: $\quad \bar{\gamma} = \left(\frac{1}{L}\right) [\theta_R]^T (\{d_\beta\} - \{d_\alpha\}) + \left(\frac{1}{2L^2}\right) ([d_\beta^T]$

$$- [d_\alpha^T]) \ (\{d_\beta\} - \{d_\beta\})$$

Elongation: $\quad\quad\quad\quad \lambda_{\bar{R}} = \sqrt{1 + 2\bar{\gamma}} \ \lambda_R$

Direction: $\quad\quad\quad \{\theta_{\bar{R}}\} = \left(\frac{\lambda_R}{\lambda_{\bar{R}}}\right) \{\theta_R\} + \left(\frac{1}{L}\right) (\{d_\beta\} - \{d_\alpha\})$

Tension: $\quad\quad\quad\quad T_{\bar{R}} = T_R + E_R A_0 \ (\lambda_{\bar{R}} - \lambda_R)$

Length: $\quad\quad\quad\quad\quad \bar{L} = \lambda_{\bar{R}} \ L$

Then an additional load can be applied. Note that the stiffness matrix $[_{\bar{e}}K]$ and load matrix $\{_{\bar{e}}Q\}$ have to be reformed before applying the additional load. In this fashion the nonlinear response can be simulated by a sequence of piecewise linear problems.

Example problems. To illustrate the assembly and solution process for cable finite element modeling, consider the simple problem of a cable cross shown in Fig. 3.2.5. In that structure the two crossing cables prestress one another. An arbitrary magnitude of horizontal force H_R (equal in both cables if the spans ℓ and sag ratios f_R are equal) can be specified. We will determine the added deflection w due to an in-service load P at the intersection of the two cables.

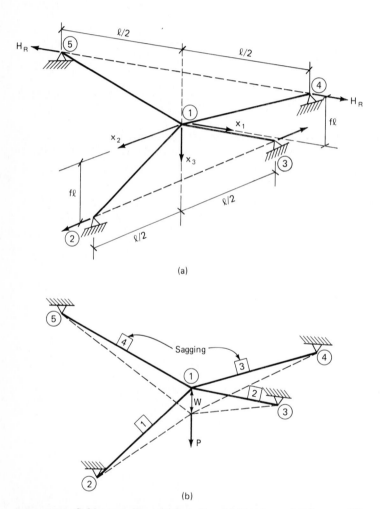

(a)

(b)

Figure 3.2.5 Cable cross linearized statics. (a) Prestressed reference; (b) in-service phase.

We will place the origin of the cartesian coordinate system at node 1 in Fig. 3.2.5 and direct each element away from that node, i.e., $_e\alpha = 1$ for all elements e and $_e\beta = e + 1$ where a preceding subscript denotes element number as shown in Fig. 3.2.5b.

The tension, elongation ratio, and length of each element e in the pre-stressed reference configuration shown in Fig. 3.2.5a are the same. Let

$$\rho = \sqrt{1 + 4f_R^2}$$

Then,

$$_eT_R = H_R\rho$$

$$_e\lambda_R = 1 + \frac{_eT_R}{E_RA_0} = 1 + \frac{H_R\rho}{E_RA_0}$$

$$_eL = \frac{\ell\rho}{2}$$

where E_R is taken as constant for a linearly elastic material.

The matrices of direction cosines for each member are different. Based on the coordinates of each node

$$[_1\theta_R]^T = \left[0 \; \frac{1}{\rho} \; \frac{2f_R}{\rho} \right] \tag{3.2.30a}$$

$$[_2\theta_R]^T = \left[0 \; -\frac{1}{\rho} \; \frac{2f_R}{\rho} \right] \tag{3.2.30b}$$

$$[_3\theta_R]^T = \left[\frac{1}{\rho} \; 0 \; -\frac{2f_R}{\rho} \right] \tag{3.2.30c}$$

$$[_4\theta_R]^T = \left[-\frac{1}{\rho} \; 0 \; -\frac{2f_R}{\rho} \right] \tag{3.2.30d}$$

The submatrix $[_eB]$ for each element is given by Eq. (3.2.26) as

$$[_eB] = H_R\rho \, [I] + \left[E_RA_0 \left(1 + \frac{H_R\rho}{E_RA_0} \right) - H_R\rho \right] \{_e\theta_R\} \, [_e\theta_R]^T$$

$$= H_R\rho \begin{bmatrix} 1 & 0 & 0 \\ 0 & 1 & 0 \\ 0 & 0 & 1 \end{bmatrix} + E_RA_0 \begin{bmatrix} _e\theta_1^2 & _e\theta_1 \, _e\theta_2 & _e\theta_1 \, _e\theta_3 \\ _e\theta_2 \, _e\theta_1 & _e\theta_2^2 & _e\theta_2 \, _e\theta_3 \\ _e\theta_3 \, _e\theta_{11} & _e\theta_3 \, _e\theta_2 & _e\theta_3^2 \end{bmatrix}$$

and therefore,

$$
[_1B] = \begin{bmatrix} H_R\rho & 0 & 0 \\[2mm] 0 & \left(H_R\rho + \dfrac{E_R A_0}{\rho^2}\right) & \left(\dfrac{E_R A_0 2 f_R}{\rho^2}\right) \\[3mm] 0 & \left(\dfrac{E_R A_0\, 2 f_R}{\rho^2}\right) & \left(H_R\rho + \dfrac{E_R A_0 4 f_R^2}{\rho^2}\right) \end{bmatrix}
$$

$$(3.2.31a)$$

$$
[_2B] = \begin{bmatrix} H_R\rho & 0 & 0 \\[2mm] 0 & \left(H_R\rho + \dfrac{E_R A_0}{\rho^2}\right) & -\left(\dfrac{E_R A_0 2 f_R}{\rho 2}\right) \\[3mm] 0 & -\left(\dfrac{E_R A_0 2 f_R}{\rho^2}\right) & \left(H_R\rho + \dfrac{E_R A_0 4 f_R^2}{\rho^2}\right) \end{bmatrix}
$$

$$(3.2.31b)$$

$$
[_3B] = \begin{bmatrix} \left(H_R\rho + \dfrac{E_R A_0}{\rho^2}\right) & 0 & -\left(\dfrac{E_R A_0 2 f_R}{\rho^2}\right) \\[3mm] 0 & H_R\rho & 0 \\[3mm] -\left(\dfrac{E_R A_0 2 f_R}{\rho^2}\right) & 0 & \left(H_R\rho + \dfrac{E_R A_0 4 f^2{}_R}{\rho^2}\right) \end{bmatrix}
$$

$$(3.2.31c)$$

$$
[_4B] = \begin{bmatrix} \left(H_R\rho + \dfrac{E_R A_0}{\rho 2}\right) & 0 & +\left(\dfrac{E_R A_0 2 f_R}{\rho^2}\right) \\[3mm] 0 & H_R\rho & 0 \\[3mm] +\left(\dfrac{E_R A_0 2 f_R}{\rho^2}\right) & 0 & \left(H_R\rho + \dfrac{E_R A_0 4 f^2{}_R}{\rho^2}\right) \end{bmatrix}
$$

$$(3.2.31d)$$

Assembling the system stiffness matrix with

$$
[_e K] = \frac{1}{_e L}\begin{bmatrix} _e B & -_e B \\ -_e B & _e B \end{bmatrix}
$$

for each element, we have ($_e L$ constant)

$$
\left(\frac{1}{_eL}\right)
\left[
\begin{array}{c:cccc}
(_1B + {_2}B + {_3}B + {_4}B) & (-{_1}B)(-{_2}B)(-{_3}B)(-{_4}B) \\
\hdashline
(-{_1}B) & (_1B) & \varnothing & \varnothing & \varnothing \\
(-{_2}B) & \varnothing & (_2B) & \varnothing & \varnothing \\
(-{_3}B) & \varnothing & \varnothing & (_3B) & \varnothing \\
(-{_4}B) & \varnothing & \varnothing & \varnothing & (_4B)
\end{array}
\right]
\left[
\begin{array}{c}
D_1 \\ \hdashline D_2 \\ D_3 \\ D_4 \\ D_5
\end{array}
\right]
=
\left[
\begin{array}{c}
P_1 \\ \hdashline P_2 \\ P_3 \\ P_4 \\ P_5
\end{array}
\right]
$$

$$(3.2.32)$$

where $[\varnothing]$ denotes a 3×3 null matrix. The displacements $\{D_2\}$, $\{D_3\}$, $\{D_4\}$, and $\{D_5\}$ of the support nodes 2 to 5 are zero. The external loads $\{P_2\}$, $\{P_3\}$, $\{P_4\}$, and $\{P_5\}$ nodes are unknown changes in reactions. Thus, we partition the matrix as shown by dashed lines in Eq. (3.2.32) and decouple the problem, i.e., solve

$$[(_1B + {_2}B + {_3}B + {_4}B)]\{D_1\} = \{P_1\} \qquad (3.2.33)$$

for $\{P_1\}$ given $\{D_1\}$ and then determine reactions from

$$
\left[
\begin{array}{c}
P_2 \\ P_3 \\ P_4 \\ P_5
\end{array}
\right]
= -
\left[
\begin{array}{c}
{_1}B \\ {_2}B \\ {_3}B \\ {_4}B
\end{array}
\right]
\{D_1\}
\qquad (3.2.34)
$$

Substituting Eq. (3.2.31d) into Eq. (3.2.33) and setting

$$
\{P_1\} =
\left\{
\begin{array}{c}
0 \\ 0 \\ P
\end{array}
\right\}
$$

we obtain

$$
\left(\frac{2}{\ell\rho}\right)
\left[
\begin{array}{ccc}
\left(4H_R\rho + \dfrac{2E_R A_0}{\rho^2}\right) & 0 & 0 \\
0 & \left(4H_R\rho + \dfrac{2E_R A_0}{\rho^2}\right) & 0 \\
0 & 0 & \left(4H_R\rho + \dfrac{16E_R A_0 f_R^2}{\rho^2}\right)
\end{array}
\right]
\left\{
\begin{array}{c}
d_{11} \\ d_{12} \\ d_{13}
\end{array}
\right\}
=
\left\{
\begin{array}{c}
0 \\ 0 \\ P
\end{array}
\right\}
$$

$$(3.2.35)$$

where d_{1i} is the displacement of node 1 in the x_i direction. The solution to the simultaneous equations in Eqs. (3.2.35) is

$$d_{11} = d_{12} = 0 \tag{3.2.36}$$

$$w = d_{13} = \frac{P\ell\rho}{8(H_R\rho + 4E_RA_0f_R^2/\rho^2)}$$

The changes in reactions are obtained by back-substitution of Eq. (3.3.36) into Eq. (3.2.34). For $i = 2,3,4,5$,

$$\{P_i\} = \begin{Bmatrix} \pm\Delta H \text{ or } 0 \\ \pm \Delta H \text{ or } 0 \\ -P/4 \end{Bmatrix}$$

where ΔH is the change in the horizontal forces:

$$\Delta H = \frac{f_R P/2}{4f^2\,(1 + H_R\rho/EA) + H_R/E_RA_0}$$

and the $\pm$ signs (or zero) depend on the value of i. The tension increases in the top members and decreases in the bottom members. Plotted in Fig. 3.2.6 (as a function of sag ratio f_R) are nondimensionalized values of w for two ratios of E_RA_0/H_R. We see that the sag ratio has a more pronounced effect on the response than does the modulus of rigidity. Also shown in

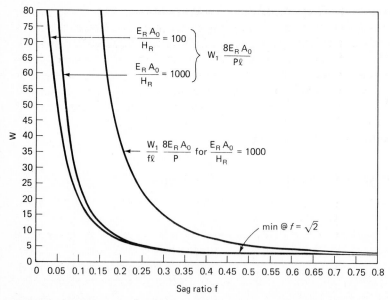

Figure 3.2.6 Deflections vs. sag ratio of cable cross.

Fig. 3.2.6 is the ratio $(w_0/f_R\ell)$ for a particular value $E_R A_0/H_R = 1000$. It can be seen for small sag ratios that the ratio increases dramatically. We have assumed small deflections, i.e., $w/f_R\ell \ll 1$. Thus, for small sag ratios, the ratio of added load to rigidity $E_R A_0$ must be small. For example, if $E_R A_0/H_R = 1000$ and if $f_R = 0.1$, then $(w_0/f_R\ell) \approx 32.3P/EA$ and hence we must limit $P/EA \ll 0.031$, which is not severe. However, if $f_R = 0.05$, $w/f_R\ell = 230P/EA$, and we must limit $P/EA \ll 0.0043$ to enable our assumption of small deflections.

Hyperbolic paraboloid network. Figure 3.2.7 shows a hyperbolic paraboloid network of cables with a rise-to-span ratio of 1:4. Each cable is assumed to be a ⅜-in (9.5-mm) diameter 6×7 wire strand core (WSC) bridge rope for which $A_0 = 0.063$ in^2 (42 mm^2), weight $= 0.24$ lb/ft (0.36 kg/m), $E_R = 20 \times 10^3$ ksi (140×10^3 MN/m^2). Each cable is prestressed to 100 ksi (690 MN/m^2).

As an illustration of the capability of the finite element model described herein to treat nonlinear static problems, this network was loaded in the $-x_3$ direction at node 1 (Fig. 3.2.7a) with concentrated forces of varying magnitudes. Different load increment sizes were considered. Because of the nearly flat initial shape of the network, the static response was very sensitive to the size of load increments, especially at the beginning of loading where the response is highly nonlinear. Figure 3.2.7b shows the static stress response to different rates of loading increments [39]. The stress was much more sensitive than the displacement response. In both instances, the predominant effect is in the initial stages of loading. The curves parallel each other after sufficient stiffness has developed in the network. This implies that in modeling *incrementally* a nonlinear static problem with an initially slack, or nearly flat, network, care should be taken to apply extremely small increments for the first few steps. The increment sizes can be increased thereafter. Alternatively, an iterative process can be used for nonlinear problems. However, those iteration processes have their own set of difficulties in modeling small stiffness systems, as will be seen subsequently.

3.3   Nonlinear Cable Finite Element Analysis

In this section we will discuss solution methods and equation formulations for nonlinear finite element models of cable networks. We will attempt to retain the full nonlinearity of the problem, including finite displacements, nonlinear constitutive equations, and nonconservative loadings. The virtual work statements for total finite displacement from the unstressed state, Eq. (3.2.12b), and for incremental finite displacements from a prestressed reference configuration, Eq. (3.2.15), will be used to develop the finite element models. We will limit our discussion here to a straight ele-

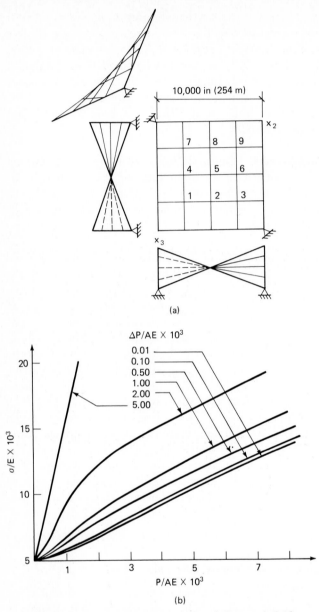

Figure 3.2.7 Incremental analysis of hyperbolic paraboloid network. (*a*) Hyperbolic paraboloid network; (*b*) static stress (nodes 1 to 2) of hyperbolic paraboloid network.

ment approximation with linear interpolating functions, as in Eqs. (3.2.18) to (3.2.22) for the linearized in-service response of cables. Before attempting to formulate the nonlinear algebraic equations for nodal displacements, it is worthwhile to consider the solution methods that will be used to solve those equations.

3.3.1   Solutions of nonlinear algebraic equations

Let us postulate a set of N nonlinear algebraic equations in N variables, y_i, $i = 1, N$:

$$g_i(y_j) = 0 \quad i, j = 1, N \tag{3.3.1}$$

An approximate solution $^P y_i$ is sought to Eq. (3.3.1), where we use a superscript to denote a successive approximation to y_i, i.e., $^P y_i$ is a supposedly better approximation to y_i than $^{P-1} y_i$.

There are two basic classes of approximate solution methods for nonlinear equations, i.e., iterative and incremental methods. Each class has relative strengths and weaknesses compared to the other class. A weakness in the iterative methods is that an initial trial solution $^0 y_i$ is required in the neighborhood of the exact solution y_i (within the radius of convergence). If the trial solution is within that radius of convergence, the convergence is very rapid. The strength of the iterative method is that the accuracy of the solution at a particular value of the forcing function embedded in Eq. (3.3.1) is not dependent on the accuracy obtained at any other value of the forcing function.

In the incremental methods of solution, no initial trial solution is required. Instead, the solution at a particular value of the incremented forcing function is taken as the solution at the end of the previous increment plus the solution to the locally linearized equations. However, the weakness in the incremental method is that unless the equations are exactly linear, small errors generated within each increment will accumulate, making the total solution progressively less accurate as succeeding values of the forcing functions are considered.

To combine the strengths of both methods, a combined approach can be used. In this approach the incremental method is used to obtain an initial trial solution, which is then iterated upon to converge within a prescribed error bound of the exact solution. The initial trial obtained by the incremental solution should be sufficiently close to the exact solution that few iterations would be required.

Methods to initiate the nonlinear solution methods are of concern for both classes of methods, i.e., trial solutions for the iterative methods and first increment solutions for the incremental methods. If the nonlinear equations are a slight perturbation of a set of equivalent linear equations, the initial trial solution for the iterative methods could be taken as the solution to that equivalent linear set. Unfortunately, for tension structures the nonlinear equations are usually not slight perturbations of a linear problem. In fact, often there is no equivalent linear problem.

To initiate the incremental solution process, a stable reference state is required. Often, the unstressed states of tension structures have no initial stiffness with regard to the first increment of load applied, no matter how small that first increment of load is. Special methods have to be used to start the solution process in that case, e.g., iterative method or viscous relaxation [16, 89, 90]. In the viscous relaxation method an artificial dynamic problem is solved in which inertial or viscous resistance to motion provides initial stability to the solution process.

Iteration methods.   There are several iteration methods for the solution of Eq. (3.3.1). The simplest of these is the Gauss-Seidel method of successive substitution. In the total-step Gauss-Seidel approach Eq. (3.3.1) is rewritten as

$$y_i = g_i'(y_j) \tag{3.3.2a}$$

and thus successive approximations can be based on

$$^{p+1}y_i = g_i'(^py_j) \tag{3.3.2b}$$

Convergence is presumed when a suitable norm of the unknowns is within a prescribed tolerance, e.g., the euclidean norm is often used:

$$\sqrt{(^{p+1}y_i - \ ^py_i)\,(^{p+1}y_i - \ ^py_i)} \leqslant \varepsilon$$

The descriptive phrase "total-step" implies that all of $^{p+1}y_i$ are based on values of ^py_i only. A more efficient scheme is the single-step Gauss-Seidel method in which most recent estimates of y_i are used to develop a better estimate $^{p+1}y_i$, i.e.,

$$^{p+1}y_i = g_i'\,(^{p+1}y_j, j < i; \ ^py_j, j \geqslant i) \tag{3.3.3}$$

The advantage of the Gauss-Seidel methods is their simplicity. But convergence is not guaranteed in that the equations must be diagonally dominant.

A more stable and efficient scheme is the Newton-Raphson method discussed previously in Chap. 2. Define the error between the true solution y_i and the most recent approximation ^py_i as

$$\Delta^p y_i = y_i - \ ^py_i \tag{3.3.4}$$

Taking a Taylor expansion of Eq. (3.3.1) about the most recent estimate ^Py_i, we have

$$0 = g_i(y_j) = g_i(^Py_j) + \left(\left.\frac{\partial g_i}{\partial y_j}\right|_{y_i\ =\ ^Py_j}\right)(\Delta^Py_j)$$

Thus, we compute Δ^Py_i from

$$\left(\left.\frac{\partial g_i}{\partial y_j}\right|_{y_i\ =\ ^Py_j}\right)\Delta^Py_j = -g_i(^Py_j) \tag{3.3.5}$$

which is a linear set of equations given a prior estimate ^Py_i. Classic methods of linear algebraic solutions can be used to calculate Δ^Py_i. Thus, an improved estimate is given by

$$^{p+1}y_i = {}^Py_i + \Delta^Py_i$$

Subsequent approximations can be calculated from Eqs. (3.3.4) and (3.3.5). Convergence is defined by either a chosen norm of $g_i(^{p+1}y) \leqslant \varepsilon$, or by a chosen norm of $\Delta^{p+1}y \leqslant \varepsilon_2$. The advantage of the Newton-Raphson method is that its convergence rate is quadratic near the true solutions. A one-dimensional analog of the method is shown in Fig. 3.3.1. It can be seen that the choice of initial trial solution plays an important role in the convergence of the Newton-Raphson method.

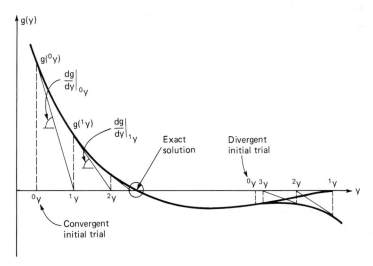

Figure 3.3.1   Newton-Raphson method.

The gradient matrix in Eq. (3.3.5) is called the tangent stiffness matrix:

$$^pJ_{ij} = \left[\frac{\partial g_i}{\partial y_j} \Big|_{y_j = ^py_j} \right]$$

The evaluation of $^pJ_{ij}$ is required in every iteration, and that process is potentially very time consuming in a finite element analysis. In situations where the nonlinearities are not severe and good initial trial solutions are available, the tangent stiffness matrix can be left unchanged. This is called the modified Newton-Raphson method and is illustrated in Fig. 3.3.2a. Its convergence rate is only linear, but it is more computationally efficient since J_{ij} does not have to be recomputed. It has difficulties in converging for highly nonlinear or singular problems. A perturbation of the modified Newton-Raphson method is to change the tangent stiffness matrix only at intervals in the iteration process as shown in Fig. 3.3.2b.

Incremental methods. Incremental schemes are extensions of linear analysis. In a finite element model the loads are applied in small increments and the structure is assumed to respond linearly during each load increment. Rewrite Eq. (3.3.1) as

$$f_i (y_i) = P_i (y_i) \tag{3.3.6}$$

where $f_i(y_j)$ represents the internal forces (from strain energy) and $P_i(y_j)$ represents the external forces (from external work). Both f_i and P_i are nonlinear functions of y_i, e.g., P_i could represent nonconservative forces.

Let us assume that we have obtained a solution ^qy_i after q increments have been applied and that we seek an increment $^q\Delta y_i$ due to an increment $^q\Delta P_i$ of load.

$$^{q+1}y_i = {}^qy_i = {}^q\Delta y_i \tag{3.3.7a}$$

$$^q\Delta P_i = {}^{q+1}P_i ({}^{q+1}y_j) - {}^qP_i ({}^qy_j) \tag{3.3.7b}$$

Equation (3.3.6) must be satisfied for every $^{q+1}y_i$. Thus, taking a Taylor expansion of Eq. (3.3.6) about the solutions ^qy_i, we have

$$f_i ({}^{q+1}y_j) = f_i ({}^qy_j + {}^q\Delta y_j) = P_i ({}^{q+1}y_j)$$

$$= f_i ({}^qy_j) + \left(\frac{\partial f_i}{\partial y_j} \Big|_{y_j = ^qy_j} \right) {}^q\Delta y_i = P_i ({}^{q+1}y_j) \tag{3.3.8}$$

Subtracting Eq. (3.3.6) evaluated at ^qy_i from Eq. (3.3.8), we have

$$\left(\frac{\partial f_i}{\partial y_j} \Big|_{y_j = ^qy_j} \right) {}^q\Delta y_j = {}^q\Delta P_i \tag{3.3.9}$$

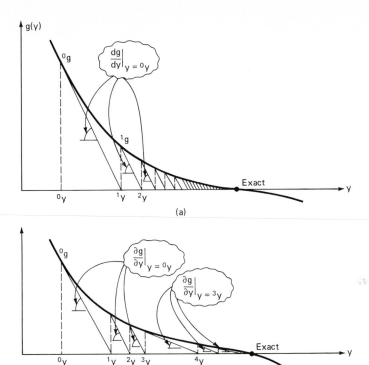

Figure 3.3.2 Modified Newton-Raphson methods. (a) Modified Newton-Raphson method; (b) periodic updates of J_{ij}.

Equation (3.3.9) is a set of linear equations for $^q\Delta y_i$, and again we have obtained a tangent stiffness matrix of gradients of f_i evaluated at the end of the previous increment $^q y_i$. If $P_i (y_j)$ is a nonconservative set of forces, approximations must be used to evaluate $^q\Delta P_i$ since, by Eq. (3.3.7b), it is a function of $^q\Delta y_i$. If $^q\Delta P_i$ is small enough, Eq. (3.3.9) provides a good approximation to Δy_i and the total displacement y_i is the vector sum of incremental displacement vectors.

The drawback to the incremental method is that the error tends to be cumulative, as shown in Fig. 3.3.3 for the one-dimensional case. This error can grow unacceptably large. An improvement on the purely incremental scheme is the so-called self-corrective incremental scheme. In that method, instead of using an increment of load $^q\Delta P_i$ as defined by Eq. (3.3.7b), $^q P_i(^q y_j)$ is replaced by $^q f_i(^q y_j)$ from Eq. (3.3.6) and we have

$$\left(\frac{\partial f_i}{\partial y_j}\bigg|_{y_j \,=\, ^q y_j}\right) {}^q\Delta y_j = {}^{q+1}P_i \left({}^{q+1}y_j\right) - {}^q f_i \left({}^q y_j\right) \qquad (3.3.10)$$

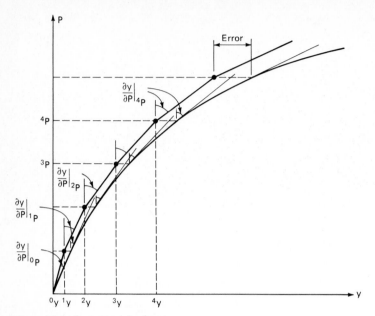

Figure 3.3.3   Incremental scheme.

This provides better error control than the purely incremental scheme, but extra computational effort is required to compute the internal forces ^qf_i after every increment. The extra effort has been shown to be worthwhile in terms of accuracy.

Combined incremental-iterative method.   A combination of the incremental and iterative methods is recommended. The incremental method provides good initial trial solutions for an iterative method. Let a superscript p denote an iteration and a superscript q denote an increment. Equation (3.3.1) can be written at the end of the $q+1$st increment as

$$^{q+1}g_i\,(^{q+1}y_j^*) = 0 \qquad (3.3.11)$$

where $^{q+1}y_j^*$ is the true solutions at the end of the $q+1$st increment. Assume that

$$^{q+1}y_j^* = {}^qy_j + {}^0\Delta y \qquad (3.3.12)$$

in which ^qy_i is the computed solution after q increments and $^0\Delta y$ is the error after p iterations. Expand Eq. (3.3.11) by Taylor series to obtain

$$^{q+1}g_i\,(^{q+1}y_j^*) = {}^{q+1}g_i\,(^qy_j) + \left(\left.\frac{\partial g_i}{\partial y_j}\right|_{y_j={}^qy_j}\right){}^0\Delta y_j = 0 \qquad (3.3.13)$$

which can be solved for $^0\Delta y_i$. Thus, the initial iterate for a Newton-Raphson scheme on $^{q+1}y_i^p$ is

$$^{q+1}y_i^0 = {}^qy_i + {}^0\Delta y_i \qquad (3.3.14)$$

The $p+1$st iterate is

$$^{q+1}y_i^{p+1} = {}^qy_i + \sum_{p=0}^{p+1} {}^p\Delta y_i \qquad (3.3.15)$$

in which $^p\Delta y_i$ is obtained by solving

$$\left(\frac{\partial g_i}{\partial y_j}\bigg|_{y_j = {}^{q+1}y_j}\right) {}^p\Delta y_j = - {}^{q+1}g_i \left({}^{q+1}y_j^p\right) \qquad (3.3.16)$$

See Fig. 3.3.4 for a one-dimensional analog, where Newton-Raphson iterations are used after three increments.

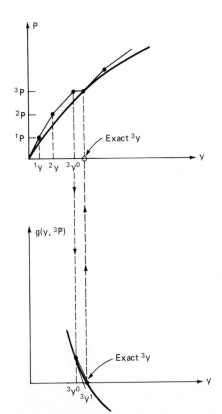

Figure 3.3.4 Combined incremental-iterative scheme.

3.3.2 Nonlinear solutions for cable elements

The solution methods described in Sec. 3.3.1 can be applied to finite element models of cables. Our task in this subsection is to derive the tangent stiffness matrix $[^P J]$ for an iterative method and $[^q J]$ for an incremental method. The tangent stiffness matrices for a system of cable elements are assembled from the tangent stiffness matrices for the individual elements in the assemblage in the same fashion described previously for the linear stiffness matrix. We will focus our attention on an individual element.

The first step in the derivation of a tangent stiffness matrix for an element is to develop an expression of nodal equilibrium of forces analogous to Eq. (3.3.1) for an iterative method or to Eq. (3.3.6) for an incremental method. To that end we adopt the virtual work expression Eq. (3.2.14b) in terms of added displacements u_i from some reference state R. That reference state could be the unstretched configuration if we were considering the nonlinear problem of the prestressing phase, in which case $T_R = 0$ and $\lambda_R = 1$. Alternatively, the reference state could be the prestressed configuration, or any other intermediate state, with stipulated T_R, λ_R, and E_R. In matrix terminology, the virtual work, Eq. (3.2.14b), is

$$\int_{S_R} \left[\frac{\partial \{\delta u\}^T}{\partial s_R} \{\theta\}(T) - \frac{1}{\lambda_R} \{\delta u\}^T \{q\} \right] ds_R - \{\delta D\}^T \{P_c\} = 0 \quad (3.3.17)$$

where $\{P_c\}$ are concentrated nodal forces, and we have used the fact that the direction cosines $\{\theta\}$ and tension T in the additionally deformed state are

$$\{\theta\} = \left(\{\theta_R\} + \frac{\partial \{u\}}{\partial s_R} \right) \frac{\lambda_R}{\lambda} \quad (3.3.18a)$$

$$T = T_R + (\lambda - \lambda_R)(E_R A_0) \quad (3.3.18b)$$

with the relative strain γ and elongation ratio λ given by

$$\gamma = \{\theta_R\}^T \frac{\partial \{u\}}{\partial s_R} + \frac{1}{2} \left[\frac{\partial u}{\partial s_R} \right]^T \left[\frac{\partial u}{\partial s_R} \right] \quad (3.3.18c)$$

$$\lambda = \sqrt{1 + 2\gamma}\, \lambda_R \quad (3.3.18d)$$

Note that $\{q\}$ in Eq. (3.3.17) is the total distributed load on the additionally deformed cable element per unit length of the unstressed element. The term $E_R A_0$ in Eq. (3.3.18b) is the tangent modulus of elasticity at the reference state ($T = T_R$, $\lambda = \lambda_R$). Thus, we are adopting here a piecewise linear representation of the stress-strain relation, i.e., between T_R and T, T is linear in λ. We do this in order to treat general stress-strain curves and to enable elastoplastic analysis. A completely analogous derivation could be pursued with any other form of the constitutive relations, e.g., using the unstressed state as the reference state we could assume a nonlinear curve

$$T = C_1(\lambda - 1)^{C_2}$$

with C_1 and C_2 being experimentally determined constants. In our piecewise linear representation we must be cautious to update the reference state R sufficiently often to maintain a proper tangent modulus $E_R A_0$, e.g., if unloading occurs in a plastic region. Also, if upon application of a load $\{q\}$ we calculate a relative displacement $\{u\}$ which significantly changes EA_0 (the instantaneous tangent modulus) from $E_R A_0$, we should instead apply a fraction of that load and update the reference state and $E_R A_0$ before applying the remainder of the load.

Let us adopt the straight-element approximating functions used in Sec. 3.2.2.:

$$\{u\} = [N_\alpha \ N_\beta] \begin{Bmatrix} d_\alpha \\ d_\beta \end{Bmatrix} \tag{3.3.19a}$$

$$\{\delta u\} = [N_\alpha \ N_\beta] \begin{Bmatrix} \delta d_\alpha \\ \delta d_\beta \end{Bmatrix} \tag{3.3.19b}$$

$$\{q\} = [N_\alpha \ N_\beta] \begin{Bmatrix} p_\alpha \\ p_\beta \end{Bmatrix} \tag{3.3.19c}$$

$$\frac{\partial \{u\}}{\partial s_R} = \frac{1}{L} (\{d_\beta\} - \{d_\alpha\}) \tag{3.3.19d}$$

$$\frac{\partial \{\delta u\}}{\partial s_R} = \frac{1}{L} (\{\delta d_\beta\} - \{\delta d_\alpha\}) \tag{3.3.19e}$$

with

$$[N_\alpha] = (1 - \xi)[I] \tag{3.3.19f}$$

$$[N_\beta] = \xi[I] \tag{3.3.19g}$$

Therefore, the relative strain γ and direction cosines $\{\theta\}$ are

$$\gamma = \frac{1}{L} \{\theta_R\}^T (\{d_\beta\} - \{d_\alpha\}) + \frac{1}{2L^2} (\{d_\beta\}^T - \{d_\alpha\}^T)(\{d_\beta\} - \{d_\alpha\}) \tag{3.3.20a}$$

$$\{\theta\} = \frac{1}{\sqrt{1 + 2\gamma}} \{\theta_R\} + \frac{1}{L} (\{d_\beta\} - \{d_\alpha\}) \tag{3.3.20b}$$

Some useful expressions of derivatives of Eqs. (3.3.18) with respect to $\{d_\alpha\}$ and $\{d_\beta\}$ are

$$\frac{1}{\sqrt{1+2\gamma}}\frac{\partial\gamma}{\partial\{d_\alpha\}} = -\frac{1}{L}\{\theta\}^T \qquad (3.3.21a)$$

$$\frac{1}{\sqrt{1+2\gamma}}\frac{\partial\gamma}{\partial\{d_\beta\}} = +\frac{1}{L}\{\theta\}^T \qquad (3.3.21b)$$

$$\frac{\partial\{\theta\}}{\partial\{d_\alpha\}} = -\frac{1}{L\sqrt{1+2\gamma}}([I] - \{\theta\}\{\theta\}^T) \qquad (3.3.21c)$$

$$\frac{\partial\{\theta\}}{\partial\{d_\beta\}} = +\frac{1}{L\sqrt{1+2\gamma}}([I] - \{\theta\}\{\theta\}^T) \qquad (3.3.21d)$$

$$\frac{\partial T}{\partial\{d_\alpha\}} = E_R A_0 \frac{\partial\lambda}{\partial\{d_\alpha\}} = -\frac{E_R A_0 \lambda_R}{L}\{\theta\}^T \qquad (3.3.21e)$$

$$\frac{\partial T}{\partial\{d_\beta\}} = +\frac{E_R A_0 \lambda_R}{L}\{\theta\}^T \qquad (3.3.21f)$$

We can now substitute Eqs. (3.3.18) to (3.3.20) into Eq. (3.3.17) and integrate over the nondimensional length $0 \leq \xi \leq 1$ of each element. Note that T and $\{\theta\}$ are independent of ξ and that Eq. (3.2.20b) can be used. We obtain the virtual work as

$$\sum_e \left[(\delta d_\alpha^T \; \delta d_\beta^T) \left(T \begin{Bmatrix} -\theta \\ \theta \end{Bmatrix} - \frac{L}{6\lambda_R} \begin{Bmatrix} 2p_\alpha + p_\beta \\ p_\beta + 2p_\beta \end{Bmatrix} \right) \right]$$

$$- [\delta D^T]\{P_c\} = 0 \qquad (3.3.22)$$

Since the virtual displacements δD are not zero, we have upon assemblage of the total force contributions from all elements

$$\{G\} = \{F\} - \{P\} = 0 \qquad (3.3.23a)$$

where $\{F\}$ is the internal force contributions of each element incident at a node

$$F = \sum_e \left(T \begin{Bmatrix} -\theta \\ \theta \end{Bmatrix} \right) \qquad (3.3.23b)$$

and $\{P\}$ is the external force contributions of each element incident at a node plus the concentrated force at the node

$$\{P\} = \sum_e \left(\frac{L}{6\lambda_R} \begin{bmatrix} 2p_\alpha + p_\beta \\ p_\alpha + 2p_\beta \end{bmatrix} \right) + \{P_c\} \qquad (3.3.23d)$$

Equations (3.3.20) are now of the form of Eqs. (3.3.1) and (3.3.6) in Sec. 3.3.1. Here we will consider only conservative forces $\{P\}$. In the next chapter on dynamics we will consider nonconservative forces. For the present we take

$$\frac{\partial \{P\}}{\partial D} = \{0\}$$

In an iterative method the tangent stiffness matrix, according to Eq. (3.3.5) for a Newton-Raphson method, is from Eq. (3.3.23):

$$[{}^pJ] = \frac{\partial \{G\}}{\partial \{D\}} \bigg|_{\{D\} = \{{}^pD\}}$$

$$= \sum_e \frac{\partial}{\partial \begin{Bmatrix} {}^pd_\alpha \\ {}^pd_\beta \end{Bmatrix}} \left({}^pT \begin{bmatrix} {}^p\theta \\ {}^p\theta \end{bmatrix} \right) \qquad (3.3.24a)$$

where the superscript p denotes the most recent trial solutions for $\{{}^pD\}$. In an incremental method the tangent stiffness matrix, according to Eq. (3.3.9), is likewise

$$[{}^qJ] = \frac{\partial \{F\}}{\partial \{D\}} \bigg|_{\{D\} = \{{}^qD\}}$$

$$= \sum_e \frac{\partial}{\partial \begin{Bmatrix} {}^qd_\alpha \\ {}^qd_\beta \end{Bmatrix}} \left({}^qT \begin{bmatrix} {}^q\theta \\ {}^q\theta \end{bmatrix} \right) \qquad (3.3.24b)$$

where the superscript q denotes the solution $\{{}^qD\}$ after q increments of load.

We can take the indicated derivatives in Eqs. (3.3.20) with the assistance of Eqs. (3.3.21) to obtain an assembled $[{}^pJ]$ or $[{}^qJ]$ of element tangent stiffness matrices:

$$[J] = \sum_e [{}_eJ] \qquad (3.3.25a)$$

$$[{}_eJ] = \frac{1}{L\sqrt{1 + 2\gamma}} \begin{bmatrix} B & -B \\ -B & B \end{bmatrix}$$

where

$$[B] = (T)[I] + (E_R A_0 \lambda_R - T_R)\{\theta\}\{\theta\}^T \qquad (3.3.25b)$$

For an iterative Newton-Raphson solution we have to solve for $\{^P\Delta D\}$ from

$$[^PJ]\{^P\Delta D\} = -\{^PG\} = \{P\} - \{^PF\} \qquad (3.3.26)$$

where $\{^P\Delta D\}$ is the correction to a previous trial solution $\{^PD\}$, and $[^PJ]$ and $\{^PF\}$ are the tangent stiffness matrix and internal force vector, respectively, evaluated at $\{^PD\}$, i.e., use ^PT and $^P\{\theta\}$ in Eqs. (3.3.25) and (3.3.23). Then an improved trial solution is given by Eqs. (3.3.18) and (3.3.20) as

$$\{^{p+1}D\} = \{^PD\} + \{^P\Delta D\} \qquad (3.3.27a)$$

$$^{p+1}\gamma = \frac{1}{L}\{\theta_R\}^T(\{^{p+1}d_\beta\} - \{^{p+1}d_\alpha\}^T)(\{^{p+1}d_\beta\} - \{^{p+1}d_\alpha\}) \qquad (3.3.27b)$$

$$^{p+1}\lambda = \lambda_R\sqrt{1 + 2^{p+1}\gamma} \qquad (3.3.27c)$$

$$\{^{p+1}\theta\} = \frac{\lambda_R}{^{p+1}\lambda}\left[\{\theta_R\} + \frac{1}{L}(\{^{p+1}d_\beta\} - \{^{p+1}d_\alpha\})\right] \qquad (3.3.27d)$$

$$^{p+1}T = T_R + (^{p+1}\lambda - \lambda_R)E_RA_0 \qquad (3.3.27e)$$

Equation (3.3.26) is then used to calculate a new correction.

For an incremental self-corrective method, as in Eq. (3.3.10), we have that after $(q+1)$ load increments the displacement $\{^{q+1}D\}$ is

$$\{^{q+1}D\} = \{^qD\} + \{^q\Delta D\} \qquad (3.3.28a)$$

where $\{^q\Delta D\}$ is the increment of displacement to be added to the accumulated relative displacement $\{^qD\}$ after q increments. The increment $\{^q\Delta D\}$ is determined from solution of

$$[^qJ]\{^q\Delta D\} = \{^{q+1}P\} - \{^qF\} \qquad (3.3.28b)$$

in which $[^qJ]$ and $\{^qF\}$ are the tangent stiffness matrix and internal force vector, respectively, evaluated at $\{^qD\}$, and $\{^{q+1}P\}$ is the accumulated external load through $q+1$ increments. Having determined $\{^{q+1}D\}$, we can advance the solutions through another increment of load by using Eqs. (3.3.18) and (3.3.20):

$$^{q+1}\gamma = \frac{1}{L}\{\theta_R\}^T(\{^{q+1}d_\beta\} - \{^{q+1}d_\alpha\}) + \frac{1}{2L^2}(\{^{q+1}d_\beta\} - \{^{q+1}d_\alpha\}) \qquad (3.3.29a)$$

$$\cdot (\{^{q+1}d_\beta\} - \{^{q+1}d_\alpha\})$$

$$^{q+1}\lambda = \lambda_R\sqrt{1 + 2^{q+1}\lambda} \qquad (3.3.29b)$$

$$\{^{q+1}\theta\} = \frac{\lambda_R}{^{q+1}\lambda} \left[\{\theta_R\} + \frac{1}{L} (\{^{q+1}d_\beta\} - \{^{q+1}d_\alpha\}) \right] \tag{3.3.29c}$$

$$^{q+1}T = T_R + (^{q+1}\lambda - \lambda_R) E_R A_0 \tag{3.3.29d}$$

and by then reapplying Eq. (3.3.28).

3.3.3 Examples of nonlinear problems

To illustrate the solution methods for nonlinear cable analysis using finite elements we will consider three examples. The first example will be a simple problem that can be solved by hand calculations. The other examples require digital computer simulation.

Cable cross example. We considered previously under "Example Problems" the simple linear problem of two cables intersecting as shown in Fig. 3.2.5. Let us now develop a nonlinear solution to that same problem. As observed under "Example Problems" there is only one degree of freedom, the vertical deflection of node 1, because of the symmetry of geometry and loading. The two top cables, 3 and 4, are called sagging cables and will have the same tensions, relative strains, and elongation ratios for all values of the deflections w. The two bottom cables, 1 and 2, are called hogging cables and will have tensions, relative strains, and elongation ratios different from the sagging cables but equal for each of the hogging cables.

We shall define some nondimensional variables to facilitate the description of the nonlinear solution. For the prestressed configuration, let the sag ratio be f_R and

$$\rho_R = \sqrt{1 + 4f_R^2} \qquad \beta = \frac{H_R}{E_R A_0}$$

and thus

$$T_R = H_R \rho_R$$

$$\lambda_R = 1 + \beta \rho_R$$

$$L = \rho_R \frac{\ell}{2}$$

For the hogging cables:

$$\{_1\theta_R\}^T = \frac{1}{\rho_R} [0 \quad 1 \quad 2f_R]$$

$$\{_2\theta_R\}^T = \frac{1}{\rho_R} [0 \quad -1 \quad 2f_R]$$

For the sagging cables:

$$\{_3\theta_R\}^T = \frac{1}{\rho_R} [1 \quad 0 \quad -2f_R]$$

$$\{_4\theta_R\}^T = \frac{1}{\rho_R} [-1 \quad 0 \quad -2f_R]$$

Since we have assumed a linear elastic material, we have for all cables

$$T_R - \lambda_R E_R A_0 = -E_R A_0$$

To calculate a tangent stiffness matrix, we assume a nondimensional deflection of node 1 as

$$f = \frac{w}{\ell}$$

where w is either a trial solution for a Newton-Raphson method or the solution after one increment of load. Then for the hogging cables we have

$$f_h = f_R - f$$

$$r_h = \frac{1}{\sqrt{\rho_R^2 - 8f \, (f_R - \frac{1}{2} f)}}$$

$$\gamma_h = -\frac{4}{\rho_R^2} f \, (f_R - \frac{1}{2} f)$$

$$\lambda_h = \frac{1 + \beta \rho_R}{\rho_R r_h}$$

$$\sqrt{1 + 2\gamma_h} = \frac{1}{\rho_R r_h}$$

$$T_h = E_R A_0 \left(\frac{1 + \beta \rho_R}{\rho_R r_h} - 1 \right)$$

$$\frac{1}{L \sqrt{1 + 2\gamma_h}} = \frac{2r_h}{\ell}$$

$$\{_1\theta\}^T = r_h [0 \quad 1 \quad 2f_h]$$

$$\{_2\theta\}^T = r_h [0 \quad -1 \quad 2f_h]$$

$$\{_1F\} + \{_2F\} = -T_h(\{_1\theta\} + \{_2\theta\}) = -E_RA_0\left(\frac{1 + \beta\rho_R}{\rho_R r_h} - 1\right) r_h \begin{Bmatrix} 0 \\ 0 \\ 4f_h \end{Bmatrix}$$

$$\{_3\theta\}^T = r_s[1 \quad 0 \quad -2f_s]$$

$$\{_4\theta\}^T = r_s[-1 \quad 0 \quad -2f_s]$$

$$\{_3F\} + \{_4F\} = -T_s(\{_3\theta\} + \{_4\theta\}) = -E_RA_0\left(\frac{1 + \beta\rho_R}{\rho_R} - r_s\right) \begin{Bmatrix} 0 \\ 0 \\ -4f_s \end{Bmatrix}$$

$$\frac{1}{L\sqrt{1 + 2\gamma_s}}[_3B + {}_4B] = \frac{4E_RA_0}{\ell}$$

$$\times \begin{bmatrix} \dfrac{1 + \beta\rho_R}{\rho_R} - r_s + r^3{}_s & 0 & 0 \\ 0 & \dfrac{1 + \beta\rho_R}{\rho_R} - r_s & 0 \\ 0 & 0 & \dfrac{1 + \beta\rho_R}{\rho_R} - r_s + 4r^3{}_sf^2{}_s \end{bmatrix}$$

Assembling the tangent stiffness matrices for the sagging and hogging cables and the residual force vectors, we see that since all deflections but those at node 1 are zero,

$$[J]\{^P\Delta D\} = \{P\} - \{^PF\}$$

for a Newton-Raphson method becomes

$$\begin{bmatrix} \left(\begin{aligned} 2\left(\dfrac{1 + \beta\rho_R}{\rho_R}\right) \\ - r_h - r_s + r^3{}_s \end{aligned}\right) & 0 & 0 \\ 0 & \left(\begin{aligned} 2\left(\dfrac{1 + \beta\rho_R}{\rho_R}\right) \\ - r_h - r_s + r^3_h \end{aligned}\right) & 0 \\ 0 & 0 & \left(\begin{aligned} 2\left(\dfrac{1 + \beta\rho_R}{\rho_R}\right) \\ - r_h - r_s + 4 \\ \cdot(r^3_hf^2_h + r^3_sf^2_s) \end{aligned}\right) \end{bmatrix}$$

$$
\cdot \left\{ \begin{array}{c} 0 \\ 0 \\ \Delta f \end{array} \right\} = \left\{ \begin{array}{c} 0 \\ 0 \\ \dfrac{1}{4}\beta \left(\dfrac{P}{H_R}\right) \end{array} \right\} + \left\{ \begin{array}{c} 0 \\ 0 \\ \left(\dfrac{1 + \beta\rho_R}{\rho_R} - r_h\right) f_h \\ -\left(\dfrac{1 + \beta\rho_R}{\rho_R} - r_s\right) f_s \end{array} \right\}
$$

Thus, the problem reduces to the solution of

$$
\left[2\left(\frac{1 + \beta\rho_R}{\rho_R}\right) - r_h - r_s + 4(r_h^3 f_h^2 + r_s^3 f_s^2) \right] \Delta f
$$

$$
= \left[\frac{1}{4}\beta\frac{P}{H_R} + \left(\frac{1 + \beta\rho_R}{\rho_R} - r_h\right) f_h - \left(\frac{1 + \beta\rho_R}{\rho_R} - r_s\right) f_s \right]
$$

for $\Delta f = (\Delta w/\ell)$, the nondimensional correction to a previous nondimensional trial $f = w/\ell$ for a total nondimensional load (P/H_R). For an incremental method (self-corrective) the same equation is obtained but with Δf interpreted as the increment to be added to the previously accumulated f and with (P/H_R) interpreted as the total accumulated load.

Two numerical examples are considered for a Newton-Raphson solution. Let us assume a prestress H_R and material properties $E_R A_0$ such that $\beta = 0.001$. Also, assume a total load such that $(P/H_R) = 10$. Tables 3.3.1 and 3.3.2 show the successive iterations to f for two different initial sag ratios, $f_R = 0.1$ and 0.05.

For the initial sag ratio of 0.1 the behavior is nearly linear. With an initial trial of $f = 0$, the result of the first iteration is the equivalent linear solution. We see that convergence is rapid. After three iterations a correction, $\Delta f/f$, of 0.014 percent was needed. Note that the nonlinear solution only differs from the linear solution by -3.7 percent.

For the initial sag ratio of 0.05 the behavior is more nonlinear. With an initial trial of $f = 0$, five iterations were required to reduce the corrections

TABLE 3.3.1   Cable Cross with Larger Sag

		$f_R = 0.1$	$\beta = 0.001$	$\dfrac{P}{H_R} = 10$	
f	f_h	f_s	r_h	r_s	Δf
0.0	0.1	0.1	0.980508	0.980508	$+0.0322945$
0.0322945	0.0677055	0.1322945	0.990956	0.966733	-0.00114375
0.03115078	0.06884922	0.1311508	0.990652	0.9672781	-0.00000340
0.031146477	...	...	...	...	...

TABLE 3.3.2 Cable Cross with Smaller Sag

	$f_R = 0.05$		$\beta = 0.001$	$\dfrac{P}{H_R} = 10$	
f	f_h	f_s	r_h	r_s	Δf
0.0	0.05	0.05	0.995031	0.995037	+0.1151877
0.1151877	−0.0651877	0.1651877	0.991608	0.9495226	−0.033999
0.0811886	−0.03118866	0.13118866	0.9980603	0.9672602	−0.0130447
0.06814399	−0.01814399	0.11814399	0.9993423	0.97320115	−0.0019082
0.06623575	−0.01623575	0.11623528	0.9994733	0.9740267	−0.00003775
0.066197998	...	...	...	...	...

$\Delta f / f = 0.006$ percent. Note that in this case the nonlinear solution differs from the linear solutions (result of first iteration) by −74 percent.

Incremental example. The second example [20] is the incremental analysis of the in-service response to an added 3 kips (13.3-kN) load applied to node 2 of a plane suspended cable prestressed by an initial dead load of 4 kips (17.8 kN) at each internal node (see Fig. 3.3.5). This same example was considered in a paper [163] by F. Baron and M. S. Vendatesan, "Nonlinear Analysis of Cable and Truss Structures," *Journal of the Structural Division, ASCE,* vol. 97, no. ST2, February 1971, pp. 679–710. The area of the cable is 0.227 in² (1.46 cm²) and E = 12,000 ksi (82,800 MPa).

In this treatment of the example [20] the external load of 3 kips (13.3 kN) was divided into 10 and 20 uniform increments for comparison purposes. Table 3.3.3 gives the incremental displacements of nodes 1 and 2 and stress in member 2 at the end of each increment for the 10-increment

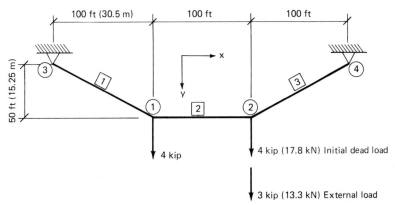

Figure 3.3.5 Plane suspended cable.

TABLE 3.3.3 Incremental Displacements of Plane Suspended Cable

Increment no.	Displacement of node 1, in		Stress in member 2, ksi	Displacement of node 2, in	
	x	y		x	y
1	3.36	−6.40	36.52	3.49	7.37
2	3.20	−6.15	37.81	3.18	6.65
3	3.03	−5.89	39.11	2.90	6.02
4	2.88	−5.64	40.43	2.65	5.47
5	2.73	−5.39	41.76	2.43	4.99
6	2.58	−5.14	43.09	2.23	4.57
7	2.45	−4.91	44.44	2.05	4.19
8	2.32	−4.68	45.79	1.89	3.87
9	2.20	−4.46	47.14	1.75	3.75
10	2.08	−4.25	48.50	1.62	3.31
Sum	26.83	−52.91	...	24.19	50.01

NOTE: 1 in = 2.54 cm, 1 ksi = 6.9 MPa.

case. Baron and Vendatesan's results are given for comparison purposes in Tables 3.3.4 and 3.3.5, where the final displacements and stresses for both cases are shown. Good agreement was achieved. It should be noted that more accurate results are achieved with the greater number of increments.

TABLE 3.3.4 Displacement Comparison for Suspended Cable

Node no.	Component	Ref. 163, in	Current study Ref. 20, in			
			10 increments	% of difference	20 increments	% of difference
1	x	26.28	26.83	2.10	26.00	−0.84
	y	−52.08	−52.91	1.57	−52.36	0.54
2	x	23.40	24.19	−3.37	−23.80	1.71
	y	48.48	50.01	3.13	49.11	1.30

NOTE: 1 in = 2.54 cm.

TABLE 3.3.5 Stress Comparison for Suspended Cable

Member no.	Ref. 163, ksi	Current study Ref. 20, ksi			
		10 increments	% of difference	20 increments	% of difference
1	53.32	52.95	−0.69	52.96	−0.67
2	48.91	48.50	−0.84	48.52	−0.80
3	55.52	55.17	−0.63	55.18	−0.61

NOTE: 1 ksi = 6.9 MPa.

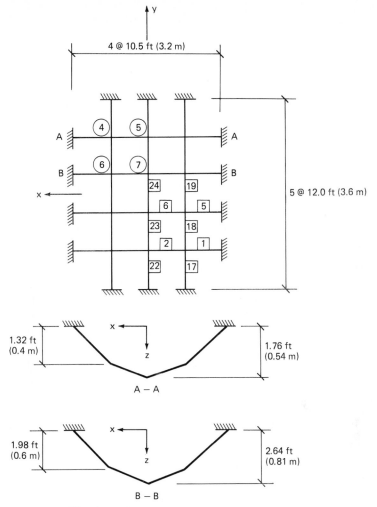

Figure 3.3.6   Hanging net.

Combined incremental-iterative example. The third example involves results obtained by A. Lo [90] for a combined incremental and iterative scheme in which after every increment of load is applied, Newton-Raphson iterations are performed to obtain improved convergence. A hanging net is shown in Fig. 3.3.6. That prestressed configuration is due to self-weight of the cable elements. The in-service problem considered is the superposition of an additional vertical dead load of 13.608 kips applied at every interior node. For each cable element $A_0 = 0.01$ ft^2 (0.30 cm^2), E_R = 3,600,000 ksf (172,440 MPa). The same example was considered by Baron and Vendatesan [163].

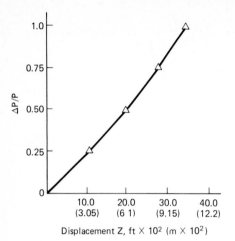

Figure 3.3.7 Displacement of node 7 of hanging net [90].

Displacement Z, ft × 10² (m × 10²)

In the present study [90] the in-service loads were applied in four equal-sized increments and a Newton-Raphson iteration was performed after each increment to achieve a convergence tolerance of 0.0025. The first increment required five iterations, while three iterations are sufficient for the subsequent increments. The results are shown in Fig. 3.3.7 and Tables 3.3.6 and 3.3.7.

The displacement of nodes 5 and 7 is plotted in Fig. 3.3.7 to demonstrate the nonlinear load displacement behavior of the hanging net. The displacements and element forces are shown in Tables 3.3.6 and 3.3.7, respectively, together with Baron and Vendatesan's [163] results. It can be seen that there is virtually no difference between the two sets of results. The slight discrepancies can be explained by the residual errors in the iterations and by the different definitions of strain used; nonlinear strain

TABLE 3.3.6   Hanging Net—Comparison of Displacements

Model	Nodal points (see Fig. 3.3.7)	Displacement components, ft × 10²		
		x	y	z
Present study, Ref. 90	4	−2.35	−2.73	34.0
	5	0.00	−3.66	35.1
	6	−2.25	−0.73	23.4
	7	0.00	−0.96	23.9
Ref. 163	4	−2.36	−2.73	34.0
	5	0.00	−3.67	35.2
	6	−2.26	−0.73	23.4
	7	0.00	−0.96	24.0

NOTE: 1 ft = 0.305 m.

TABLE 3.3.7   Hanging Net—Comparison of Member Forces

	Member forces, kips									
	Member (see Fig. 2.3.7)									
Model	1	2	5	6	17	18	19	22	23	24
Present study, Ref. 90	83.6	82.5	80.0	78.5	44.4	43.8	43.8	58.9	58.0	57.9
Ref. 163	83.5	82.4	80.0	78.5	44.4	43.9	43.8	58.9	58.0	57.8

NOTE: 1 kip = 4.45 kN.

is used here, whereas engineering strains are used by Baron and Vendatesan.

Since the deformations in this problem are not excessive, it might have been more efficient to apply the loads in a single step and then perform iterations. However, such a decision is purely problem-dependent, and, in general, the magnitude of deformations in a large displacement problem cannot be predicted beforehand.

Dynamics of Cable Systems

Linearized Dynamics of Cable Systems

If time-varying forces shake a structure at less than one-third the lowest natural frequency of vibration for the structure, the response is essentially static in nature and dynamic motions and stresses can be considered negligible [94]. If greater frequencies of time variation are present in the spectrum of external loads applied, then inertial forces on the structural components are mobilized and dynamic motions become important. For extremely high frequencies of load variation greater than four times the highest natural frequency of vibration for the structure, there is insufficient time for the structure to respond to the load before the direction of load reverses and there are negligible dynamic motions due to those frequency components of the load [94].

In tension structures, the stiffness is relatively small compared to other structural components, and the mass may be large because of attached components or cladding. Since the natural frequencies are proportional to the square root of the stiffness-to-mass ratio, the natural frequencies of tension structures can be expected to be smaller than most other structural types. Dynamic responses to time-varying loads are more significant for tension structures in that dynamic stresses superposed on static prestresses may lead to failure of members as a result of overstressing or fatigue. The dynamic stresses are particularly large near resonant conditions, i.e., when the frequency of load is near a natural frequency of the system. In cable systems an increase in mass of the system, e.g., snow loads, may be accompanied by a decrease in stiffness of counterstressing cables (stiffness is proportional to cable tension) and the natural frequencies may be significantly decreased [95]. Thus, under combined load systems, e.g., static snow load and dynamic wind loads, the dynamic response of the system may become more pronounced if the natural frequencies are decreased to within three times the frequency of load.

The dynamic loads that should be considered for cable structures include blast, seismic, wave, and wind forces [10]. Blast loadings from diffraction and refraction of shock waves about the structure consist of an abrupt rise in pressures and suctions which then rapidly decay. Seismic loads due to ground motions have a frequency range of 3 to 10 Hz. Cable structures have low natural frequencies (<4 Hz), whereas supporting structures have much higher natural frequencies. Since seismic loads are transferred to the cable system through the supports, high-frequency components of the load will be amplified by the supports and the low-frequency components will be attenuated. Thus, seismic effects on the cable may not be significant [10]. Hydrodynamic wave pressures from the relative motion of the cables and the fluid have a frequency range from 0.05 to 0.5 Hz [96]. Also drift forces with long periods and fluctuations in currents have significant effects on extremely compliant systems used as mooring systems [96]. In a deterministic model of wind velocities, loads due to wind pressures or suctions can be considered to consist of a slowly varying component plus a rapidly varying gust component with a frequency range of 0.1 to 1.0 Hz [97]. It is the low-frequency gust components that are most important for cable structures. Because of the flexible nature of cable systems, an aeroelastic model of their response to winds is preferable to a rigid structural model of the wind forces. Aerodynamic flutter may occur because of coupling between modes of vibration, causing energy to be drawn from the airstream to excite instabilities.

Transient dynamic motions in structural systems eventually decay when the forcing functions are removed. This damping is due to conversion of mechanical energy into energy radiated from the structure by waves or heat. There are three sources of damping: material, connection, and environmental. In addition to the usual hysteresis effects, material damping in cables is attributable to friction forces developed between the separate strands laid into a cable rope. Connection damping is associated with loss of energy from friction in joints and support connections. Environmental damping is associated with interaction of the cable system with the medium in which it is embedded, e.g., aerodynamic damping, real fluid effects, and radiation of waves in a fluid. Typical damping factors [10] for unclad cable systems are between 0.5 and 3 percent of a critical damping factor (the value of damping at which no oscillation occurs and transient motion decays in a purely exponential fashion). When fluid damping is included, the damping factors range between 3 and 5 percent [11, 102]. For cable systems with cladding, damping factors may become as high as 20 percent [98, 108]. It has been found that damping is considerably greater for slack cables than for taut cables [11]. Damping is a desirable aspect in that the dynamic stresses at resonant peaks of the dynamic response are considerably reduced. It is very difficult to establish the form of damping in structural systems and to estimate percentages of damping. Experimental work is required in models or prototypes to estimate the damping present in the system.

There are three sources of nonlinearity in cable behavior: material, loads, and geometry. Material nonlinearity is of importance in the prediction of the dynamic response of cable systems in the calculation of the instantaneous stiffness of the system. For elastoplastic behavior, the reduction in strain upon reversal of load when beyond the elastic limit requires the use of the original elastic modulus during the unloading and subsequent reloading to the new elastic limit. Hysteretic damping occurs. Load nonlinearities occur primarily in hydrodynamic applications, wherein the dynamic pressure always remains normal to the moving and rotating cable axis. The geometric nonlinearities due to large deflections change the stiffness and, hence, the natural frequencies of the system.

One factor to remember in the dynamic analysis of cable systems is that superposition of loads and displacements is not strictly valid for nonlinear problems. Thus, the modal superposition method of forced vibration analysis in which results from decoupled analyses of the separate modes are superposed is not a true representation of the nonlinear behavior. However, if a prestressed static configuration is established which has significant stiffness and if small in-service dynamic loads are then added, it is reasonable to assume linearity of the superposed dynamic response. In that case, methods of linear structural dynamics can be used [93, 94, 110], as described in this chapter. The most common linear methods adopt the approach of transforming the problem from the time domain to the frequency domain, wherein harmonic analysis of each mode of natural vibration is performed and results appropriately recombined to obtain histories of motions and stress. This approach is particularly useful in the analysis of stochastic loads which reflect the random nature of wind and waves.

In this chapter, we will review methods of linear structural dynamics beginning with a simple cable example with a single degree of freedom. Some aspects of vibration of continuous cable segments will be noted. The main emphasis of the chapter will be on multiple-degree-of-freedom systems. Methods of nodal analysis of free vibration and modal superposition will be reviewed. The dynamic equations of motion for a linearized finite element model of cable systems will be developed. The forms that the mass matrix can take will be discussed, and results for an example cable will be used to illustrate the difference in results obtained with different mass matrices. A straight cable element will be adopted in the formulation of the equations of motion.

4.1 Single-Degree-of-Freedom Example

To introduce the subject of dynamics of cable systems and to review some concepts of dynamics, let us consider the response of the simple cable problem presented in Sec. 2.1.1 of a single cable segment transversely loaded by a central concentrated force (see Fig. 4.1.1). We will assume that all mass M of the system is concentrated at the central load point and, furthermore, that the point is constrained to move only vertically. (A more realistic representation would include the possibility of horizontal motion of the mass.) These assumptions enable us to completely characterize the dynamic motions of, and stresses in, the system by the single degree of freedom of vertical motion at the mass point.

The determination of the static equilibrium state of the system prestressed by the force P (due to the weight of the mass M and any other dead load) was the subject of Sec. 2.1.1. We consider here the in-service behavior of the system wherein either an additional time-dependent load

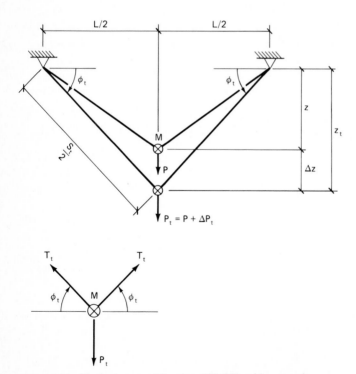

Figure 4.1.1 Single-degree-of-freedom (SDOF) cable example.

ΔP_t is applied, or the system is disturbed by an initial displacement or velocity of the mass. To that end, we draw a free-body diagram of the mass at any arbitrary time t as in Fig. 4.1.1 and write the equation of motion for the mass by summing the vertical external and internal forces on the mass and equating that sum to the inertial force of mass times vertical acceleration.

$$M\ddot{z}_t = P_t - 2T_t \sin \phi_t \tag{4.1.1}$$

where a dot used as a superscript denotes differentiation with respect to time ($\dot{z}_t = \partial z_t/\partial t$, $\ddot{z}_t = \partial^2 z_t/\partial t^2$), and a subscript t denotes the fact that the quantity is time-varying and is evaluated at time t.

From geometric considerations, the total stretched length is

$$\frac{S_t}{L} = \sqrt{1 + e_t^2} \tag{4.1.2}$$

and the angle of inclination ϕ_t of each section is given by

$$\sin \phi_t = \frac{e_t}{\sqrt{1 + e_t^2}} \tag{4.1.3}$$

where we have defined a nondimensional coordinate

$$e_t = \frac{2z_t}{L} \tag{4.1.4}$$

We will assume here a linear elastic material as in Eq. (2.1.1) of Sec. 2.1.1, such that the total stretched length is related to the total unstressed length S_0 by

$$\frac{S_t}{S_0} = 1 + \frac{T_t}{AE} \tag{4.1.5}$$

Thus, by combining Eqs. (4.1.2) and (4.1.5), the tension is

$$T_t = AE \left(\frac{L}{S_0} \sqrt{1 + e_t^2} - 1 \right) \tag{4.1.6}$$

Finally, by substitution of Eqs. (4.1.3) and (4.1.6) into Eq. (4.1.1), we obtain the nonlinear equation of motion of the mass as a function of the forcing function P_t:

$$\left(\frac{LM}{2} \right) \ddot{e}_t = P_t - 2AEe_t \left(\frac{L}{S_0} - \frac{1}{\sqrt{1 + e_t^2}} \right) \tag{4.1.7}$$

Now, assume that P_t consists of a steady prestressing force P and a transient in-service force ΔP_t:

$$P_t = P + \Delta P_t \qquad (4.1.8a)$$

Likewise, the nondimensional deflection e_t will consist of a static deflection e and a transient deflection Δe_t:

$$e_t = e + \Delta e_t \qquad (4.1.8b)$$

The prestressing phase was considered in Sec. 2.1.1 where, in the present notation, it was determined that e and P were related by (see Eq. 2.1.8)

$$0 = P - 2AEe\left(\frac{L}{S_0} - \frac{1}{\sqrt{1 + e^2}}\right) \qquad (4.1.9)$$

Since $\ddot{e}_t = (\ddot{e} + \Delta \ddot{e}_t) = \Delta \ddot{e}_t$, we can subtract Eq. (4.1.9) from Eq. (4.1.7) to obtain an equation of motion for the increment Δe_t of in-service deflection as

$$\left(\frac{LM}{2}\right) \Delta \ddot{e}_t = \Delta P_t - 2AE$$

$$\times \left\{ \frac{L}{S_0} \Delta e_t - \frac{e}{\sqrt{1 + e^2}} \left[\frac{1 + \Delta e_t/e}{\sqrt{\dfrac{1 + (e + \Delta e_t)^2}{1 + e^2}}} - 1 \right] \right\} \qquad (4.1.10)$$

wherein e is assumed known a priori from the solution to Eq. (4.1.9).

Equation (4.1.10) is still nonlinear in Δe_t. We can obtain a linearized version of that equation by assuming that Δe_t is small compared to unity, by taking Taylor expansions of the radical term, and by retaining only first-order terms in Δe_t. This gives

$$M \Delta \ddot{e}_t + K \Delta e_t = \frac{2 \Delta P_t}{L} \qquad (4.1.11a)$$

where

$$K = \frac{4AE}{L}\left[\frac{L}{S_0} - \frac{1}{(1 + e^2)^{3/2}}\right] \qquad (4.1.11b)$$

is the same stiffness coefficient derived in Sec. 2.1.1 for small static in-service deflections from a prestressed state (see Eq. 2.1.10).

Equation (4.1.11) is a second-order linear differential equation describing the small time-dependent deflections Δe_t to be superposed on the static prestressed deflection e. The stiffness coefficient K is dependent on the magnitude of the prestressed deflection and, thus, will affect the in-service response. Note that we have neglected the higher-order terms in Δe_t in the transition from Eq. (4.1.10) to (4.1.11). In our linearized analysis of the in-service dynamic response we will assume that e is invariant with time and K can, therefore, be assumed constant. The effect of these assumptions will be examined subsequently when the nonlinear solution to Eq. (4.1.7) is considered.

4.1.1   Free vibrations

If we assume that ΔP_t is zero in Eq. (4.1.11), we can rewrite that equation as

$$\Delta \ddot{e}_t + \frac{K}{M} \Delta e_t = 0 \qquad (4.1.12)$$

which is the classic equation of a linear undamped oscillator [93] and admits a nontrivial solution for Δe_t in the form

$$\Delta e_t = \overline{A} \sin \omega t + \overline{B} \cos \omega t \qquad (4.1.13a)$$

or in the form

$$\Delta e_t = \overline{R} \sin (\omega t + \theta) \qquad (4.1.13b)$$

where ω = natural circular frequency of free vibration

$\overline{A}$ and $\overline{B}$ (or $\overline{R}$ $\sqrt{\overline{A}^2 + \overline{B}^2}$) = amplitudes of that vibration

θ = a phase angle ($\tan-1$ $\overline{B}/\overline{A}$)

Upon substitution of Eq. (4.1.13) into Eq. (4.1.12) we obtain

$$\left(\frac{K}{M} - \omega^2 \right) (\overline{R} \sin (\omega t + \theta) = 0$$

which must be satisfied for all t. Thus, the natural circular frequency is

$$\omega = \sqrt{\frac{K}{M}} = \left\{ \frac{4AE}{ML} \left[\frac{L}{S_0} - \frac{1}{(1 + e^2)^{3/2}} \right] \right\}^{1/2} \qquad (4.1.14a)$$

The natural circular frequency ω is related to the period $\overline{T}$ of one cycle of the response by

$$\overline{T} = \frac{2\pi}{\omega} \qquad (4.1.14b)$$

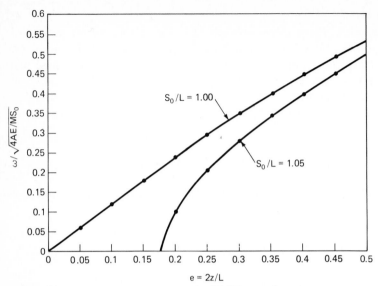

Figure 4.1.2   Natural frequency vs. sag for SDOF example.

The natural circular frequency for the transversely loaded cable is plotted in Fig. 4.1.2 as a function of the prestressed deflection e for two different choices of S_0/L. The nonlinear nature of ω can be seen in that figure. Also, note the transition to a nearly linear dependence as the magnitude of e increases. See Fig. 4.1.3 for a representative plot of the free vibration response of the system.

The amplitudes $\overline{A}$ and $\overline{B}$ (or $\overline{R}$ and θ) remain to be determined. As far as Eq. (4.1.12) is concerned, their magnitudes are unconstrained. They depend instead on the conditions imposed on the system at $t = 0$. Suppose that an initial in-service displacement Δe_0 and velocity $\Delta \dot{e}_0$ are prescribed. Then, evaluation of Eq. (4.1.13) and its first time derivative at $t = 0$ gives

$$\overline{A} = \frac{\Delta \dot{e}_0}{\omega}$$

$$\overline{B} = \Delta e_0$$

and thus

$$\overline{R} = \sqrt{\overline{A}^2 + \overline{B}^2} = \sqrt{(\Delta e_0)^2 + \left(\frac{\Delta \dot{e}_0}{\omega}\right)^2}$$

$$\theta = \tan^{-1}\left(\frac{\omega \, \Delta e_0}{\Delta \dot{e}_0}\right)$$

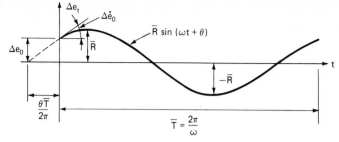

Figure 4.1.3 Harmonic motion of SDOF example.

The equation of free undamped vibrations during the in-service phase of behavior of the transversely loaded cable is, therefore,

$$\Delta e_t = \left(\frac{\Delta\dot{e}_0}{\omega}\right) \sin \omega t + \Delta e_0 \cos \omega t \qquad (4.1.15a)$$

$$e_t = e + \Delta e_t \qquad (4.1.15b)$$

where ω is given by Eq. (4.1.14a).

4.1.2 Free damped vibrations

As noted previously, all real structural systems are subject to damping of natural vibrations. We can represent viscous damping in our example by addition of a restraining force proportional to the velocity of additional displacement $\Delta\dot{e}_t$, i.e., Eq. (4.1.1) becomes

$$M \Delta\ddot{e}_t + C \Delta\dot{e}_t + K \Delta e_t = \frac{2 \Delta P_t}{L} \qquad (4.1.16)$$

where C is the proportional damping factor, which we will assume constant in our linearized analysis. Nonlinear representations of C will be considered in a subsequent chapter when we treat hydrodynamic forces on cable systems.

For free underdamped vibrations we write

$$\Delta\ddot{e}_t + \beta\omega \Delta\dot{e}_t + \omega^2 \Delta e_t = 0 \qquad (4.1.17a)$$

where we have defined a damping ratio

$$\beta = \frac{C}{C_{\text{cr}}} = \frac{C}{M\omega} \qquad (4.1.17b)$$

where C_{cr} = the critical damping factor, i.e., the smallest damping factor for which oscillation occurs in the real response. By "underdamped" we

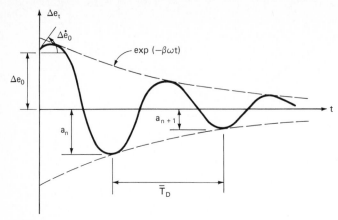

Figure 4.1.4 Exponentially damped harmonic motion of SDOF example.

mean $\beta < 1$ and therefore oscillations do occur. The term β is often expressed as a percentage of critical damping.

The solution to Eq. (4.1.17a) is

$$\Delta e_t = \exp(-\beta\omega t)(\overline{A} \sin \omega_D t + \overline{B} \cos \omega_D t) \qquad (4.1.18a)$$

where

$$\omega_D = \omega\sqrt{1 - \beta^2} \qquad (4.1.18b)$$

is the natural circular frequency of the exponentially decaying motion. Initial conditions on displacement Δe_0 and velocity $\Delta\dot{e}_0$ are again used to evaluate $\overline{A}$ and $\overline{B}$. Thus

$$\Delta e_t = \exp(-\beta\omega t)\left(\frac{\Delta\dot{e}_0 + \Delta e_0 \beta\omega}{\omega_D} \sin \omega_D t + \Delta e_0 \cos \omega_D t\right) \qquad (4.1.18c)$$

A representative plot of a damped Δe_t is shown in Fig. 4.1.4.

The damped natural frequency does not differ significantly from the undamped natural frequency for lightly damped systems ($\beta < 0.2$) [93], as usually occurs in structural problems. The damping characteristics of real responses are difficult to quantify. It is common to express damping in terms of viscous damping ratios determined from free vibration tests. Consider two successive peaks a_n and a_{n+1} in the response curve of Fig. 4.1.4. The ratio of those peaks [from Eq. (4.1.18) evaluated at $t = t_1$ and $t = t_2 = t_1 + \overline{T}_D$] is

$$\frac{a_n}{a_{n+1}} = \exp\left(\frac{2\pi\beta\omega}{\omega_D}\right) \qquad (4.1.19a)$$

Taking the natural logarithm of both sides of Eq. (4.1.19a), we obtain the logarithmic decrement δ as

$$\delta = \log_e\left(\frac{a_n}{a_{n+1}}\right) = 2\pi\beta\,\frac{\omega}{\omega_D} = \frac{2\pi\beta}{\sqrt{1-\beta^2}} \qquad (4.1.19b)$$

Equation (4.1.19b) can be used to estimate β from measured peaks a_n and a_{n+1} in a free vibration test:

$$\beta = \frac{\delta}{\sqrt{\delta^2 + 4\pi^2}} \qquad (4.1.19c)$$

4.1.3 Dynamic response [93]

For cases in which ΔP_t is nonzero in Eq. (4.1.11), we can construct a linearized solution from the free vibration solutions, Eqs. (4.1.15) or (4.1.18), plus a particular solution to Eq. (4.1.11). For example, if a sudden small increase in load

$$\Delta P_t = \alpha P \qquad 0 < t < \infty$$

were applied to the prestressed cable, the undamped solution to Eq. (4.1.11) would be

$$\Delta e_t = \overline{A}\sin\omega t + \overline{B}\cos\omega t + \frac{2\alpha P}{KL} \qquad (4.1.20a)$$

and $\overline{A}, \overline{B}$ would be evaluated from initial conditions on Δe_0 and $\Delta \dot{e}_0$, e.g., if $\Delta e_0 = \Delta \dot{e}_0 = 0$,

$$\Delta e_t = \frac{2\alpha P}{KL}\,(1 - \cos\omega t)$$

$$e_t = e + \Delta e_t \qquad (4.1.20b)$$

as plotted in Fig. 4.1.5. Recall that the superposed deflection must be small enough that our assumption of constant K from Eq. (4.1.11b) remains valid.

If an oscillatory load

$$\Delta P_t = \alpha P \sin \Omega t$$

were superposed, the total undamped solution for Δe_t would be

$$\Delta e_t = \overline{A}\sin\omega t + \overline{B}\cos\omega t + \frac{2\alpha P/L}{M(\omega^2 - \Omega^2)}\sin\Omega t \qquad (4.1.21a)$$

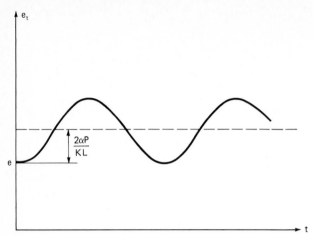

Figure 4.1.5 Forced vibration of SDOF example with $\Delta P = \alpha P$.

and if $\Delta e_0 = \Delta \dot{e}_0 = 0$,

$$\Delta e_t = \frac{2\alpha P}{ML(\omega^2 - \Omega^2)} \left(\sin \Omega t - \frac{\Omega}{\omega} \sin \omega t \right) \qquad (4.1.21b)$$

For those same initial conditions, the damped solution is

$$\Delta e_t = \frac{2\alpha P}{ML[(\omega^2 - \Omega^2)^2 + 4\beta^2\omega^2\Omega^2]} \left\{ (\omega^2 - \Omega^2) \sin \Omega t + 2\beta\omega\Omega \cos \Omega t \right.$$

$$\left. + \exp(-\beta\omega t) \left[2\omega\Omega\beta \cos \omega_D t - (\omega^2 - \Omega^2 - 2\beta\omega^2) \frac{\Omega}{\omega_D} \sin \omega_D t \right] \right\}$$

$$(4.1.21c)$$

In Eq. (4.1.21b) we can observe the possibility of unbounded resonant deflection if $\omega = \Omega$. The presence of damping bounds that resonance, as can be seen in Eq. (4.1.21c), but large deflections still occur in lightly damped systems if $\omega = \Omega$. The resonance is not accurately predicted by either of Eqs. (4.1.21) because as the deflection increases, the assumption of small Δe_t and, hence, constant K is invalidated. The structure stiffens and the natural frequency shifts so as to inhibit that resonance. A nonlinear analysis is required to predict the magnitudes of displacement and stress near a resonant condition accurately.

For other loading functions superposed on the prestressed state, we can use the Duhamel integral technique [93] to write the total solution as (undamped):

$$\Delta e_0 = \frac{\Delta \dot{e}_0}{\omega} \sin \omega t + \Delta e_0 \cos \omega t$$

$$+ \frac{2}{ML\omega} \int_0^t \Delta P_t (\tau) \sin [\omega(t - \tau)] \, d\tau \quad (4.1.22)$$

given $\Delta P_t (\tau)$ for $0 \leqslant \tau \leqslant t$. An equivalent solution to Eq. (4.1.22) can be written for the case in which damping is included.

4.1.4 Nonlinear effects

To estimate the effects of the linearization assumption of $\Delta e_t \ll 1$ on the free vibration of a transversely loaded cable and to illustrate a simplified nonlinear solution technique, let us reconsider the nonlinear equation for total deflection, Eq. (4.1.7), here rewritten as

$$\ddot{e}_t = \frac{2P_t}{LM} - \left(\frac{4AE}{LM} \right) e_t \left(\frac{L}{S_0} - \frac{1}{\sqrt{1 + e_t^2}} \right) \quad (4.1.23)$$

As a particular example, consider the case of an initially straight unstressed cable ($L/S_0 = 1$) to which a mass of weight W is attached at the center ($M = W/g$, $P_t = W$, g = gravitational constant) and suddenly released. Assume $W/AE = 0.006$. Then Eq. (4.1.23) is

$$\frac{L}{g} \ddot{e}_t = 2 - 666.67 \left(1 - \frac{1}{\sqrt{1 + e_t^2}} \right) e_t \quad (4.1.24)$$

The time history of e_t can be determined approximately by numerical integration of Eq. (4.1.24). The relative merits of various explicit and implicit techniques for integration of second-order nonlinear equations will be considered in a subsequent chapter. For now we will adopt a simple explicit scheme [61] in which we propagate e_t through a small time interval Δt from a discrete time t_1, to time $t_2 = t_1 + \Delta t$ by assuming constant acceleration during the interval, i.e. (see Fig. 4.1.6),

$$\ddot{e}_1 = 2 - 666.67 \left(1 - \frac{1}{\sqrt{1 + e_1^2}} \right) e_1 \quad (4.1.25a)$$

$$\dot{e}_2 = \dot{e}_1 + \int_{t_1}^{t_2} \ddot{e}_t \, dt = \dot{e}_1 + \ddot{e}_1 \, \Delta t \quad (4.1.25b)$$

$$e_2 = e_1 + \int_{t_1}^{t_2} \dot{e}_t \, dt = e_1 + \dot{e}_1 \, \Delta t + \frac{\ddot{e}_1 \, \Delta t^2}{2} \quad (4.1.25c)$$

$$\ddot{e}_2 = 2 - 666.67 \left(1 - \frac{1}{\sqrt{1 + e_2^2}} \right) e_2 \quad (4.1.25d)$$

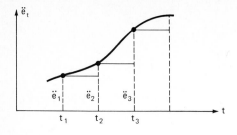

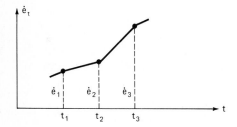

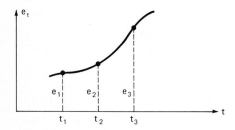

Figure 4.1.6 Explicit numerical integration with constant acceleration.

Equations (4.1.25) provide a recursive scheme in which e_t, $\dot{e}_t$, and $\ddot{e}_t$ at $t = t_2$ are calculated from values at $t = t_1$ starting with initial values e_0 and $\dot{e}_0$. Care must be taken in using an explicit scheme such as this [61] in that the time interval must be small enough to guarantee stability of results for progressively large times. In this example, a small interval $\Delta t = (0.001)\sqrt{L/g}$ was used to obtain the results displayed in Fig. 4.1.7.

We see in Fig. 4.1.7 that the nonlinear response is periodic but not harmonic. The static response to the load $P/AE = 0.006$ is shown as a horizontal line at $e = 0.183$ and was calculated using the inverse method described in Sec. 2.1. For a deflection less than that of the static value, the system has a smaller stiffness and, hence, a longer period than when the deflection is greater than the static value. We can observe this effect by comparing the partial period of $0.443\sqrt{L/g}$ when the deflection is greater than 0.183 to the partial period of $0.896\sqrt{L/g}$ when the deflection is less than 0.183. Also, the maximum deflection of 0.294 is less than twice the static value.

For comparison purposes the linearized solution of free vibration about

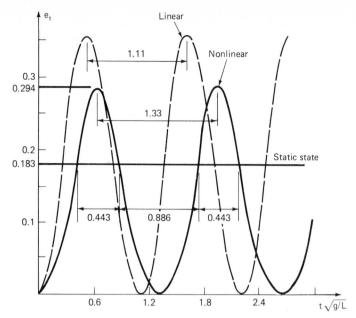

Figure 4.1.7 Nonlinear free response of SDOF example.

the static state with an initial deflection $\Delta e_0 = -0.183$ is also plotted in Fig. 4.1.7. That solution was obtained by calculating ω from Eq. (4.1.14a) as $\omega = 5.669 \sqrt{L/g}$ (with a corresponding period of $1.11 \sqrt{L/g}$) and using Eqs. (4.1.15) with $\Delta \dot{e}_0 = 0$ and $\Delta e_0 = -0.183$. We see that the period is underestimated by the linear solution and that the maximum deflection is overestimated (0.366). An alternative linearized solution could be obtained by assuming a partially prestressed state at a lesser value of $\overline{P}/AE$ and then superposing as a dynamic load the remainder $(P - \overline{P})/AE$ of load. In other words, we would assume a prestressing deflection $\bar{e} < 0.0183$ and calculate a corresponding $\overline{P}/AE$ by the inverse method. Then we would add a sudden load $(P - \overline{P})$ with initial conditions $\Delta e_0 = -\bar{e}$. This would lead to a reduced stiffness and frequency and, hence, an increased period. However, this increase in period would be accompanied by an increase in $e_{t,\max}$. For example, if we selected $\bar{e} = 0.15$, we would calculate a period of $1.34 \sqrt{L/g}$, which is close to the nonlinear period of $1.33 \sqrt{L/g}$, but the resulting maximum deflection would be 0.38.

The discrepancy between the linear and nonlinear solutions is due to two factors. First, we are underestimating the frequency over part of the period and overestimating it over the remainder of the period. Second, our initial deflection of 0.183 is not very small compared to unity and our basic assumption of small superposed deflections is questionable.

4.2   Free Vibration of Continuous Cable Segments

In Sec. 4.2, we examined the dynamic behavior of a cable structure which was represented as a single-degree-of-freedom system. Even for that simple structure the assumption was an approximation of the true dynamic behavior of the separate segments. The individual segments have distributed mass and will in fact have spatial and temporal change in dynamic response along their length, i.e., an infinite number of degrees of freedom.

The dynamic response of single suspended cables with distributed mass and elasticity is considered in Ref. 11. There it is noted that for taut cables

1. The swinging motion out of the plane of the suspended cable is uncoupled from the in-plane motion and has natural frequencies

$$\omega_n = n\pi \sqrt{\frac{H}{mL^2}} \qquad (4.2.1)$$

where n = number of nth harmonic mode of vibration
m = distributed mass per unit length of cable
H = horizontal component of pretension

2. The amplitude of in-line motion is considerably less than the amplitude of transverse in-plane motion, and the frequency of in-line motion (stress wave velocity) is considerably higher than the frequency of transverse motions (transverse disturbance wave velocity), implying that temporal changes occur mainly in unison with fluctuations in transverse deflection and that spatial changes in tension can reasonably be assumed to occur instantaneously.

3. The antisymmetric modes of in-plane transverse deflection generate no additional tension in the cable and have natural frequencies

$$\omega_n = 2n\pi \sqrt{\frac{H}{mL^2}} \qquad (4.2.2)$$

4. The symmetric modes of in-plane transverse deflections do generate additional tension in the cable and have natural frequencies embedded in the roots of the transcendental equation

$$\tan\frac{\tilde{\omega}}{2} = \frac{\tilde{\omega}}{2} - \frac{4}{\lambda^2}\left(\frac{\tilde{\omega}}{2}\right)^3 \qquad (4.2.3a)$$

where

$$\tilde{\omega} = \omega \sqrt{\frac{mL^2}{H}} \qquad (4.2.2b)$$

$$\lambda^2 = \left(\frac{mgL}{H}\right)^2 \frac{EA}{H(1 + 8f^2)} \qquad (4.2.3c)$$

$$f = \text{sag ratio}$$

5. Modal crossover between the first symmetric and first antisymmetric modes occurs at $\lambda^2 = 4\pi^2$ where the two frequencies are equal; crossover between the corresponding second modes occurs at $\lambda^2 = 16\pi^2$.

A. G. Pugsley [8, 164] presented a semiempirical approach for determining the lower frequencies of free vibration of an inextensible suspension chain with a span L and a sag ratio f which is valid for a range $1/10 \leqslant f \leqslant 1/4$. The first three frequencies are

$$\omega_1 = \frac{\pi}{\sqrt{2}} \sqrt{\frac{g}{L}} \sqrt{\frac{1 - 3f^2}{f}} \qquad (4.2.4a)$$

$$\omega_2 = \pi \sqrt{\frac{g}{L}} \sqrt{\frac{1 - 1.5f^2}{f}} \qquad (4.2.4b)$$

$$\omega_3 = \sqrt{2}\,\pi \sqrt{\frac{g}{L}} \sqrt{\frac{1 - 0.7f^2}{f}} \qquad (4.2.4c)$$

Krishna [10] presented an approximate analysis technique for determining natural frequencies of shallow-rise cable nets based on modeling the discrete cable system as a continuous taut membrane. Under that assumption, the system can be analyzed by the Rayleigh-Ritz or Galerkin's numerical procedure, whereby mode shapes are assumed which are compatible with boundary conditions and corresponding modal amplitudes and frequencies are calculated numerically.

4.3 Multiple-Degree-of-Freedom Systems

The dynamic response of multiply connected cable systems can be described in terms of dynamic motions of a finite number of discrete points in the system. The external forces and the inertial forces due to accelerations of discrete and distributed masses are counterbalanced by internal tensions in the cable segments interconnecting those discrete points.

Even the simple problem presented in Sec. 4.1 can be considered as a multiple-degree-of-freedom system. If we were to remove the constraint there imposed of no horizontal motion of the mass, we would have two degrees of freedom with the discrete motions being the horizontal and ver-

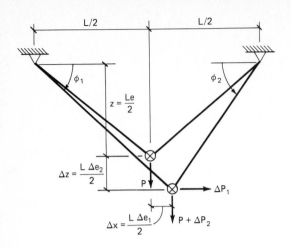

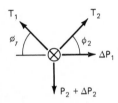

Figure 4.3.1 Two-degree-of-freedom cable example.

tical motions of the mass. With the possibility of horizontal motion allowed, the tensions in the two segments may be different (see Fig. 4.3.1).

If we follow the same procedure as in Sec. 4.1 (i.e., equilibrium in two directions, geometry of deformation, constitutive equation for each segment, and retain only first-order terms) but consider each segment separately, we arrive at the two linearized equations of motion:

$$M \, \Delta\ddot{e}_1 + K_1 \, \Delta e_1 = \frac{2 \, \Delta P_1}{L} \tag{4.3.1a}$$

$$M \, \Delta\ddot{e}_2 + K_2 \, \Delta e_2 = \frac{2 \, \Delta P_2}{L} \tag{4.3.1b}$$

where Δe_2 = nondimensional additional vertical motion of mass
  $\qquad$ = $e_2 - e$
  Δe_1 = nondimensional horizontal motion of mass
  ΔP_2 = added vertical load = $P_2 - P$
  ΔP_1 = added horizontal load
  K_2 = vertical stiffness coefficient (as in Sec. 4.1)

$$= \frac{4AE}{L} \left[\frac{L}{S_0} - \frac{1}{(1 + e^2)^{3/2}} \right] \tag{4.3.2a}$$

K_1 = horizontal stiffness coefficient

$$= \frac{4AE}{L} \left[\frac{L}{S_0} - \frac{1}{(1 + e)^{1/2}} + \frac{1}{(1 + e)^{3/2}} \right] \qquad (4.3.2b)$$

In this case the two equations of motion are decoupled. Under free vibrations, Eqs. (4.3.1) have separate solutions:

$$\Delta e_1 = \overline{A}_1 \sin \omega_1 t + \overline{B}_1 \cos \omega_1 t$$

$$\Delta e_2 = \overline{A}_2 \sin \omega_2 t + \overline{B}_2 \cos \omega_2 t$$

where ω_1 and ω_2 are two different natural frequencies:

$$\omega_i = \sqrt{\frac{K_i}{M}} \qquad i = 1, 2$$

The natural frequency ω_1 of sway motion is much higher than that of vertical motion ω_2 for small sag e. As e increases, however, ω_1 decreases, while ω_2 increases until, for $e > 1$, ω_1 is less than ω_2.

In Sec. 4.3.1, we will derive the matrix equations of motion of cable systems in terms of a finite number of degrees of freedom of motion at discrete points and will present linearized solutions to those equations. In this section, we consider solution techniques that can be used to determine the free and forced vibration responses of linear structural systems with a finite number of degrees of freedom, i.e., we will review solutions to second-order ordinary differential equations of the form

$$[M]\{\ddot{y}\} + [C]\{\dot{y}\} + [K]\{y\} = \{F\} \qquad (4.3.3)$$

where $\{y\}$, $\{\dot{y}\}$, $\{\ddot{y}\}$ are vectors of time-varying motions, velocities, and accelerations; $[M]$, $[C]$, $[K]$ are time- and motion-invariant mass, damping, and stiffness matrices of the interconnected degrees of freedom; and $\{F\}$ is the vector of time-varying external forces.

The linearized equation of motion, Eq. (4.3.3), represents dynamic motion about an equilibrium configuration. In the case of cable systems, that configuration will be the prestressed state. The forces $\{F\}$ are the added in-service forces. The motions $\{y\}$ are the added in-service components of deflection of the discrete points. The symmetric stiffness matrix $[K]$ will, in general, provide coupling between the degrees of freedom and in the case of cable systems will be a function of the prestressed deflections and tensions as well as elastic properties. The symmetric mass matrix $[M]$ will include diagonal entries from discrete masses corresponding to each degree of freedom and could also include either approximate diagonal entries only or coupling off-diagonal entries to represent the distributed mass of the cables. That choice of mass representation will be discussed subsequently. The damping matrix $[C]$ is difficult to quantify, and often, for lack of a better representation, approximate values proportional to the mass and stiffness matrices are used for computational convenience.

4.3.1   Free vibrations

We will first consider instances in which $\{F\}$ and $[C]$ are zero in Eq. (4.3.3), i.e., free undamped motion:

$$[M]\{\ddot{y}\} + [K]\{y\} = \{0\} \tag{4.3.4}$$

The free vibration solution to Eq. (4.3.4) will be used subsequently in the construction of solutions of Eq. (4.3.3) when $\{F\}$ and $[C]$ are nonzero.

Equation (4.3.4) admits a homogeneous harmonic solution of the form

$$\{y(t)\} = \{Y\} \exp{(i\omega t)} \tag{4.3.5}$$

under conditions on amplitudes $\{Y\}$ and circular frequencies ω. Upon substitution of Eq. (4.3.5) into Eq. (4.3.4) we obtain

$$([K] - \omega^2[M])\{Y\} = \{0\} \tag{4.3.6}$$

which is an eigenvalue problem possessing nonzero solutions for the eigenvector $\{Y\}$ at only N (N = number of degrees of freedom) values of the eigenvalues ω_n^2, $n = 1$ to N, i.e., when the determinant $p(\omega^2)$ of the matrix in Eq. (4.3.6) is zero:

$$p(\omega_n^2) = \det{([K] - \omega_n^2[M])} = 0 \tag{4.3.7}$$

The eigenvectors $\{Y_n\}$ corresponding to each ω_n have arbitrary magnitude but can be determined to within a constant of proportionality A_n by

$$\{Y_n\} = A_n\{\phi\}_n \tag{4.3.8a}$$

$$([K] - \omega_n^2[M])\{\phi\}_n = \{0\} \tag{4.3.8b}$$

where $\{\phi_n\}$ is the nth "mode" shape of $\{Y_n\}$. The mode shapes are orthogonal to one another with respect to both the mass and stiffness matrices, i.e., for two modes r and s with distinct frequencies ω_r and ω_s:

$$\{\phi\}_r^T[M]\{\phi\}_s = 0 \tag{4.3.9a}$$

$$\{\phi\}_r^T[K]\{\phi\}_s = 0 \tag{4.3.9b}$$

if $\omega_r^2 \neq \omega_s^2$. Note that the frequencies must be distinct. This may cause difficulties in cable systems where, as noted previously, equal frequencies can occur for certain selections of prestressed configurations. These orthogonality conditions will prove useful subsequently when forced motions are considered. One immediate use is to generate values for the normalizing factors A_n for each mode. Often A_n is selected such that the maximum entry in $\{\phi\}_n$ is unity. Alternatively, it is convenient for digital computer programs to define A_n such that

$$\{\phi\}_n^T [M] \{\phi\}_n = 1 \tag{4.3.10a}$$

in which case, we can define a square $N \times N$ matrix of all mode shapes

$$[\phi] = \left[\{\phi\}_1 \cdot \{\phi\}_2 \cdot \cdot \cdot \cdot \{\phi\}_N \right] \tag{4.3.10b}$$

such that

$$[\phi]^T [M] [\phi] = [I] \tag{4.3.10c}$$

Before elaborating on numerical procedures for determining the natural frequencies and mode shapes from Eqs. (4.3.7) and (4.3.8) for structural systems with large numbers of degrees of freedom, let us illustrate the free vibration solution process for a cable problem with three degrees of freedom. Consider the tightly stretched (zero sag) cable shown in Fig. 4.3.2a

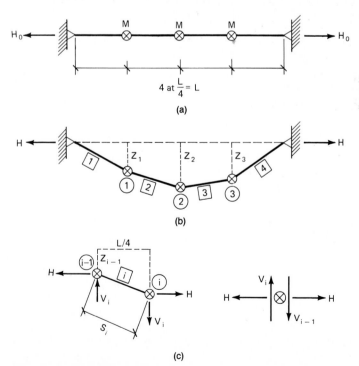

Figure 4.3.2   Three-DOF cable example.

with three equally spaced masses. Recalling from Sec. 4.2 that in-line motions are much less than transverse motions, we will assume large H_0 and neglect horizontal motions of the masses. This reduces the number of degrees of freedom to the three vertical motions z_1, z_2, z_3 of the masses.

Equilibrium of each mass i, shown in the additionally deformed position in Fig. 4.3.2b, requires

$$\frac{ML}{4} \ddot{e}_i = V_{i+1} - V_i \qquad i = 1, 2, 3 \qquad (4.3.11)$$

where $e_i = 4z_i/L \ll 1$ and V_i is the vertical component of tension T_i in the ith segment:

$$V_i = T_i \frac{L}{4S_i} (e_i - e_{i-1}) \qquad (4.3.12a)$$

with S_i = stretched length of the ith segment:

$$S_i = \frac{L}{4} \sqrt{1 + (e_i - e_{i-1})^2} \qquad (4.3.12b)$$

For an elastic material

$$T_i = AE \left(\frac{S_i}{S_{i0}} - 1 \right) \qquad (4.3.12c)$$

where S_{i0} = initial unstressed length of each segment (total length = S_0 = $4S_{i0} < L$ in this case).

Upon substitution of Eq. (4.3.12) into Eq. (4.3.11) and retaining only first-order terms, we obtain

$$\frac{ML}{4} \ddot{e}_i = AE \left(\frac{L}{S_0} - 1 \right) (e_{i-1} - 2e_i + e_{i+1}) \qquad i = 1, 2, 3 \quad (4.3.13a)$$

or, in matrix terms

$$\left(\frac{ML}{4} \right) \begin{bmatrix} 1 & 0 & 0 \\ 0 & 1 & 0 \\ 0 & 0 & 0 \end{bmatrix} \begin{Bmatrix} \ddot{e}_1 \\ \ddot{e}_2 \\ \ddot{e}_3 \end{Bmatrix}$$

$$+ AE \left(\frac{L}{S_0} - 1 \right) \begin{bmatrix} 2 & -1 & 0 \\ -1 & 2 & -1 \\ 0 & -1 & 2 \end{bmatrix} \begin{Bmatrix} e_1 \\ e_2 \\ e_3 \end{Bmatrix} = \begin{Bmatrix} 0 \\ 0 \\ 0 \end{Bmatrix}$$

which is in the form of Eq. (4.3.4). Following Eqs. (4.3.5) to (4.3.9), we have

$$\left\{ \begin{array}{c} e_1 \\ e_2 \\ e_3 \end{array} \right\} = \{e\} = \{\phi\} \exp{(i\omega t)}$$

$$\begin{bmatrix} (2\Lambda - \omega^2) & -\Lambda & 0 \\ -\Lambda & (2\Lambda - \omega^2) & -\Lambda \\ 0 & -\Lambda & (2\Lambda - \omega^2) \end{bmatrix} \left\{ \begin{array}{c} \phi_1 \\ \phi_2 \\ \phi_3 \end{array} \right\} = \left\{ \begin{array}{c} 0 \\ 0 \\ 0 \end{array} \right\} \qquad (4.3.13b)$$

where

$$\Lambda = 4AE/ML \ (L/S_0 - 1)$$

Thus, the three frequencies of free vibration ω_1, ω_2, ω_3 are determined from the zeros of the determinant of the matrix in Eq. (4.3.13b):

$$p(\omega^2) = (2\Lambda - \omega^2)^3 - 2\Lambda^2(2\Lambda - \omega^2) = 0$$

Those three zeros occur when

$$(2\Lambda - \omega^2) = 0$$

$$(2\Lambda - \omega^2)^2 = 2\Lambda^2$$

implying

$$\omega_1 = \sqrt{(2 - \sqrt{2}) \Lambda}$$

$$\omega_2 = \sqrt{2\Lambda}$$

$$\omega_3 = \sqrt{(2 + \sqrt{2}) \Lambda}$$

The mode shapes $\{\phi\}_1$, $\{\phi\}_2$, and $\{\phi\}_3$ are determined from Eq. (4.3.13b). Let us normalize by setting $\phi_1 = 1$ in each mode. For $\omega_2 = \sqrt{2\Lambda}$ we have from Eq. (4.3.13)

$$\Lambda \begin{bmatrix} 0 & -1 & 0 \\ -1 & 0 & -1 \\ 0 & -1 & 0 \end{bmatrix} \left\{ \begin{array}{c} 1 \\ \phi_2 \\ \phi_3 \end{array} \right\}_2 = \left\{ \begin{array}{c} 0 \\ 0 \\ 0 \end{array} \right\}$$

Thus

$$\left\{ \begin{array}{c} \phi_1 \\ \phi_2 \\ \phi_3 \end{array} \right\}_2 = \left\{ \begin{array}{c} 1 \\ 0 \\ -1 \end{array} \right\}$$

For $\omega_1 = \sqrt{(2 - \sqrt{2})}\,\Lambda$ we have

$$\Lambda \begin{bmatrix} \sqrt{2} & -1 & 0 \\ -1 & \sqrt{2} & -1 \\ 0 & -1 & \sqrt{2} \end{bmatrix} \begin{Bmatrix} \phi_1 \\ \phi_2 \\ \phi_3 \end{Bmatrix}_1 = \begin{Bmatrix} 0 \\ 0 \\ 0 \end{Bmatrix}$$

Thus

$$\begin{Bmatrix} \phi_1 \\ \phi_2 \\ \phi_3 \end{Bmatrix}_1 = \begin{Bmatrix} 1 \\ 2 \\ 1 \end{Bmatrix}$$

Similarly, for $\omega_3 = \sqrt{(2 + \sqrt{2})}\,\Lambda$ we have

$$\begin{Bmatrix} \phi_1 \\ \phi_2 \\ \phi_3 \end{Bmatrix}_3 = \begin{Bmatrix} 1 \\ \sqrt{2} \\ 1 \end{Bmatrix}$$

The three mode shapes are displayed in Fig. 4.3.3, where we see the first and third modes are symmetric modes and the second is antisymmetric.

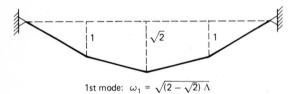

1st mode: $\omega_1 = \sqrt{(2 - \sqrt{2})}\,\Lambda$

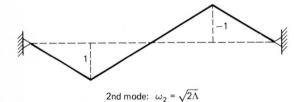

2nd mode: $\omega_2 = \sqrt{2}\Lambda$

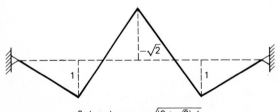

3rd mode: $\omega_3 = \sqrt{(2 + \sqrt{2})}\,\Lambda$

Figure 4.3.3 Mode shapes of three-DOF cable example.

Note that the nth mode has $n - 1$ antinodes, i.e., the cable crosses the zero axis $n - 1$ times. The orthogonality of the three modes can be demonstrated by forming the products $\{\phi\}_1^T\{\phi\}_2$, $\{\phi\}_1^T\{\phi\}_3$, and $\{\phi\}_2^T \{\phi\}_3$, since the mass matrix is an identity matrix multiplied by a scalar.

Let us now return to the subject of determining the natural frequencies (eigenvalues) and mode shapes (eigenvectors) of Eq. (4.3.6) for large systems of equations using a digital computer. For extensive coverage of the subject see Chaps. 10 and 11 of Ref. 94. We only consider a few aspects of the problem here. There are three groups of efficient methods of eigenvalue extraction, each group corresponding to a different basic property of the eigenvalue problem:

1. Vector iteration methods

2. Transformation methods

3. Polynomial iteration methods

The extraction of eigenvalues and eigenvectors of Eq. (4.3.6) is intrinsically an iterative process, so all three groups involve iteration. The selection of which method to use depends on the form and size of the problem at hand and how many and which eigenvalues are desired. Often it is desirable to combine the methods.

Vector iteration methods.   Use Eq. (4.3.8b) in the form

$$[K]\{\phi\}_n = \omega_n^2[M]\{\phi\}_n$$

to obtain successive improvements to a starting vector approximation $\{\Psi_1\}$ to $\{\phi\}_1$:

$$[K]\{\Psi_2\} = \{R_1\} = [M]\{\Psi_1\}$$

which is solved for $\{\Psi_2\}$. Then successive improvements are given by

$$[K]\{\Psi_{i+1}\} = \{R_i\} = [M]\{\Psi_i\}$$

which converges linearly to the eigenvalue with the smallest frequency. The Rayleigh quotient is used to determine ω_1:

$$\lim_{i \to \infty} \frac{\{\Psi_{i+1}\}^T[K]\{\Psi_{i+1}\}}{\{\Psi_{i+1}\}^T[M]\{\Psi_{i+1}\}} = \omega_1^2$$

The matrix $[K]$ must be positive definite, but $[M]$ need not be, i.e., $[M]$ may possess zeroed entries on the diagonal. The rate of convergence is dependent on the quality of the starting iteration vector $\{\Psi_1\}$. Inverse iteration (operating on $[M]\{\phi\}_m = 1/\omega^2[K]\{\phi\}_n$) will converge on the eigenvector with the largest frequency. To determine other eigenvectors the

Gram-Schmidt method of making starting vectors orthogonal to all previously determined eigenvectors can be used to converge to the next highest eigenvalue. Only those eigenvalues desired need to be determined, and
the vector iteration method is best for calculating a few of the lower frequency eigenvectors. The accuracy of the mode shapes for the higher frequencies deteriorates because of accumulating errors in the Gram-
Schmidt method. The vector iteration methods are simple to program and
make efficient use of computer storage.

Transformation methods. Use the orthogonality properties of the
eigenvectors

$$[\phi]^T[K][\phi] = \text{diag}(\omega_n^2)$$

$$[\phi]^T[M][\phi] = [I]$$

The matrices $[K]$ and $[M]$ are reduced into diagonal form by successive
pre- and postmultiplication by orthogonal matrices, i.e., with $[K_1] = [K]$,
$[M_1] = [M]$:

$$\lim_{i \to \infty} ([K_{i+1}] = [P_i]^T[K_i][P_i]) = \text{diag}(\omega_n^2)$$

$$\lim_{i \to \infty} ([M_{i+1}] = [P_i]^T[M_i][P_i]) = [I]$$

Then $[\phi] = \Pi_{i=1}^m([P_i])$, and $m =$ last iteration. The matrices $[P_i]$ are
selected differently for different transformation methods, e.g., by the
Jacobi method or the Householder-QR method. In the Jacobi method, a
simple and stable method with quadratic convergence characteristics, $[P_i]$
are rotation matrices such that off-diagonal terms in $[K_i]$ are zeroed. In
the transformation methods all the eigenvalues are determined, and no
restrictions on the forms of $[M]$ and $[K]$ are made. The disadvantages of
the transformation methods are that they require extensive operations
and off-line storage for large problems.

Polynomial iteration methods. Use Eq. (4.3.7) for the zero roots of the
determinant to first determine the frequencies. Then, by appropriate
shifting, the vector iteration method can be used to rapidly calculate the
mode shapes. The roots of Eq. (4.3.7) can be determined explicitly by calculating $p(\omega^2)$ as a polynomial:

$$p(\omega^2) = \sum_{i=1}^{n} a_i(\omega^2)^i$$

and using a root finding method, e.g., Newton's method, to calculate ω_n^2.
However, for large problems, the expansion of the determinant into poly-

nomial form requires extensive operations, and also small errors in the coefficients of the polynomial lead to large errors in the roots. In an implicit polynomial iteration solution, $p(\omega^2)$ is evaluated directly without expansion into polynomial form by decomposing $[K] - \omega^2[M]$ into a lower unit triangular matrix $[L]$ and an upper triangular matrix $[S]$:

$$[K] - \omega^2[M] = [L][S]$$

and then, for an assumed ω^2,

$$p(\omega^2) = \prod_{i=1}^{n} (S_{ii})$$

is solved for $p(\omega^2) = 0$ by iteration. An advantage of the polynomial iteration method is that each eigenvalue is determined independently and cumulative errors do not occur. However, we are never sure which ω_n^2 has been calculated without recourse to other methods, e.g., Sturm sequence checks.

4.3.2   Forced vibrations

There are two basic approaches to the solution of Eq. (4.3.3), here rewritten as

$$[M]\{\ddot{y}\} + [C]\{\dot{y}\} + [K]\{y\} = \{F(t)\}$$

in cases in which $\{F(t)\}$ and/or $[C]$ are nonzero:

1. Direct integration
2. Modal superposition

Direct integration relies on explicit or implicit numerical integration of Eq. (4.3.3) directly in time starting with initial values of $\{y_0\}$ and $\{\dot{y}_0\}$. In those methods, the equations are not restricted to be linear. Therefore, we will defer discussion of direct integration until we consider nonlinear dynamics of cable systems.

The modal superposition method [93, 94] requires linearized equations and relies upon prior calculation of natural frequencies ω_n and mode shapes $\{\phi\}_n$, as discussed in Sec. 4.3.1. Since the mode shapes are orthogonal, they constitute a set of linearly independent base vectors in terms of which any other vector can be expressed. Specifically, we will assume that

$$\{y(t)\} = \sum_{n=1}^{n} x_n(t)\{\phi\}_n \tag{4.3.14}$$

where each $x_n(t)$ is a scalar, unknown, time-dependent, "participation" factor for the nth mode $\{\phi\}_n$.

Substituting Eq. (4.3.14) into Eq. (4.3.3), we have

$$\sum_{n=1}^{n} (\ddot{x}_n[M]\{\phi\}_n + \dot{x}_n[C]\{\phi\}_n + x_n[K]\{\phi\}_n) = \{F(t)\} \qquad (4.3.15)$$

Convenient use of the orthogonality principle, Eq. (4.3.9), can now be made to decouple Eq. (4.3.15) into equations for each mode. Premultiply Eq. (4.3.15) by each of $\{\phi\}_m^T$, $m = 1, N$, and use Eq. (4.3.9) to eliminate summations. Thus,

$$\ddot{x}_m\{\phi\}_m^T[M]\{\phi\}_m + x_m\{\phi\}_m^T[K]\{\phi\}_m$$

$$+ \{\phi\}_m^T \sum_{n=1}^{n} (\dot{x}_n[C]\{\phi\}_n) = \{\phi\}_m^T\{F(t)\} \qquad (4.3.16)$$

Also, premultiplying Eq. (4.3.8b) by $\{\phi\}_n^T$ we see that

$$\{\phi\}_m^T[K]\{\phi\}_m = \omega_m^2\{\phi\}_m^T[M]\{\phi\}_m \qquad (4.3.17)$$

Note that the damping term in Eq. (4.3.16) has not yet been decoupled.

Since $[C]$ is ill-defined in most cases, let us assume Rayleigh damping [94] in which $[C]$ is proportional to $[M]$ and $[K]$ by

$$[C] = \alpha_1[M] + \alpha_2[K] \qquad (4.3.18)$$

where α_1 and α_2 are proportionality constants. Then, the damping term in Eq. (4.3.16) becomes [using orthogonality and Eq. (4.3.17)]

$$\{\phi\}_m^T \sum_{n=1}^{n} (\dot{x}_n[C]\{\psi\}_n) = (\alpha_1 + \alpha_2\omega_m^2)\dot{x}_m\{\phi\}_m^T[M]\{\phi\}_m$$
$$= 2\omega_m\beta_m\dot{x}_m\{\phi\}_m^T[M]\{\phi\}_m \qquad (4.3.19)$$

where β_m = modal damping factor for the mth mode, i.e., the percent of critical damping in the mth mode. Given any two β_m from estimates or experiments, we can evaluate α_1 and α_2 from Eq. (4.3.19), and the rest of β_m are also estimated from Eq. (4.3.19).

Let $\{\phi\}_m$ be factored according to Eq. (4.3.10a), i.e.,

$$\{\phi\}_m^T[M]\{\phi\}_m = 1 \qquad (4.3.20)$$

Then, with Eqs. (4.3.17), (4.3.19), and (4.3.20), Eq. (4.3.16) becomes

$$\ddot{x}_m + 2\omega_m\beta_m\dot{x}_m + \omega_m^2 x_m = \Gamma_m(t) \qquad (4.3.21a)$$

where

$$\Gamma_m(t) = \{\phi\}_m^T\{F(t)\} \qquad (4.3.21b)$$

Equations (4.3.21) are decoupled equations for the mth participation factor x_m and resemble an equilibrium equation for a single-degree-of-freedom system with unit mass and stiffness ω_m^2. The initial conditions are

$$x_{m0} = \{\phi\}_m^T\{y_0\} \qquad (4.3.22a)$$

$$\dot{x}_{m0} = \{\phi\}_m^T\{\dot{y}_0\} \qquad (4.3.22b)$$

Solution techniques for single-degree-of-freedom systems can therefore be used to solve Eqs. (4.3.21) and (4.3.22). For example, using the Duhamel integral technique [93]

$$x_m = \frac{1}{\omega_m^2} \int_0^t \Gamma_m(\tau) \sin \omega_m(t - \tau)\exp(-\beta_m\omega_m(t - \tau) \, d\tau)$$

$$+ \exp(-\beta_m\omega_m t) \left(\frac{\dot{x}_{m0}}{\omega_m} \sin \omega_m + x_{m0} \cos \omega_m t \right) \quad (4.3.23)$$

Alternatively, direct integration of Eq. (4.3.21) can be performed using the techniques described subsequently for multiple-degree-of-freedom problems.

We can solve Eqs. (4.3.21) successively for x_1, x_2, x_3, etc., up to whatever mode we wish to consider and truncate the summation in Eq. (4.3.14). In most structural systems subjected to most dynamic loads, only a few of the lower modes are needed for convergence of the truncated summation. Often in cable systems, because of their flexibility and capability to deform easily in each of the modes, many modes are required to be included for convergence. Unfortunately, as indicated in Sec. 4.3.1, the higher modes become progressively less accurate.

4.4 Dynamics of Cable Finite Element Models

The emphasis in this chapter has been on linearized dynamic response of cable systems, so we will limit our discussion to the small in-service dynamics of stable prestressed networks of cables [99–103, 109], i.e., mostly to the problem of predicting frequencies and mode shapes of free vibration about the static equilibrium state. Some indication of the use of piecewise linear equations to model nonlinear responses will be given, but the introduction of more accurate modeling techniques for nonlinear dynamics will be deferred until the next chapter. We will adopt as our basic element representation the straight element approximation used in Sec. 3.2.2.

The inclusion of cable inertial forces in a finite element context will be approached by development of an incremental version of Hamilton's principle for cables. The two forms that the mass matrices for assemblages of cable elements can take will be discussed. Results from Sec. 3.2.2 for the load and stiffness matrices will be included to arrive at the linearized equations of motion in matrix form for assemblages of cable elements. Some examples will be used to demonstrate the modeling technique and to illustrate some of the behavioral aspects of the various models.

4.4.1 Hamilton's principle

Since finite element formulations of equations of motion are based on energy minimization principles, we will extend the principle of virtual work (presented in Chap. 3 for static problems) to include variational work due to inertial forces, i.e., the kinetic energy. The most generally applicable variational principle is Hamilton's principle [94], which states that the variation of kinetic energy and the potential energy of internal and external conservative forces plus the variation of work due to non-conservative forces, such as damping, during an arbitrary virtual displacement and velocity in a time interval t_1 to t_2 must be zero. The mathematical statement of Hamilton's principle is

$$\int_{t_1}^{t_2} (\delta U - \delta KE - \delta A)\, dt = 0 \qquad (4.4.1)$$

where the admissible virtual displacements $\{\delta U\}$ vanish at t_1 and t_2. The three energy terms in Eq. (4.4.1) are

δU = variation in strain energy $\qquad\qquad\qquad (4.4.2a)$

$$= \int_V \{\delta\varepsilon\}^T\{\sigma\}d(\text{vol})$$

δKE = variation in kinetic energy

$$= +\int_V \{\delta\dot{u}\}^T m\{\dot{u}\}d(\text{vol}) + \text{conc. mass effects} \qquad (4.4.2b)$$

$$= -\int_V \{\delta u\}^T m\{\ddot{u}\}d(\text{vol})$$

(by integration by parts)

δA = virtual work of all conservative and

nonconservative external forces $\qquad$ (4.4.2c)

$$= + \int_V \{\delta u\}^T \{f\} d(\text{vol}) + \text{conc. load effects}$$

The terms in Eqs. (4.4.2) are

$\{\delta\varepsilon\}$ = virtual small strain due to virtual displacement $\{\delta u\}$

$\{\sigma\}$ = stresses in dynamic equilibrium at a point in V

$\{\delta\ddot{u}\}$ = accelerations of an arbitrary point in V

m = mass density at a point in V

$\{f\}$ = external body force density at a point in V

We will disregard the concentrated load and mass effects. They can be simply added to the system matrices after assembly of the finite element equations. Note that if inertial terms are suppressed, i.e., $\delta KE = 0$, Eqs. (4.4.1) and (4.4.2) reduce to the statement of the principle of virtual work given by Eq. (3.1.5) in Chap. 3.

For cables, we will again adopt the concept of two configurations, the additionally (dynamically) deformed state S and a reference state S_R from which all further displacements are measured. We will include the effects of possible velocities $\{v_R\}$ and accelerations $\{a_R\}$ of the reference state. If the reference state is selected as a static equilibrium state, then $\{v_R\}$ and $\{a_R\}$ will be zero.

The constitutive, topological, and kinematic relations developed in Sec. 3.2.1 for static problems still apply for dynamic motions; i.e., we have

Direction cosines in the reference state:

$$\theta_{Ri} = \frac{dx_{Ri}}{dS_R} \qquad (4.4.3a)$$

Direction cosines in the additionally deformed state:

$$\theta_i = \frac{\theta_{Ri} + \dfrac{\partial u_i}{\partial s_R}}{\sqrt{1 + 2\gamma}} \qquad (4.4.3b)$$

Relative strain:

$$\gamma = \theta_{Ri} \frac{\partial u_i}{\partial s_R} + \tfrac{1}{2} \frac{\partial u_i}{\partial s_R} \frac{\partial u_i}{\partial s_R} \qquad (4.4.3c)$$

Tension in the additionally deformed state:

$$T = T_R + E_R A_o T_R (\sqrt{1 + 2\gamma} - 1) \qquad (4.4.3d)$$

Small virtual strains of the two states:

$$\delta\varepsilon_R = \theta_{Ri} \frac{\partial(\delta u_i)}{\partial s_R} \qquad (4.4.3e)$$

$$\delta\varepsilon = \theta_i \frac{\partial(\delta u_i)}{\partial s} \qquad (4.4.3f)$$

Elongation ratios of the two states to initial length ds_0:

$$\lambda_R = \frac{ds_R}{ds_0} \qquad (4.4.3g)$$

$$\lambda = \frac{ds}{ds_0} = \frac{ds}{ds_R} \lambda_R = \lambda_R \sqrt{1 + 2\gamma} \qquad (4.4.3h)$$

where T_R = tension in reference state
E_R = instantaneous modulus of elasticity in reference state
A_0 = unstressed area

Now we write the Hamilton's principle for both the reference and additionally deformed states of the differential cable length:

$$\int_{t_1}^{t_2} \left[\int_{S_R} \left(\frac{\partial\delta u_i}{\partial s_R} \theta_{Ri} T_R - \delta u_i \, q_{Ri}^* + \delta u_i \, m_R A_R a_{Ri} \right) ds_R \right] dt = 0 \qquad (4.4.4a)$$

$$\int_{t_1}^{t_2} \left[\int_{S} \left(\frac{\partial\delta u_i}{\partial s} \theta_i T - \delta u_i q_i^* + \delta u_i \, m A \ddot{u}_i \right) ds \right] = 0 \qquad (4.4.4b)$$

where q_{Ri}^*, q_i^* = distributed loads per unit length in S_R and S
m_R, m = mass densities per unit volume in S_R and S respectively

To restate Eqs. (4.4.4) in terms of the known geometry and stress of the reference state we define load intensities q_{Ri}, q_i per unit unstressed length:

$$q_{Ri} = \lambda_R q_{Ri}^* \qquad (4.4.5a)$$

$$q_i = \lambda q_i^* \qquad (4.4.5b)$$

and apply the principle of conservation of mass

$$m_R A_R \, ds_R = m A \, ds = m_0 A_0 \, ds_0$$

which implies

$$m_R A_R \, ds_R = \frac{m_0 A_0}{\lambda_R} \, ds_R \qquad (4.4.6a)$$

$$mA \, ds = \frac{m_0 A_0}{\lambda} \, ds_R \qquad (4.4.6b)$$

where m_0 is mass density per unit volume in the unstressed state.
Thus, Eqs. (4.4.4) become

$$\int_{t_1}^{t_2} \left[\int_{S_R} \left(\frac{\partial \delta u_i}{\partial s_R} T_R \theta_{Ri} - \frac{\delta u_i \, q_{Ri}}{\lambda_R} + \frac{\delta u_i \, m_0 A_0 a_{Ri}}{\lambda_R} \right) ds_R \right] dt = 0 \qquad (4.4.7a)$$

$$\int_{t_1}^{t_2} \left(\int_{S_R} \left\{ \frac{\partial \delta u_i}{\partial s_R} \left[T_R + E_R A_0 \lambda_R \left(\sqrt{1 + 2\gamma} - 1 \right) \right] \frac{\theta_{Ri} + \dfrac{\partial u_i}{\partial s_R}}{\sqrt{1 + 2\gamma}} \right. \right. \qquad (4.4.7b)$$

$$\left. \left. - \frac{\delta u_i \, q_i}{\lambda_R} + \frac{\delta u_i \, m_0 A_0 \ddot{u}_i}{\lambda_R} \right\} ds_R \right) dt = 0$$

The incremental version of Hamilton's principle is obtained by subtracting Eq. (4.4.7a) from (4.4.7b):

$$\int_{t_1}^{t_2} \left[\int_{S_R} \left(\frac{\partial \delta u_i}{\partial s_R} \left\{ \frac{E_R A_0 \lambda_R - T_R}{\sqrt{1 + 2\gamma}} \left[(\sqrt{1 + 2\gamma} - 1)\theta_{Ri} - \frac{\partial u_i}{\partial s_R} \right] \right. \right. \right.$$

$$\left. \left. + E_R A_0 \lambda_R \frac{\partial u_i}{\partial s_R} - \frac{\delta u_i \, \Delta q_i}{\lambda_R} + \frac{\delta u_i \, m_0 A_0 (\ddot{u}_i - a_{Ri})}{\lambda_R} \right\} ds_R \right) dt = 0 \qquad (4.4.8)$$

which holds for large additional displacements u_i. The linearized version of Eq. (4.4.8) for small additional displacements u_i is obtained by retaining only first-order terms in u_i, i.e.,

$$\int_{t_1}^{t_2} \left(\int_{S_R} \left\{ \frac{\partial \delta u_i}{\partial s_R} \left[T_R \, \delta_{ij} + (E_R A_0 \lambda_R - T_R)\theta_{Ri}\theta_{Rj} \right] \frac{\partial u_j}{\partial s_R} \right. \right.$$

$$\left. \left. - \frac{\delta u_i (\Delta q_i + m_0 A_0 a_{Ri})}{\lambda_R} + \delta u_i \frac{m_0 A_0}{\lambda_R} \ddot{u}_i \right\} ds_R \right) dt = 0 \qquad (4.4.9)$$

Equation (4.4.9) is the basis for the finite element model for the linearized dynamic response of cable systems. Equations (4.4.7) or Eq. (4.4.8) will be used subsequently to develop nonlinear finite element models.

4.4.2   Mass matrix representation for straight elements

The formulation in Sec. 3.2.2 in which straight cable elements were adopted to obtain a finite element model of the static equivalent to Eq. (4.4.9) will now be extended to include the added inertial effects displayed in Eq. (4.4.9). We define linear polynomial approximating functions for the displacements u_i:

$$u_i = \eta_\alpha \, d_{i\alpha} + \eta_\beta \, d_{i\beta} \qquad (4.4.10a)$$

where

$$\eta_\alpha = 1 - \xi \qquad (4.4.10b)$$

$$\eta_\beta = \xi \qquad (4.4.10c)$$

in which $d_{i\alpha}$, $d_{i\beta}$ = deflections of nodes α and β of a typical element with nondimensional arc length coordinate ξ ($0 \leqslant \xi = S_R/L \leqslant 1$), L being the length of the element.

There is no a priori reason that the velocities and accelerations need be approximated by the same polynomial interpolation functions as for the displacements. So assume for the moment that

$$\dot{u}_i = \bar{\eta}_\alpha \dot{d}_{i\alpha} + \bar{\eta}_\beta \dot{d}_{i\beta} \qquad (4.4.11a)$$

$$\delta \dot{u}_i = \bar{\eta}_\alpha \delta \dot{d}_{i\alpha} + \bar{\eta}_\beta \delta \dot{d}_{i\beta} \qquad (4.4.11b)$$

where $\bar{\eta}_\alpha(\xi)$, $\bar{\eta}_\beta(\xi)$ are as yet unspecified polynomial interpolating functions; $\dot{d}_{i\alpha}$, $\dot{d}_{i\beta}$, and $\delta \dot{d}_{i\alpha}$, $\delta \dot{d}_{i\beta}$ are the real and virtual velocities of nodes α and β. Then, since

$$\delta KE = \int_{t_1}^{t_2} [\textstyle\int \delta \dot{u}_i^T \, m \dot{u}_i \, d(\text{vol})] \, dt$$

$$= - \int_{t_1}^{t_2} [\textstyle\int \delta u_i^T \, m \ddot{u}_i \, d(\text{vol})] \, dt$$

we obtain upon substitution of Eqs. (4.4.10) and (4.4.11) into Eq. (4.4.9) and elimination of the virtual displacement

$$\sum_e ([_eM]\{d\} + [_eK]\{d\} - \{_eQ\}) = \{0\} \qquad (4.4.12a)$$

where the summation Σ_e, denotes assemblage of contributions from each element, and

$[_eK]$ = element stiffness matrix derived in Sec. 3.2.2 as

$$= \frac{1}{L} \begin{bmatrix} B & -B \\ -B & B \end{bmatrix} \qquad (4.4.12b)$$

with

$$[B] = T_R[I] + (E_R A_0 \lambda_R - T_R)\{\theta_R\}\{\theta_R\}^T \qquad (4.4.12c)$$

$\{_eQ\}$ = element contributions from incremental distributed loads plus distributed inertial forces due to reference accelerations

$$= \frac{L}{6\lambda_R} \begin{Bmatrix} 2p_\alpha + p_\beta \\ p_\alpha + 2p_\beta \end{Bmatrix} + [_eM] \begin{Bmatrix} a_{R\alpha} \\ a_{R\beta} \end{Bmatrix} \qquad (4.4.12d)$$

with $\{p_\alpha\}$, $\{p_\beta\}$ and $\{a_{R\alpha}\}$, $\{a_{R\beta}\}$ being the intensities of added loads and of reference acceleration at nodes α and β.

$[^eM]$ = element mass matrix

$$= \frac{m_0 A_0 L}{\lambda_R} \int_0^1 \begin{bmatrix} \bar{\eta}_\eta^2 I & \bar{\eta}_\alpha \bar{\eta}_\beta I \\ \bar{\eta}_\alpha \bar{\eta}_\beta I & \bar{\eta}_\beta^2 I \end{bmatrix} d\xi \qquad (4.4.12e)$$

wherein we have left $\bar{\eta}_\alpha$, $\bar{\eta}_\beta$ as undefined interpolating functions. The element stiffness, load, and mass matrices can be assembled into system matrices, as outlined in Chap. 3, and then concentrated loads and masses at nodes can be added.

There are two predominant choices for inertial interpolation functions $\bar{\eta}_\alpha$, $\bar{\eta}_\beta$, each leading to a different form of the mass matrix [94]. If $\bar{\eta}_\alpha$, $\bar{\eta}_\beta$ are taken equal to η_α, η_β, i.e., accelerations are "consistent" with our definition of displacement interpolations, we arrive at the so-called consistent mass matrix [26]:

$$[_eM] = \frac{m_0 A_0 L}{6\lambda_R} \begin{bmatrix} 2I & I \\ I & 2I \end{bmatrix} \qquad (4.4.13a)$$

in which there is coupling between nodal accelerations due to the off-diagonal terms.

If $\bar{\eta}_\alpha$, $\bar{\eta}_\beta$ are taken as Heaviside functions, defined as unity over the first and last halves of the element, respectively, i.e.,

$$\bar{\eta}_\alpha = 1 \qquad 0 \leqslant \xi < \tfrac{1}{2}$$

$$\bar{\eta}_\alpha = 0 \qquad \tfrac{1}{2} \leqslant \xi \leqslant 1$$

$$\bar{\eta}_\beta = 0 \qquad 0 \leqslant \xi < \tfrac{1}{2}$$

$$\bar{\eta}_\beta = 1 \qquad \tfrac{1}{2} < \xi \leqslant 1$$

and thus the accelerations of elements in the neighborhood of a node are assumed equal to the accelerations of the node, then we arrive at the so-called lumped mass matrix:

$$[_eM]_e = \frac{m_0A_0L}{6\lambda_R} \begin{bmatrix} 3I & 0 \\ 0 & 3I \end{bmatrix}$$

$$= \left(\frac{m_0A_0L}{2}\right) \frac{1}{\lambda_R} \begin{bmatrix} I & 0 \\ 0 & I \end{bmatrix} \qquad (4.4.13b)$$

in which there is no coupling between nodal accelerations because the matrix is diagonal. Note that half the mass of each element is assigned to the nodes at either end of the element.

It can be seen that the consistent mass matrix $[M]_e$ given by Eq. (4.4.13a) is in fact more consistent with the fact that accelerations are continuously distributed over the length of an element and there is no acceleration discontinuity at the center of the element as assumed in the lumped mass matrix. The use of a consistent mass matrix leads to more accurate results in comparison to the use of a lumped mass matrix if the same mesh of elements is used. However, the increased accuracy is often only achieved at greater computational expense. We might, in fact, achieve the same accuracy at less expense by using a lumped mass matrix in a more refined mesh of elements.

The consistent mass matrix is a positive-definite banded matrix of the same size as the stiffness matrix. The lumped mass matrix is also positive-definite if at least one element adjoining a node has nonzero mass.* Because the lumped mass is diagonal and thus requires much less storage space than does the consistent mass matrix, the lumped mass matrix is cheaper to formulate and manipulate on a digital computer. The computational time in an eigenvalue analysis is considerably faster using a lumped mass matrix if $[M]^{-1}$ is required. Consistent mass matrices lead to upper bounds on the exact frequencies of free vibration, whereas lumped mass matrices tend to be lower bounds.

In direct time integration solutions, the two alternatives lead to comparable overall computation time if implicit integration is used in that an inverse of a modified stiffness matrix is typically required rather than $[M]^{-1}$. On the other hand, if an explicit integration technique is used, $[M]^{-1}$ is required. Thus, explicit integration with a lumped mass matrix

---

* There are interesting implications and special techniques required for flexural problems in finite element modeling if rotational degrees of freedom are included. In that case, the lumped mass matrix is not positive-definite because no mass is associated with the rotational degrees of freedom. See Chap. 11 of Ref. 165 for a discussion of those implications. In our instance, we suppress rotational degrees of freedom because of the assumption of negligible bending rigidity in cables.

is much faster per time step. However, as we will see subsequently, explicit integration requires smaller time steps in the integration process. So the choice of whether to use a lumped or consistent mass matrix is dependent on the type of problem at hand.

The assemblage of contributions from each element as described by Eqs. (4.4.12) leads to a system of matrix equations of motion:

$$[M]\{\ddot{D}\} + [K]\{D\} = \{P(t)\}$$

which is a linear equation of stresses in that reference state.

4.4.4   Free vibrations of cable systems

Let us illustrate the solution process for free vibrations of cable systems by two examples: the cable cross problem considered in Chap. 3, and the vibration of an orthogonal network of cables.

Cable cross.   In Chap. 3, we considered the simple problem of two crossing cables as shown in Fig. 3.2.4, with equal spans ℓ and sag ratios f_R. Initially, let us assume a crude mesh of four elements as shown in Fig. 3.2.4b with the only degrees of freedom being the three components of motion at node 1. The stiffness matrix for the assemblage is given by Eq. (3.2.32) rewritten as

$$[K] = \frac{1}{L}
\begin{bmatrix}
(_1B + {}_2B + {}_3B + {}_4B) & -_1B-_2B-_3B-_4B \\
\hline
-_1B & {}_1B & 0 & 0 & 0 \\
{}_2B & 0 & {}_2B & 0 & 0 \\
-_3B & 0 & 0 & {}_3B & 0 \\
-_4B & 0 & 0 & 0 & {}_4B
\end{bmatrix}
\tag{4.4.14}$$

where [0] denotes a 3×3 null matrix, $[_eB]$ are given by Eqs. (3.2.31), and $L = $ length of each element:

$$L = \frac{\ell}{2}\sqrt{1 + 4f_R^2}
\tag{4.4.15}$$

The mass matrix for each element is

$$[_eM] = \frac{m_0 A_0 L a}{6\lambda_R}
\begin{bmatrix}
I & bI \\
bI & I
\end{bmatrix}
\tag{4.4.16}$$

where for a lumped mass matrix $a = 3$, $b = 0$ and for a consistent mass matrix $a = 2$, $b = 0.5$. Also

$$\lambda_R = 1 + \frac{H_R}{E_R A_0} \sqrt{1 + 4f_R^2} \qquad (4.4.17)$$

where H_R = horizontal component of pretension. The assembled mass matrix is therefore

$$[M] = \frac{m_0 A_0 L}{6\lambda_R} (a) \begin{bmatrix} 4I & bI & bI & bI & bI \\ \hline bI & I & 0 & 0 & 0 \\ bI & 0 & I & 0 & 0 \\ bI & 0 & 0 & I & 0 \\ bI & 0 & 0 & 0 & I \end{bmatrix} \qquad (4.4.18)$$

Applying the boundary conditions $\{D_2\} = \{D_3\} = \{D_4\} = \{D_5\} = 0$, we arrive at the equations of free vibration of node 1 as

$$\left[-\left(\frac{2m_0 A_0 L^2 a\omega^2}{3\lambda_R} \right) [I] + \left(\sum_e [{}_eB] \right) \right] \{d\} = \{0\} \qquad (4.4.19a)$$

where

$$\left(\sum_e [{}_eB] \right) = 2\kappa \begin{bmatrix} (2c + 1) & 0 & 0 \\ 0 & (2c + 1) & 0 \\ 0 & 0 & (2c + 8f_R^2) \end{bmatrix} \qquad (4.4.19b)$$

with

$$\kappa = \frac{E_R A_0}{1 + 4f_R^2} \qquad (4.4.19c)$$

$$c = \frac{H_R}{E_R A_0} (1 + 4f_R^2)^{3/2} \qquad (4.4.19d)$$

We see that since $\sum_e [{}_eB]$ is a diagonal matrix, Eq. (4.4.19a) decouples into three single-degree-of-freedom equations for each component d_{1i}, $i = 1$, 3, of displacement at node 1, i.e.,

$$\left(g_i - \frac{m_0 A_0 L^2 a\omega^2}{3\kappa\lambda_R} \right) d_{1i} = 0 \qquad (4.4.20)$$

where $g_1 = g_2 = 2c + 1$ and $g_3 = 2c + 8f_R^2$. Thus, the three frequencies of vibrations are [substituting Eqs. (4.4.15), (4.4.17), and (4.4.19d) into Eq. (4.4.20)]

$$\omega_1 = \left[\frac{1}{a} \frac{24E_R}{m_0\ell^2} \left(\frac{1}{1 + 4f_R^2} \right)^2 \left(1 + \frac{c}{1 + 4f_R^2} \right) (c + 4f_R^2) \right]^{1/2} \tag{4.4.21a}$$

$$\omega_2 = \omega_3 = \left[\frac{1}{a} \frac{12E_R}{m_0\ell^2} \left(\frac{1}{1 + 4f_R^2} \right)^2 \left(1 + \frac{c}{1 + 4f_R^2} \right) (2c + 1) \right]^{1/2} \tag{4.4.21b}$$

with mode shapes

$$\{\phi\}_1 = \begin{Bmatrix} 0 \\ 0 \\ 1 \end{Bmatrix} \qquad \{\phi\}_2 = \begin{Bmatrix} 0 \\ 1 \\ 0 \end{Bmatrix} \qquad \{\phi\}_3 = \begin{Bmatrix} 1 \\ 0 \\ 0 \end{Bmatrix}$$

We see that we have two repeated frequencies ω_2 and ω_3 for horizontal plane vibration. Since f_R is usually much less than unity, ω_1 usually is less than ω_2 and ω_3. If $8f_R^2 > 1$, then $\omega_1 > \omega_2 = \omega_3$. See Figs. 4.4.1 and 4.4.2 for a sensitivity study of ω_1 and ω_2 as a function of f_R and $H_R/E_R A_0$.

In Eqs. (4.4.21) the term a reflects our choice of a consistent or lumped mass. For a lumped mass at the node 1 equal to one-half the total mass of the four cables, $a = 3$. For a consistent mass at the node 1 equal to one-third the total mass of the four cables, $a = 2$. Thus, comparing frequencies of free vibration based on lumped or consistent mass, we see that

$$\frac{\omega_1(\text{lumped})}{\omega_i(\text{consistent})} = \sqrt{\frac{a(\text{consistent})}{a(\text{lumped})}} = \sqrt{\frac{2}{3}}.$$

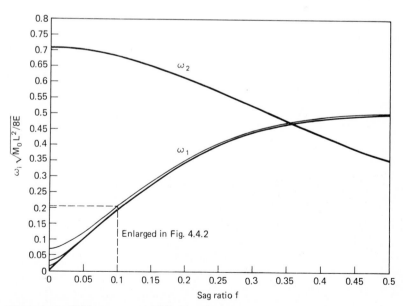

Figure 4.4.1 Cable cross natural frequencies.

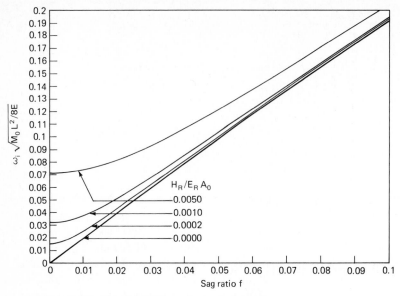

Figure 4.4.2 Cable cross natural frequencies (enlarged).

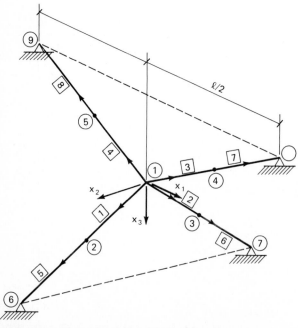

Figure 4.4.3 Eight-element model of cable cross.

The frequency based on a lumped mass representation is 81.65 percent of that based on a consistent mass.

The difference between the frequencies as predicted by a lumped or consistent mass representation can be reduced by considering a finer mesh of elements. Let us further subdivide the cable segments in the cable cross such that each segment is represented by two elements as in Fig. 4.4.3. We now have a system with 15 degrees of freedom if all three components of displacement at each of the five internal nodes 1 to 5 are unconstrained. Because of our arrangement of elements,

$$[_eB] = [_{4+e}B]$$

and Eqs. (3.2.31) can still be used for $[_eB]$, $e = 1, 4$. Now, however, the length of each element is

$$\mathring{L} = \frac{\ell}{4} \sqrt{1 + 4f_R^2} \tag{4.4.22}$$

The assembled stiffness matrix is

$$[K] = \frac{1}{\mathring{L}}
\begin{bmatrix}
\left(\sum_1^e B\right) & -_1B & -_2B & -_3B & -_4B & 0 & 0 & 0 & 0 \\
-_1B & 2_1B & 0 & 0 & 0 & -_1B & 0 & 0 & 0 \\
-_2B & 0 & 2_2B & 0 & 0 & 0 & -_2B & 0 & 0 \\
-_3B & 0 & 0 & 2_2B & 0 & 0 & 0 & -_3B & 0 \\
-_4B & 0 & 0 & 0 & 2_4B & 0 & 0 & 0 & -_4B \\
0 & -B_1 & 0 & 0 & 0 & _1B & 0 & 0 & 0 \\
0 & 0 & -_2B & 0 & 0 & 0 & _2B & 0 & 0 \\
0 & 0 & 0 & -_3B & 0 & 0 & 0 & _3B & 0 \\
0 & 0 & 0 & 0 & -_4B & 0 & 0 & 0 & _4B
\end{bmatrix} \tag{4.4.23a}$$

and the assembled mass matrix is

$$[M] = \frac{m_0 A_0 \mathring{L}a}{6\lambda_R}
\begin{bmatrix}
4I & bI & bI & bI & bI & 0 & 0 & 0 & 0 \\
bI & 2I & 0 & 0 & 0 & bI & 0 & 0 & 0 \\
bI & 0 & 2I & 0 & 0 & 0 & bI & 0 & 0 \\
bI & 0 & 0 & 2I & 0 & 0 & 0 & bI & 0 \\
bI & 0 & 0 & 0 & 2I & 0 & 0 & 0 & bI \\
0 & bI & 0 & 0 & 0 & I & 0 & 0 & 0 \\
0 & 0 & bI & 0 & 0 & 0 & I & 0 & 0 \\
0 & 0 & 0 & bI & 0 & 0 & 0 & I & 0 \\
0 & 0 & 0 & 0 & bI & 0 & 0 & 0 & I
\end{bmatrix} \tag{4.4.23b}$$

TABLE 4.4.1. Stiffness Matrix for Cable Cross*

							Columns								
Rows	**1**	**2**	**3**	**4**	**5**	**6**	**7**	**8**	**9**	**10**	**11**	**12**	**13**	**14**	**15**
1	$2(2c+1)$	0	0	$-c$	0	0	$-c$	0	0	$-(c+1)$	0	$2f_R$	$-(c+1)$	0	$-2f_R$
2		$2(c+1)$	0	0	$-(c+1)$	$-2f_R$	0	$-(c+1)$	$2f_R$	0	$-c$	0	0	$-c$	0
3			$4(c+4f_R^2)$	0	$-2f_R$	$-(c+4f_R^2)$	0	$2f_R$	$-(c+4f_R^2)$	$2f_R$	0	$-(c+4f_R^2)$	$-2f$	0	$-(c+4f_R^2)$
4				$2c$	0	0	$2c$	0	0						
5					$2(c+1)$	$4f_R$									
6						$2(c+4^2)$									
7							$2c$	0	0						
8				(Symmetric)				$2(c+1)$	$-4f_R$						
9									$2(c+4f_R^2)$						
10										$2(c+1)$	0	$-4f_R$			
11											$2c$	$2c$			
12												$2(c+4f_R^2)$			
13													$2(c+1)$	0	$4f_R$
14														$2c$	$2c$
15															$2(c+4f_R^2)$

* All coefficients multiplied by $[\kappa/\bar{L}]$.

162

TABLE 4.4.2   Mass Matrix for Cable Cross*

Rows	1	2	3	4	5	6	7	8	9	10	11	12	13	14	15
							Columns								
1	4	0	0	b	0	0	b	0	0	b	0	0	b	0	0
2		4	0	0	b	0	0	b	0	0	b	0	0	b	0
3			4	0	0	b	0	0	b	0	0	b	0	0	b
4				2	0	0									
5					2	0		0			0			0	
6						2									
7							2	0	0						
8								2	0		0			0	
9									2						
10										2	0	0			
11						(Symmetric)					2	0		0	
12												2			
13													2	0	0
14														2	0
15															2

* All coefficients multiplied by $(m_0 A_0 \tilde{L} a / 6\lambda_R)$.

with boundary conditions, $\{D_6\} = \{D_7\} = \{D_8\} = \{D_9\} = \{0\}$. If Eqs. (3.2.31) are used in conjunction with our definition of κ and c by Eqs. (4.4.19c and d), we arrive at an assembled stiffness matrix where coefficients are given in Table 4.4.1. The corresponding mass matrix is given by coefficients in Table 4.4.2. The equation of free vibration is then given for the mth modal frequency ω_m and mode slope $\{\phi\}_m$ by

$$([K] - \omega_m^2[M])\{\rho\}_m = \{0\} \tag{4.4.24}$$

Let us examine cases in which the cables are very taut and the sag ratio $f_R \ll 1$. The major motion will be in the x_3 direction. We will constrain the system such that $u_1 = u_2 = 0$ at all nodes and further seek out the symmetric modes of vibration in which u_3 are equal at nodes 2, 3, 4, and 5. So we delete rows and columns corresponding to u_1 and u_2 degrees of freedom (i.e., rows and columns numbered 1, 2, 4, 5, 7, 8, 10, 11, 13, 14) and add rows and columns corresponding to vertical motion at nodes 2, 3, 4, and 5 (i.e., add rows and columns 6, 9, 12, 15). Thus we obtain for the remaining two-degree-of-freedom system (vertical motions at nodes 1 and 2) the following equation:

$$\left(\begin{bmatrix} 1 & -1 \\ -1 & 2 \end{bmatrix} - \omega^2 \frac{m_0 A_0 \tilde{L}^2 a}{6\lambda_R \kappa (c + 4f_R^2)} \begin{bmatrix} 1 & b \\ b & 2 \end{bmatrix} \right) \begin{Bmatrix} \phi_1 \\ \phi_2 \end{Bmatrix} = \begin{Bmatrix} 0 \\ 0 \end{Bmatrix} \tag{4.4.25a}$$

Let

$$\Omega = \omega^2 \frac{m_0 A_0 \tilde{L}^2 a}{6\lambda_{RK}(c + 4f_R^2)}$$

Then Eq. (4.4.25a) is

$$\begin{bmatrix} (1 - \Omega) & -(1 + b\Omega) \\ -(1 + b\Omega) & 2(1 - \Omega) \end{bmatrix} \begin{Bmatrix} \phi_1 \\ \phi_2 \end{Bmatrix} = \begin{Bmatrix} 0 \\ 0 \end{Bmatrix} \qquad (4.4.25b)$$

and is satisfied only if the determinate is zero, i.e.,

$$p(\Omega) = 2(1 - \Omega)^2 - (1 + b\Omega)^2 = 0 \qquad (4.4.26)$$

Thus,

$$\Omega = \left(\frac{2 + b}{2 - b^2}\right) \pm \sqrt{2}\left(\frac{1 + b}{2 - b^2}\right)$$

contains the two frequencies. For the lumped mass representation ($a = 3$, $b = 0$)

$$\Omega_1 = 1 - \frac{\sqrt{2}}{2} = 0.2929$$

$$\Omega_2 = 1 + \frac{\sqrt{2}}{2} = 1.7071$$

and therefore,

$$\omega_1 = \left[\frac{9.3728E}{m_0\ell^2}\left(\frac{1}{1 + 4f_R^2}\right)^2\left(1 + \frac{c}{1 + 4f_R^2}\right)(c + 4f_R^2)\right]^{1/2} \qquad (4.4.27a)$$

$$\omega_2 = \left[\frac{54.6272E}{m_0\ell^2}\left(\frac{1}{1 + 4f_R^2}\right)^2\left(1 + \frac{c}{1 + 4f_R^2}\right)(c + 4f_R^2)\right]^{1/2} \qquad (4.4.27b)$$

For the consistent mass representation ($a = 2$, $b = 0.5$)

$$\Omega_1 = 0.2164 \qquad \Omega_2 = 2.6408$$

and therefore

$$\omega_1 = \left[\frac{10.3872E}{m_0\ell^2}\left(\frac{1}{1 + 4f_R^2}\right)^2\left(1 + \frac{c}{1 + 4f_R^2}\right)(c + 4f_R^2)\right]^{1/2} \qquad (4.4.28a)$$

$$\omega_2 = \left[\frac{126.7584E}{m_0\ell^2}\left(\frac{1}{1 + 4f_R^2}\right)^2\left(1 + \frac{c}{1 + 4f_R^2}\right)(c + 4f_R^2)\right]^{1/2} \qquad (4.4.28b)$$

For both mass representations the mode shapes are

$$\left\{\begin{matrix} \phi_1 \\ \phi_2 \end{matrix}\right\}_1 = \left\{\begin{matrix} 1 \\ \sqrt{2} \end{matrix}\right\} \quad \left\{\begin{matrix} \phi_1 \\ \phi_2 \end{matrix}\right\}_2 = \left\{\begin{matrix} 1 \\ -\sqrt{2} \end{matrix}\right\}$$

Now we find for this two-degree-of-freedom system that the ratio of the lowest frequencies for the lumped and consistent mass representation is

$$\frac{\omega_1(\text{lumped})}{\omega_1(\text{consistent})} = \sqrt{\frac{9.3728}{10.3872}} = 95 \text{ percent}$$

and we see that both representations give nearly the same fundamental frequency. The ratio of fundamental frequencies for the two-degree-of-freedom system (2DOF) [Eq. (4.4.27a)] and the single-degree-of-freedom system (1DOF) [Eq. (4.4.21a)] is

$$\frac{\omega_1(\text{1DOF})}{\omega_1(\text{2DOF})} = \sqrt{\frac{8}{9.3728}} = 92.4 \text{ percent}$$

Orthogonal network. The second example is that of the transverse vibration of a 2 × 2 plane orthogonal grid of cables, as shown in Fig. 4.4.4a, made up of cables for which area = 0.5 in^2 (3.2 cm^2), $E_R = 20 \times 10^3$ ksi (140 × 10^3 kN/m^2), weight density = 0.29 lb/in^3, span = 10 ft (3 m), and initial stress = 150 ksi (1000 MN/m^2). The determination of the natural frequencies and mode shapes for this grid has been considered by B. S. Chandhari [104] and by A. I. Solar and H. Afshari [105] using various

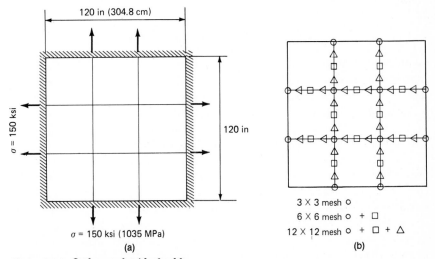

Figure 4.4.4 Orthogonal grid of cables.

TABLE 4.4.3   Frequencies of Orthogonal Grid

Method	Frequencies, cycles/s	Percent deviation from Timoshenko and Young [166]
Galerkin membrane analogy (Solar and Afshari, [105])	59.10	. . .
Membrane analogy (Chandhari, [104])	58.80	0.51
Discrete system (lumped mass) (Chandhari, [104])	56.40	4.57
Finite element method [39], 3 × 3 grid	61.59	4.21
Finite element method [39], 6 × 6 grid	59.55	0.76
Finite element method [39], 12 × 12 grid	59.05	0.08

methods limited to orthogonal plane grids. An analytic membrane analogy solution is given by S. J. Timoshento and D. H. Young [166].

Displayed in Table 4.4.3 are natural frequencies for the first mode of transverse vibration as calculated by Chandhari [104] and by Solar and Afshari [105] and as calculated using the finite element method described here [39]. Three different grids of elements were used as shown in Fig. 4.4.4b. Rapid convergence of results can be seen in Table 4.4.3 as more refined grids of elements are considered. Consistent mass matrices were adopted.

4.4.4   Piecewise linear solutions

As discussed in Chap. 3 on static problems, it is possible to treat nonlinear cable problems in a piecewise linear fashion. The dynamics of a nonlinear response can likewise be treated using Eqs. (4.4.12) in assembled form, i.e.,

$$[M]\{\ddot{D}\} + [C]\{\dot{D}\} + [K]\{D\} = \{P(t)\} \qquad (4.4.29)$$

where a damping matrix $[C]$ (proportional to the mass and stiffness matrix) has been added. This equation constitutes an initial-value problem for the small added displacements away from a reference state at time $t = t_R$. The solution to Eq. (4.4.29) can be obtained by any of a number of numerical time integration schemes [94], as discussed in Chap. 5, or by the method of modal superposition, as discussed in Sec. 4.3.2.

The solution is valid only during a small increment of loading. Thus, the solution technique for the nonlinear problem consists of the integration of the linearized initial-value problem, over small time intervals Δt between successive reference configurations separated by larger time steps Δt_u [43]. These larger time steps must be small enough that the assumption of incrementally linear behavior remains a close approximation of the actual behavior. The initial conditions at the start of the larger time steps are that the incremental displacements are zero and that the velocities and accelerations equal the velocities and accelerations at the end of the previous larger time step.

After propagation of the solution through the larger time step Δt_u to a new reference configuration, say $\overline{R}$, the matrices $[M]$, $[C]$, and $[K]$ must be recalculated in order to continue the solution process into the next increment. The finite element equivalents to Eqs. (3.2.12) are used to calculate the updated elemental values of $T_{\overline{R}}$, $E_{\overline{R}}$, $\lambda_{\overline{R}}$, $\{\theta_{\overline{R}}\}$, $L_{\overline{R}}$, and then $[_{\overline{e}}K]$ and $[_{\overline{e}}M]$ are reformulated from Eqs. (4.4.12). If the modal superposition method were being used, new frequencies and mode shapes of free vibration would then be calculated. Because of this need to recompute the eigenvalues at each reference state, the direct integration techniques are preferred.

The solution process as described above is an incremental solution technique, and all of the cautionary notes made previously for incremental statical solutions apply. This is particularly bothersome for dynamic problems in that numerous increments are necessary in time integration solution and errors can accumulate rapidly. More refined nonlinear solution techniques are given in Chap. 5.

To illustrate the incremental solution procedure, consider an initially straight cable, subdivided into four elements as shown in Fig. 4.4.5a, sub-

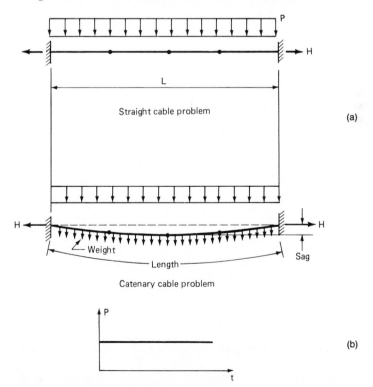

Figure 4.4.5 Straight and catenary problems.

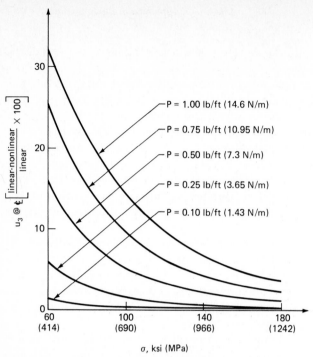

Figure 4.4.6 Midpoint peak displacement of straight cable.

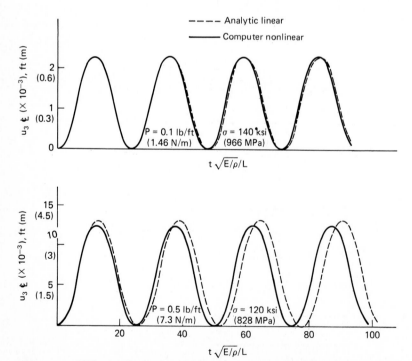

Figure 4.4.7 Midpoint displacement response curves for straight cable.

jected to a suddenly applied uniform transverse load. Results for various magnitudes of initial stress and applied load were considered [43]. The data for the cable were as follows: area = 0.065 in^2 (42 mm^2); weight = 0.24 lb/ft (3.5 N/m); E = 20 × 10^3 ksi (138 kN/m^2); m_0 = 16.5 lb·s^2/ft^4 (0.022 kg/m^3); and span = 10,000 in (259 m). In Fig. 4.4.6, the difference between the nonlinear and linear peak dynamic displacements of the midpoint as a percentage of the analytic linear displacements are shown as a function of load and initial stress. Convergence to the linear analytic results can be observed as the load decreases and the initial stress increases. Sample response curves for the midpoint displacement are shown in Fig. 4.4.7, along with the predicted linear results. A shorter

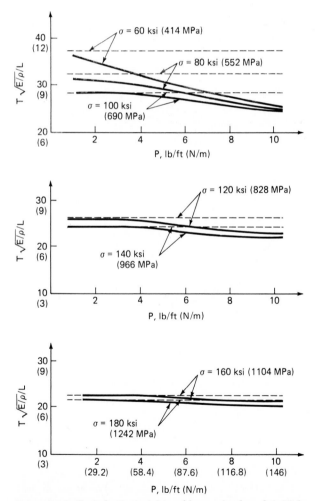

Figure 4.4.8 Period of straight cable vs. load and initial stress. Linear (---); nonlinear (———).

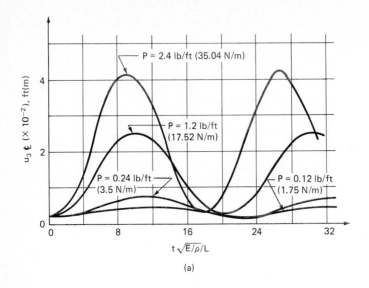

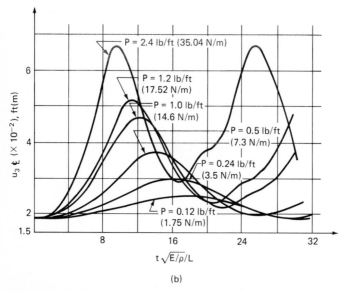

Figure 4.4.9 Midpoint displacement of catenary. (a) Initial stress = 160 ksi (approximately 1100 MPa); (b) initial stress = 20 ksi (approximately 138 MPa).

period can be seen to occur due to the nonlinear behavior. The effect of nonlinearities on the period can be better observed in Fig. 4.4.8, where the families of curves for the period as a function of load and initial stress are plotted. The corresponding periods predicted by linear theory are shown as dashed lines. According to linear theory, the period is a function of initial stress only, while the nonlinear results indicate a dependence on load magnitude also.

Another problem considered [43] was that of a cable initially at rest under its own weight (see Fig. 4.4.5b) and suddenly subjected to a uniform overload. Typical families of response curves for two sag ratios are shown in Fig. 4.4.9. For small initial sags (Fig. 4.4.9a), the period compares favorably with that for a straight cable. For large initial sags (Fig. 4.4.9b), we see appreciable nonlinearities.

Nonlinear Dynamics
of Cable Systems
and Hydrodynamic Loads

In Chap. 4 we saw that it is possible in some instances in the dynamic analysis of cable systems to use linearized equations to predict small vibrations about a prestressed reference state or to model large displacement problems incrementally by piecewise linear equations of motion. We saw further, in cases where stable prestressed reference states with significant stiffness have not been established or where large vibrations are possible, that nonlinearities in the response can have a strong effect and that sizable errors are introduced by use of linearized equations without iteration. The incrementally linear techniques used in Sec. 4.4.4 become increasingly inaccurate as successive increments are considered, and care must be taken to adopt very small time steps and to update the reference configuration relatively often.

There are three sources of nonlinearity: geometry, material, and load. When considering "soft" systems wherein the prestressed stiffness is small, e.g., when considering cases where curvature of a network is large or where large motions of supporting points are imposed, geometric nonlinearities predominate in the dynamic response [20, 39, 43]. In some cases slackening of cable segments occurs, i.e., zero pretension and stiffness, and only inertial forces can be mobilized to resist applied loads [20].

For example, in many ocean applications of mooring lines and networks the stiffness is very small and large excursions of bounding points are expected in response to changes in ocean currents, drift forces, periodic tidal currents, and sea states [16, 83, 85, 89, 90]. Also, when considering extreme forces or periodic forces close to a resonant frequency of the structure, e.g., extreme wind speeds or fluctuating wind gusts on cable roof structures, we must be concerned about momentary extreme dynamic stresses beyond the yield limit of the material and about potential resonances and flutter of the system [146, 147, 151, 152, 156].

When predicting the dynamic responses of cable systems in the ocean, the hydrodynamic forces on the cable segments due to relative motions of the cables with respect to the fluid velocity and acceleration are significant [16, 37, 83, 96]. For heavy catenary lines, e.g., chains, the principal loads on the catenary segments are due to the steady self-weight of the cable (note that frequencies are inversely proportional to the mass) and to dynamic motions of the endpoints where large compliant bodies are acted upon by hydrodynamic loads due to diffraction and radiation of waves [85, 96]. In nearly neutrally buoyant lines, e.g., polymer cables, the dynamic viscous drag forces due to relative velocity predominate over the steady self-weight and are position- and orientation-dependent as well as having nonlinear magnitudes [16, 90, 96].

In this chapter, we will develop nonlinear dynamic solution techniques for cable systems using the finite element method. Because of the importance of curvature effects and of the computational desirability of using long elements, we will present a derivation of a general isoparametric element in which curvature of the element can be included. Because of the essential nonlinearities of the dynamic response of soft cable systems we will focus on direct time integration techniques of numerical solution to the equations of motion. These same techniques can be used for linearized equations in place of using the modal superposition method presented previously. The sources of hydrodynamic loads on cables will be reviewed, and we will develop continuum and finite element formulations of loads due to ocean waves and currents.

5.1   Direct Time Integration of Nonlinear Equations

Before developing the governing matrix representation of the nonlinear equations of motion for cable systems, it is worthwhile to consider the solution technique that will be used to determine the time history of nonlinear response from those equations.

In direct time integration [93, 94], we seek the solution to the equations of motion at discrete times $t_i = (i) \Delta t$ (Δt = time increment) subject to initial conditions on displacements and velocities at $t_0 = 0$. We also assume some variation in, and self-consistent relationship between, the displacements, velocities, and accelerations to occur during Δt. The assumption made differs depending on the solution scheme selected. Thus, we seek a recursive scheme in which, having determined the displacements, velocities, and accelerations at a typical time $t = t_{i-1}$, we calculate the corresponding variables at time $t = t_i = t_{i-1} + \Delta t$, where Δt is a small time increment. Let us write the governing equations of motion at the two successive times as

$$[M]\{^{i-1}\ddot{y}\} + \{^{i-1}R\} = \{^{i-1}F\} \qquad (5.1.1a)$$

$$[M]\{^{i}\ddot{y}\} + \{^{i}R\} = \{^{i}F\} \qquad (5.1.1b)$$

where $\{^{j}R\}$ = internal restoring forces at t_j; $j = i - 1, i$
$\quad\ \{^{j}F\}$ = external forces at t_j; $j = i - 1, i$
$\quad\ \{^{j}\ddot{y}\}$ = accelerations at t_j; $i = i - 1, i$
$\quad\ [M]$ = mass matrix

For example, in a linear problem

$$\{^{j}R\} = [K]\{^{j}y\} + [C]\{^{j}\dot{y}\}$$

and we recover Eq. (4.3.3) in Sec. 4.3. In a nonlinear problem we may have that both $\{^{j}R\}$ and $\{^{j}F\}$ are nonlinear or transcendental functions of $\{^{j}y\}$ and $\{^{j}\dot{y}\}$.

Two classes of solution methods are used in structural dynamics to develop the desired recursive scheme for the numerical integration of Eqs. (5.1.1): explicit and implicit methods [61, 94]. In the explicit methods $\{^{i}y\}$ and $\{^{i}\dot{y}\}$ are assumed independent of $\{^{i}\ddot{y}\}$, $\{^{i}y\}$ is evaluated by solving Eq. (5.1.1) for $\{^{i-1}\ddot{y}\}$ at time t_{i-1} and then projecting using difference equations to predict $\{^{i}y\}$ and $\{^{i}\dot{y}\}$ at t_i. In the implicit methods $\{^{i}y\}$ and $\{^{i}\dot{y}\}$ are assumed dependent on $\{^{i}\ddot{y}\}$, $\{^{i}y\}$ is evaluated by solving Eq. (5.1.1b) for $\{^{i}\ddot{y}\}$ at time t_i as an implicit function of $\{^{i}y\}$ and $\{^{i}\dot{y}\}$, and then difference equations are used to predict $\{^{i}y\}$ $\{^{i}\dot{y}\}$. Some explicit methods are [61] Euler's method, Adams-Moulton method, Hamming's method, central difference method, and Runge-Kutta method. Some implicit methods are [61, 94] Crank-Nicholson method, Houbolt method, Wilson θ method, and Newmark method.

5.1.1 Explicit methods

The major drawback to the explicit methods is that they are conditionally stable, i.e., if the time increment Δt is greater than some critical size Δt_{cr} any errors, e.g., round-off or truncation errors, grow rapidly and eventually lead to meaningless, often unbounded, response calculations. The critical size of time increment is related to the smallest period present in

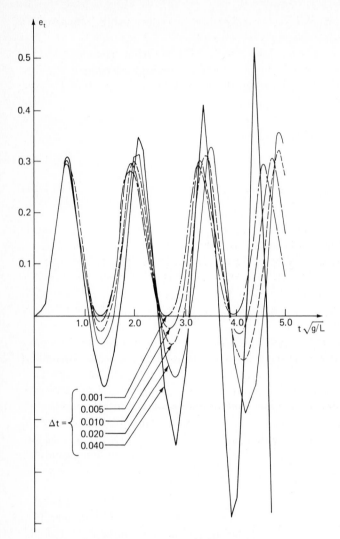

Figure 5.1.1 Explicit integration of SDOF example.

the response. This means that high-frequency fluctuations (short periods) of little consequence in the overall response would soon erroneously grow excessively large and render the overall response prediction meaningless. Thus, the time increment Δt is limited in size and may require excessive computational costs for analyses of longer ranges of time integration. For example, the scheme used in Sec. 4.1.4 is an explicit one. Displayed in Fig. 5.1.1 are response calculations using that scheme for the sample problems treated in Sec. 4.1.4 with various sizes of Δt. For $\Delta t > 0.001$, the response peak grows exponentially and the period shifts.

There is an advantage to using an explicit scheme in some instances. The solutions $\{^iy\}$, $\{^i\dot{y}\}$, and $\{^i\ddot{y}\}$ are usually predicted from values of $\{^jy\}$, $\{^j\dot{y}\}$, and $\{^j\ddot{y}\}$, $j < i$, by relatively straightforward formulas and by evaluation of the equations of motion at t_{i-1} [Eq. (5.1.1a)]. For example, in the central difference method if we neglect velocity damping effects and use the central difference expression

$$\{^{i-1}\ddot{y}\} = \frac{1}{\Delta t^2}(\{^iy\} - 2\{^{i-1}y\} + \{^{i-2}y\}) \tag{5.1.2a}$$

in conjunction with Eq. (5.1.1a) to evaluate $\{^{i-1}\ddot{y}\}$, we obtain

$$\{^iy\} = 2\{^{i-1}y\} - \{^{i-2}y\} + (\Delta t^2)[M]^{-1}(\{^{i-1}F\} - \{^{i-1}R\}) \tag{5.1.2b}$$

Thus, if the mass matrix is a diagonal matrix, i.e., a lumped mass matrix, the calculation of $\{^iy\}$ is rapid in that no inverse of the stiffness matrix embedded in $\{^{i-1}R\}$ is required and the inverse of $[M]$ is trivial. In fact, $\{F\}$ and $\{R\}$ can be assembled from contributions evaluated at the element level, and no assembly of a global stiffness matrix is required at all. With a lumped mass matrix, the computational cost per time step for an explicit scheme is very small compared to that for implicit schemes, but many more time increments are needed. Thus, for initial transient problems over short ranges of integration the explicit method may be preferable [94, 165]. However, if $[M]$ is not a lumped mass matrix, or if damping is present, the advantage is lost in that an inverse of a matrix with the same bandwidth characteristics as the stiffness matrix will be required.

5.1.2 Implicit Newmark method

We will discuss a specific implicit method, the Newmark method [94]. Other implicit methods have similar stability and convergence characteristics. The primary advantage of implicit methods is that they are unconditionally stable, i.e., there is no critical size of time step beyond which instability develops. The general form of the difference equations used in the Newmark method to propagate solutions from one time to another are

$$\{\dot{y}\} = \{\dot{z}\} + \tfrac{1}{2}(\{\ddot{z}\} + \{\ddot{y}\})\Delta t \tag{5.1.3a}$$

$$\{y\} = \{z\} + \{\dot{z}\}\,\Delta t + \tfrac{1}{2}\{\ddot{z}\}\,\Delta t^2 + \alpha(\{\ddot{y}\} - \{\ddot{z}\})\,\Delta t^2 \tag{5.1.3b}$$

where in order to avoid excess superscripts, we denote the solution at time t_i by $\{y\}$, $\{\dot{y}\}$, $\{\ddot{y}\}$ and the solution at time t_{i-1} by $\{z\}$, $\{\dot{z}\}$, $\{\ddot{z}\}$. The selection of the parameter α dictates the assumed variation of $\{\ddot{y}\}$ during the time increment. In the special case $\alpha = \tfrac{1}{4}$, we have the constant average acceleration method which has been shown to be an effective and stable solution scheme [94, 165].

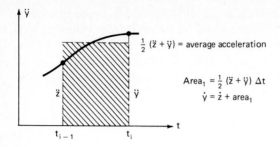

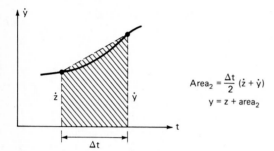

Figure 5.1.2 Constant average acceleration New-mark method.

The constant average acceleration method is depicted in Fig. 5.1.2. Since

$$\int_{t_{i-1}}^{t_i} \{d\dot{y}\} = \dot{y} - \dot{z} = \int_{t_{i-1}}^{t_i} \{\ddot{y}\} \, dt = \text{area}_1$$

$$\therefore \{\dot{y}\} = \{\dot{z}\} + \frac{\Delta t}{2} (\{\ddot{z}\} + \{\ddot{y}\}) \tag{5.1.4a}$$

Similarly,

$$\int_{t_{i-1}}^{t_i} \{dy\} = \{y\} - \{z\} = \int_{t_{i-1}}^{t_i} \{\dot{y}\} \, dt = \text{area}_2$$

$$\therefore \{\dot{y}\} = \{\dot{z}\} + \frac{\Delta t}{2} + (\{\ddot{z}\} + \{\ddot{y}\})$$

$$\{y\} = \{z\} + \Delta t \{\dot{z}\} + \frac{\Delta t^2}{4} (\{\ddot{z}\} + \{\ddot{y}\}) \qquad (5.1.4b)$$

Now, solving Eq. (5.1.4b) for $\{\ddot{y}\}$ and then substituting into Eq. (5.1.4a), we obtain (define $\mu = 2/\Delta t$)

$$\{\ddot{y}\} = \mu^2(\{y\} - \{z\}) - 2\mu\{\dot{z}\} - \{\ddot{z}\} \qquad (5.1.5a)$$

$$\{\dot{y}\} = \mu(\{y\} - \{z\}) - \{\dot{z}\} \qquad (5.1.5b)$$

by which we have now expressed the velocities and accelerations at time t_i in terms of values at t_{i-1} and of unknown $\{y\}$. Equations (5.1.5) are substituted into the equations of motion at time t_i, Eq. (5.1.1b), to give

$$\mu^2[M]\{y\} - \mu^2[M]\{z\} - 2\mu[M]\{\dot{z}\} - [M]\{\ddot{z}\}$$
$$+ \{{}^i R(z, \dot{z}, \ddot{z}, y)\} = \{{}^i F(z, \dot{z}, \ddot{z}, y)\} \qquad (5.1.6)$$

which is a nonlinear algebraic equation for $\{y\}$, and we have converted the dynamic problem into an equivalent nonlinear static problem for $\{y\}$ at each successive time t_i.

For the moment, let us consider the linear problem in order to examine convergence and stability characteristics of the Newmark method. In that case

$$\{{}^i F\} = \{{}^i F(t_i)\}$$

$$\{{}^i R\} = [K]\{y\} + [C]\{\dot{y}\}$$

$$= [K]\{y\} + \mu[C]\{y\} - \mu[C]\{z\} - [C]\{\dot{z}\}$$

and Eq. (5.1.6) becomes

$$[\hat{K}]\{y\} = \{\hat{F}\} \qquad (5.1.7a)$$

where

$$[\hat{K}] = [K] + \mu[C] + \mu^2[M]$$

$$\{\hat{F}\} = \{{}^i F\} + [C](\mu\{z\} + \{\dot{z}\}) + [M](\mu^2\{z\} + 2\mu\{\dot{z}\} + \{\ddot{z}\}) \qquad (5.1.7b)$$

in which $[\hat{K}]$ is an effective stiffness matrix at time t_i and includes mass and damping effects. The effective load vector $\{\hat{F}\}$ includes effects of displacements, velocities, and accelerations at the previous time t_{i-1}. This is an implicit method because the solution at time t_i is a function of the

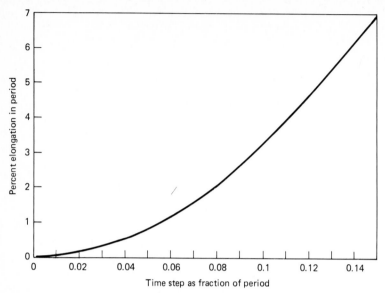

Figure 5.1.3 Period elongation in Newmark method.

stiffness $[K]$ at time t_i. Note that there is no computational advantage in using a lumped mass matrix in that $[\hat{K}]^{-1}$ is required rather than just $[M]^{-1}$ even if $[C]$ is zero.

Newmark's method is unconditionally stable [94] if $\alpha > \frac{1}{4}$ in Eqs. (5.1.3). That is, no matter how long Δt, the solution does not become unbounded. Note that if $\Delta t = \infty$ and either zero initial conditions apply or $[C] = [M] = 0$, Eqs. (5.1.7) become the static problem at t_i. Stability of results is of little value if the bounded solution is not convergent on the correct answer. Thus, we must further examine the accuracy of the Newmark method. For periodic solutions, the time increment Δt has to be a fraction of the smallest response period expected to be of importance in the overall response. Bathe [94] examined the convergence characteristics of the Newmark method for a single-degree-of-freedom system with period T and demonstrated that there was no amplitude decay during a typical period, but there was some period elongation as shown in Fig. 5.1.3. If an increment size $\Delta t = $ one-tenth the shortest period expected is used, then all longer period components of the response will have less than 3 percent period elongation.

For example, in Sec. 4.3.1 we determined the three natural frequencies of a taut cable with three attached masses as shown in Fig. 4.3.2. The corresponding periods are

$$T_1 = 4.105 \sqrt{\frac{ML/AE}{L/S_0 - 1}}$$

$$T_2 = 2.221 \sqrt{\frac{ML/AE}{L/S_0 - 1}}$$

$$T_3 = 1.700 \sqrt{\frac{ML/AE}{L/S_0 - 1}}$$

So, if we selected a time increment

$$\Delta t = 0.1 \sqrt{\frac{ML/AE}{L/S_0 - 1}}$$

the errors in period elongation for each mode participating in the response would be

$$\frac{\Delta t}{T_1} = 0.024 \rightarrow \% \text{ elongation} < 0.5 \text{ percent}$$

$$\frac{\Delta t}{T_2} = 0.045 \rightarrow \% \text{ elongation} < 0.7 \text{ percent}$$

$$\frac{\Delta t}{T_3} = 0.059 \rightarrow \% \text{ elongation} < 1.3 \text{ percent}$$

A reasonable criterion for selection of a time increment for highly irregular time series of loading can be based on a Fourier analysis of the dynamic load applied. If only frequencies below ω_0 are contained in the spectrum of the loading, then select

$$\Delta t = \frac{\pi}{40\omega_0}$$

Since Newmark's method overestimates periods and since consistent mass matrices underestimate periods, it is reasonable to use consistent mass matrices to partially counteract the period elongation errors in the Newmark method. Further, there is no computational advantage to lumped mass matrices because $[\hat{K}]^{-1}$ and not $[M]^{-1}$ is required.

For nonlinear problems there has been no conclusive proof of stability of the Newmark method except for a few specific nonlinear structural problems. However, experience by many investigators indicates that the Newmark method is a highly stable and effective scheme for integration of nonlinear equations. Since Eq. (5.1.6) is an equivalent nonlinear static problem, all the methods described previously in Sec. 3.3.1 can be applied to Eq. (5.1.6).

With $\{y\}$ being the true solution to Eq. (5.1.6), an iterative Newton-Raphson solution would proceed as follows: For an initial trial solution $\{y^P\}$ ($p = 0$), successive iterates are given by

$$\{y^{p+1}\} = \{y^p\} + \{\Delta y^p\} \tag{5.1.8}$$

where $\{\Delta y^p\}$ is the error estimate

$$\{\Delta y^p\} = \{y\} - \{y^p\} \tag{5.1.9}$$

Take a Taylor expansion of Eq. (5.1.6) about the solution of $\{y^p\}$ to obtain

$$[\hat{J}]\{\Delta y^p\} = \{^iF^p\} - \{^iR^p\} + 2\mu[M]\{\dot{z}\}$$

$$+ \mu^2[M](\{z\} - \{y^p\}) + [M]\{\ddot{z}\} \tag{5.1.10a}$$

where

$$[\hat{J}] = \mu^2[M] + \frac{\partial\{^iR^p\}}{\partial\{y^p\}} - \frac{\partial\{^iF^p\}}{\partial\{y^p\}} \tag{5.1.10b}$$

is an effective tangent stiffness matrix. Thus, solve Eq. (5.1.10a) for $\{\Delta y^p\}$ based on an iterate $\{y^p\}$ and then determine a new iterate $\{y^{p+1}\}$ from Eq. (5.1.8). Repeat this process until $\{\Delta y^p\}$ is small compared to a specified tolerance. Then the solution at time t_i is, by Eq. (5.1.5a),

$$\{y\} = \{y^p\} \tag{5.1.11a}$$

$$\{\dot{y}\} = \mu(\{y^p\} - \{z\}) - \{\dot{z}\} \tag{5.1.11b}$$

$$\{\ddot{y}\} = \mu^2(\{y^p\} - \{z\}) - 2\mu\{\dot{z}\} - \{\ddot{z}\} \tag{5.1.11c}$$

An initial estimate $\{y^0\}$ is required in the iterative process. Since this is a time-incremented solution, a projection of the solution at time t_{i-1} will give a close estimate, i.e., by Eq. (5.1.4b)

$$\{y^0\} = \{z\} + \Delta t\{\dot{z}\} + \frac{\Delta t^2}{4}\{\ddot{z}\} \tag{5.1.12}$$

Alternatively, the combined incremental-iterative technique described in Sec. 3.3.1 can be adopted, i.e., taking a Taylor expansion of Eq. (5.1.6) at time $t = t_i$ about the solution at time t_{i-1} we have

$$\{y^0\} = \{z\} + \{\Delta\bar{y}\} \tag{5.1.13a}$$

where $\{\Delta\bar{y}\}$ is the increment from t_{i-1} to t_i evaluated from

$$[\bar{J}]\{\Delta\bar{y}\} = \{^iF^0\} - \{^{i-1}R\} + [M]\{\ddot{z}\} + 2\mu[M]\{\dot{z}\} \tag{5.1.13b}$$

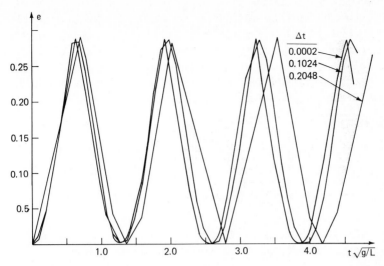

Figure 5.1.4 Implicit integration of SDOF example.

which is a residual feedback incremental equation. This scheme for predicting an initial estimate $\{y^0\}$ for an iterative method is more accurate than the first scheme of merely projecting a solution from t_{i-1} but is considerably more expensive. For small time steps commensurate with the fluctuations in the loads $\{F\}$, the first scheme predicts initial estimates $\{y^0\}$ with sufficient accuracy for rapid convergence of the iterative method. A strictly incremental scheme, or a scheme in which only one iteration is performed at every time, is not recommended because the method then begins to resemble an explicit method and errors in each increment gradually distort the response as time progresses.

As an example of application of the iterative Newmark method, let us reconsider the single-degree-of-freedom system solved in Sec. 4.1 using an explicit scheme. Displayed in Fig. 5.1.4 is the response predicted using α = ¼ and three different sizes for the time increment. We can see that the response remains stable even for a time step $\Delta t = 0.2048 \sqrt{g/L}$ and that there is period elongation for larger time increments as occurs in linear problems. Figure 5.1.4 should be compared to Fig. 5.1.1 for an explicit solution. The implicit solution yields good results for $\Delta t < 0.1 \sqrt{g/L}$, whereas the explicit solution required $\Delta t < 0.0025 \sqrt{g/L}$ for stable results.

5.1.3 Alternate form of Newmark method

In Sec. 5.1.2 we substituted expressions for the accelerations, Eq. (5.1.5a), and velocities, Eq. (5.1.5b), in terms of the displacements $\{y\}$ into the governing equations of motion to obtain an equivalent stiffness equation for

$\{y\}$. For linear problems the solution for $\{y\}$ was obtained with one inversion of $[\hat{K}]$ in Eqs. (5.1.7). For nonlinear problems, it was necessary to perform iterations and hence several inversions of $[\hat{J}]$ in Eqs. (5.1.10). Let us examine an alternate form of the Newmark method that requires iterations even for linear problems with the aim of developing a more efficient technique for nonlinear cases in which iterations are required in any case. Recall Eqs. (5.1.4):

$$\{\dot{y}\} = \{\dot{z}\} + \frac{\Delta t}{2}\left(\{\ddot{z}\} + \{\ddot{y}\}\right)$$

$$\{y\} = \{z\} - \Delta t\{\dot{z}\} + \frac{\Delta t^2}{4}\left(\{\ddot{z}\} + \{\ddot{y}\}\right)$$

which would enable calculation of $\{\dot{y}\}$ and $\{y\}$ if $\{\ddot{y}\}$ were known. Suppose we assume an initial estimate $\{\ddot{y}^p\}$ $(p = 0)$ and iterate. Define

$$\{\Delta\ddot{y}^p\} = \{\ddot{y}^{p+1}\} - \{\ddot{y}^p\}$$

Then, the new iterates to accelerations, velocities, and displacements at t_i are

$$\{\ddot{y}^{p+1}\} = \{\ddot{y}^p\} + \{\Delta\ddot{y}^p\} \tag{5.1.14a}$$

$$\{\dot{y}^{p+1}\} = \{\dot{z}\} + \frac{\Delta t}{2}\left(\{\ddot{z}\} + \{\ddot{y}^p\}\right) + \frac{\Delta t}{2}\{\Delta\ddot{y}^p\}$$

$$= \{\dot{y}^p\} + \frac{\Delta t}{2}\{\Delta\ddot{y}^p\} \tag{5.1.14b}$$

$$\{y^{p+1}\} = \{z\} + \Delta t\{\dot{z}\} + \frac{\Delta t^2}{4}\left(\{\ddot{z}\} + \{\ddot{y}^p\}\right) + \frac{\Delta t^2}{4}\{\Delta\ddot{y}^p\}$$

$$= \{y^p\} + \frac{\Delta t^2}{4}\{\Delta\ddot{y}^p\} \tag{5.1.14c}$$

Now, substitute Eqs. (5.1.14) into the equations of motion at time t_i, Eq. (5.1.16) to obtain

$$[M](\{\ddot{y}^p\} + \{\Delta\ddot{y}^p\}) + \{^iR(y^p, \dot{y}^p, \Delta\ddot{y}^p)\} - \{^iF(y^p, \dot{y}^p, \Delta\ddot{y}^p)\} = 0 \tag{5.1.15}$$

which is a nonlinear algebraic equation for $\{\Delta\ddot{y}^p\}$, the error in the acceleration estimate at time ^it. The initial estimates $\{y^0\}$, $\{\dot{y}^0\}$ are, from Eqs. (5.1.14),

$$\{\dot{y}^0\} = \{\dot{z}\} + \frac{\Delta t}{2}\{\ddot{z}\} \tag{5.1.16a}$$

$$\{y^0\} = \{z\} + \Delta t\{\dot{z}\} + \frac{\Delta t^2}{4}\{\ddot{z}\} \tag{5.1.16b}$$

with an initial estimate of $\{\Delta \dot{y}^0\} = 0$ so that

$$\{\dot{y}^0\} = \{\dot{z}\} \tag{5.1.16c}$$

If we adopt a Gauss-Seidel iteration scheme (see under "Iteration Methods" in Chap. 3), we write a recursive relation for direct iteration in Eq. (5.1.15) as

$$[M]\{\Delta \ddot{y}^p\} = -([M]\{\ddot{y}^p\} + \{^i R^p\} - \{^i F^p\}) \tag{5.1.17}$$

If we solve Eq. (5.1.17) for $\{\ddot{y}^p\}$, successive iterates are then given by Eqs. (5.1.14). Since the mass matrix is diagonally dominant, the Gauss-Seidel iteration technique will converge. Note that only the mass matrix is needed, and if a lumped mass matrix is used, the solution to Eq. (5.1.17) is very rapid and convergence is rapid because all off-diagonal terms are zero. There is no advantage to the method if a consistent matrix is used in that the Newton-Raphson method described previously has more rapid convergence properties. For linear problems, there is no advantage to this alternate approach in that iterations are required despite the linearity of the problems. This is still an implicit method in that the equations of motion are satisfied at time t_i rather than at t_{i-1}. The errors will be in $\{\ddot{y}\}$ rather than in $\{y\}$, as in the previous form of the Newmark method. Hence, smoothing of errors occurs because integration is performed to obtain $\{y\}$ rather than numerical differentiation. As in the central difference method described in Sec. 5.1.1, we need not evaluate the stiffness matrix and can assemble $\{^i R^p\}$ from element contributions. In the previous example considered in Sec. 5.1.2, convergent results were obtained for a time increment less than one-tenth the nonlinear period. If the number of iterations to convergence begins to be excessive in regions of the response curve where the accelerations are changing rapidly, a smaller (50 percent) time increment can be assumed in that region to speed convergence. The increment size can again be increased when the number of iterations again decreases.

5.2   Nonlinear Isoparametric Cable Element

For cable systems with large curvatures, numerous straight elements, as derived in Chap. 3, would be required to adequately model the curved geometry of the system. This would introduce many added degrees of freedom with a consequent increase in computer storage requirements and computational cost per time increment in the calculation of the dynamic response of such systems. In addition, spurious slope discontinuities would be introduced at nodes where no real concentrated loads or interconnections with other elements occur, thus artificially distorting the nodal equilibrium conditions at those nodes. Therefore, an element which can better model the curved geometry of cable segments subjected to distributed loads is desirable. By means of isoparametric elements [165] it is possible to develop such an element.

We will first present some aspects of isoparametric interpolating functions for one-dimensional curvilinear coordinate systems such as the arc length coordinate used in cable elements. Three definitions of element geometry will be detailed. Then, a nonlinear formulation in terms of an arbitrary isoparametric element will be derived within which any of the isoparametric approximating functions can be incorporated.

5.2.1   Isoparametric interpolation

In Chap. 3 was presented a straight element approximation in which linear polynomial approximating functions were used to represent the variation of displacements along the length of the element between the two endpoint nodes. If additional nodes were included at intermediate points on the element, higher-order interpolating polynomials for the displacements could be defined. If we simultaneously interpolate geometry, i.e., the cartesian locations of points along the arc length of the reference configuration, using the same polynomial functions in conjunction with the cartesian locations of the node points, an isoparametric formulation is obtained.

Some of the advantages of an isoparametric formulation are that [165]

1. They have been shown to meet all convergence criteria as the sizes of elements are decreased,

2. They can interpolate curved shapes and boundaries accurately with fewer elements.

3. If curvilinear coordinates are used in the development of the continuum theory, the natural coordinates of isoparametric elements provide a smooth transition from the continuous system to the discrete system.

4. The higher-order polynomial functions are more accurate predictors of equilibrium shapes.

5. They have a greater rate of convergence.

They have three disadvantages:

1. They are more costly to formulate for each element.

2. Numerical integration is required for each element to form the stiffness matrices.

3. The elements are stiffer.

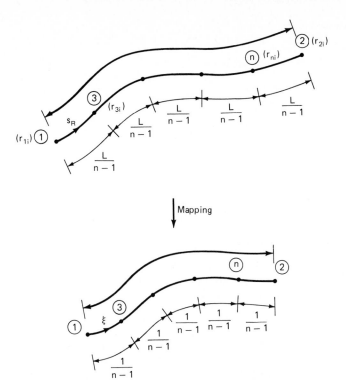

Figure 5.2.1 Isoparametric one-dimensional mapping.

However, these disadvantages are of little consequence if many fewer iso-parametric elements than straight elements can be used to model the problem.

Shown in Fig. 5.2.1 is an element of length L with n equally spaced nodes along its arc length $0 \leq s_R \leq L$. The cartesian coordinates of each node α are denoted by $r_{\alpha i}$, $\alpha = 1, n$, i.e.,

$$r_{\alpha i} = x_i(s_R = s_{R\alpha})$$

where $s_{R\alpha}$ = arc length to node α.

We can map the real element of length L into an element of unit length by defining a nondimensional coordinate $0 \leq \zeta = s_R/L \leq 1$. Let us inter-polate for the displacements $u_i(\zeta)$ at an arbitrary point along the nondi-mensional arc length ζ by a series of $(n - 1)$-order polynomials:

$$u_i(\xi) = \sum_{\alpha=1}^{n} \eta_\alpha(\xi) d_{\alpha i} \tag{5.2.1}$$

where $d_{\alpha i}$ = discrete values of u_i at node α and $\eta_\alpha(\zeta)$ are the $(n - 1)$-order polynomials. They must have the properties that $\eta_\alpha = 1$ at node α and $\eta_\alpha = 0$ at all other nodes. Similarly, we can interpolate for the cartesian

TABLE 5.2.1 Isoparametric Interpolation Polynomials

Element	No. of nodes	η_α	Interpolating polynomials	Derivatives $\partial\eta_\alpha/\partial\xi$
Linear	2	$\eta_1 =$ $\eta_2 =$	$1 - \xi$ ξ	-1 1
Parabolic	3	$\eta_1 =$ $\eta_2 =$ $\eta_3 =$	$(1 - \xi)(1 - 2\xi)$ $(-\xi)(1 - 2\xi)$ $4\xi(1 - \xi)$	$4\xi - 3$ $4\xi - 1$ $4(1 - 2\xi)$
Cubic	4	$\eta_1 =$ $\eta_2 =$ $\eta_3 =$ $\eta_4 =$	$(1 - \xi)(1 - 3\xi)(2 - 3\xi)/2$ $\xi(1 - 3\xi)(2 - 3\xi)/2$ $-9\xi(1 - \xi)(2 - 3\xi)/2$ $-9\xi(1 - \xi)(1 - 3\xi)/2$	$-1(11 - 36\xi + 27\xi^2)/2$ $(2 - 18\xi + 27\xi^2)/2$ $9(2 - 10\xi + 9\xi^2)/2$ $9(1 - 8\xi + 9\xi^2)/2$

locations $x_i(\zeta)$ at an arbitrary point along ζ in terms of the discrete values $r_{\alpha i}$ at the nodes using the same polynomials

$$x_i(\xi) = \sum_{\alpha=1}^{n} \eta_\alpha(\xi) r_{\alpha i} \tag{5.2.2}$$

The requirements on $\eta_\alpha(\zeta)$ of being unity at node α and zero at the other nodes can be satisfied by defining η_α as an $(n - 1)$-order product of linear functions of the form $(\xi - \xi_\alpha)$ where $\xi_\alpha = s_{R\alpha}/L =$ arc length coordinate of node α. In Table 5.2.1 are listed the polynomial interpolation functions for linear (2 nodes), parabolic (3 nodes), and cubic (4 nodes) elements [94, 90, 165]. Also given in Table 5.2.1 are the first derivatives $\partial\eta_\alpha/\partial\zeta$ which will be needed subsequently.

5.2.2 Element formulation

To develop our isoparametric element we will apply our isoparametric interpolation functions for displacement, Eq. (5.2.1), and location in s_r, Eq. (5.2.2), to convert the definitions of geometry, strain, stress, and constitutive relations given in Eqs. (4.4.3) for the cable continuum into discrete forms and then formulate the discrete version of the nonlinear Hamilton's principle given in Eq. (4.4.7b) for the additionally deformed state at time t_i.

First, let us rewrite Eqs. (5.2.1) and (5.2.2) in matrix form as

$$\{u\} = [N(\xi)]\{d\} \tag{5.2.3a}$$

$$\{x\} = [N(\xi)]\{r\} \tag{5.2.3b}$$

where

$$\{u\} = \begin{Bmatrix} u_1 \\ u_2 \\ u_3 \end{Bmatrix}$$

$$\{x\} = \begin{Bmatrix} x_1 \\ x_2 \\ x_3 \end{Bmatrix}$$

$$\{d\} = \begin{Bmatrix} d_1 \\ d_2 \\ \vdots \\ d_n \end{Bmatrix} \quad \{d_\alpha\} = \begin{Bmatrix} d_{\alpha 1} \\ d_{\alpha 2} \\ d_{\alpha 3} \end{Bmatrix}$$

$$\{r\} = \begin{Bmatrix} r_1 \\ r_2 \\ \vdots \\ r_3 \end{Bmatrix} \quad \{r_\alpha\} = \begin{Bmatrix} r_{\alpha 1} \\ r_{\alpha 2} \\ r_{\alpha 3} \end{Bmatrix}$$

$$[N] = [N_1 N_2 \cdots N_n] \tag{5.2.3c}$$

$$[N_\alpha] = \eta_\alpha [I] \tag{5.2.3d}$$

Thus our interpolating functions η_α given in Table 5.2.1 are embedded in the submatrix $[N_\alpha]$. The rates of change of location in s_R and of the displacements with respect to s_R are

$$\frac{\partial \{x\}}{\partial s_R} = \frac{1}{L} \frac{\partial \{x\}}{\partial \xi} = \frac{1}{L} \frac{\partial [N]\{r\}}{\partial \xi}$$

$$= \frac{1}{L} \left[\frac{\partial N_1}{\partial \xi} \frac{\partial N_2}{\partial \xi} \cdots \frac{\partial N_n}{\partial \xi} \right] \{r\}$$

$$= \frac{1}{L} [\partial N]\{r\} \tag{5.2.4a}$$

$$\frac{\partial \{u\}}{\partial s_R} = \frac{1}{L} [\partial N]\{u\} \tag{5.2.4b}$$

where

$$[\partial N] = [\partial N_1 \, \partial N_2 \cdots \partial N_n] \tag{5.2.4c}$$

$$[\partial N_\alpha] = \frac{\partial \eta_\alpha}{\partial \xi} [I] \tag{5.2.4d}$$

with $\partial \eta_\alpha / \partial \xi$ defined in Table 5.2.1.

Now, Eqs. (4.4.3) become [90]

Direction cosines in the reference state:

$$\{\theta_R\} = \begin{Bmatrix} \theta_{R1} \\ \theta_{R2} \\ \theta_{R3} \end{Bmatrix} = \frac{1}{L} [\partial N]\{r\} \tag{5.2.5a}$$

Relative strain:

$$\gamma = [\theta_r]^T \frac{\partial \{u\}}{\partial s_R} + \frac{1}{2} \frac{\partial [u]^T}{\partial s_R} \frac{\partial \{u\}}{\partial s_R}$$

$$= \frac{1}{L^2} ([r]^T [\partial N]^T [\partial N]\{d\} + [d]^T [\partial N]^T [\partial N]\{d\}) \tag{5.2.5b}$$

Direction cosines in the additionally deformed state:

$$\{\theta\} = \frac{1}{L\sqrt{1 + 2\gamma}} [\partial N](\{r\} + \{d\}) \tag{5.2.5c}$$

Tension in the additionally deformed state:

$$T = T_R + E_R A_0 \lambda_R (\sqrt{1 + 2\gamma} - 1) \tag{5.2.5d}$$

The virtual displacements and the accelerations are

$$\{\delta u\} = [N]\{\delta d\} \tag{5.2.5e}$$

$$\{\ddot{u}\} = [N]\{\ddot{d}\} \tag{5.2.5f}$$

Also,

$$\frac{\partial \{\delta u\}}{\partial s_R} = \frac{1}{L} [\partial N]\{\delta d\} \tag{5.2.5g}$$

The matrix form of Hamilton's principle, Eq. (4.4.7b), is

$$\int_{t_1}^{t_2} \int_{S_R} \left([\delta u]^T \frac{m_0 A_0}{\lambda_R} \{\ddot{u}\} + \frac{\partial [\delta u]^T}{\partial s_R} T\{\theta\} - [\delta u]^T \frac{1}{\lambda_R} \{q\} \right) ds_R \, dt = 0 \tag{5.2.6}$$

where distributed loads $\{q\} = \begin{Bmatrix} q_3 \\ q_2 \\ q_3 \end{Bmatrix}$ will also be interpolated by

$$\{q\} = [N]\{p\}$$

$$\{p\} = \begin{Bmatrix} p_1 \\ p_2 \\ p_3 \end{Bmatrix} \qquad \{p_\alpha\} = \begin{Bmatrix} q_{\alpha 1} \\ q_{\alpha 2} \\ q_{\alpha 3} \end{Bmatrix}$$

Upon substitution of Eqs. (5.2.5) into Eq. (5.2.6), we obtain

$$\int_{t_1}^{t_2} \sum_e \{\delta d\}^T \left(\int_0^1 \frac{m_0 A_0 L}{\lambda_R} [N]^T [N] \, d\xi \, \{\ddot{d}\} + \int_0^1 T \, [\partial N]^T \{\theta\} \, d\xi \right.$$

$$\left. - \int_0^1 \frac{L}{\lambda_R} [N]^T [N] \, d\xi \{p\} \right) dt = 0$$

Since the virtual displacements are arbitrary, we assemble a system of equations of motion of the form

$$[M]\{\ddot{D}\} + \{R\} = \{F\} \tag{5.2.7a}$$

where

$$[M] = [M_{\text{conc}}] + \sum_e \left([_e M] \right)$$

$$[_e M] = {}_e m \int_0^1 [N]^T [N] \, d\xi \tag{5.2.7b}$$

$$_e m = \text{total mass of element}$$

$$\{F\} = \{F_{\text{conc}}\} + \sum_e \left(\{^e F\} \right) \tag{5.2.7c}$$

$$\{_e F\} = \frac{L}{{}_e \lambda_R} \int_0^1 [N]^T [N] \, d\xi \tag{5.2.7d}$$

$$\{R\} = \sum \left(\{_e R\} \right)$$

$$\{_e R\} = \int_0^1 [\partial N]^T (T)\{\theta\} \, d\xi \tag{5.2.7e}$$

with T and $\{\theta\}$ given by Eqs. (5.2.5d and e).

Equation (5.2.7a) is the same form as Eq. (5.1.1), and thus the solution technique described in Sec. 5.1 can be applied to Eqs. (5.2.7). Using Newmark's method with iteration as described in Eqs. (5.1.8) to (5.1.10), we arrive at the governing equations for an iterative scheme for solution of Eqs. (5.2.7) at time t_i:

$$\{D^{p+1}\} = \{D^p\} + \{\Delta D^p\} \tag{5.2.8a}$$

$$[\hat{J}]\{\Delta D\} = \{{}^iF^p\} - \{{}^iR^p\} + 2\mu[M]\{\dot{z}\} + \mu^2[M](\{z\} - \{D^p\}) \tag{5.2.8b}$$

$$[\hat{J}] = \mu^2[M] + \frac{\partial\{{}^iR^p\}}{\partial\{D^p\}} \tag{5.2.8c}$$

where $\{z\}$ is the solution for $\{D\}$ at time t_{i-1}. The gradient of $\{{}^iR^p\}$ with respect to $\{D^p\}$ can be evaluated from Eq. (5.2.7d).

$$\frac{\partial\{{}^iR^p\}}{\partial\{D^p\}} = \sum_e \left(\frac{\partial\{{}^i_eR^p\}}{\partial\{d\}}\right)$$

Noting that from Eqs. (5.2.5)

$$\frac{\partial\gamma}{\partial\{d\}} = \frac{1}{L}\sqrt{1+2\gamma}\,[\theta]^T\,[\partial N]$$

$$\frac{\partial\{\theta\}}{\partial\{d\}} = \frac{1}{L\sqrt{1+2\gamma}}\,([I] - \{\theta\}[\theta]^T)\,[\partial N]$$

$$\frac{\partial T}{\partial\{d\}} = \frac{EA\lambda_R}{L}\,[\theta]^T\,[\partial N]$$

we obtain

$$\frac{\partial\{{}^iR^p\}}{\partial\{d^p\}} = \sum_e \left(\frac{1}{L}\int_0^1 \frac{1}{\sqrt{1+2\gamma}}\,[\partial N]^T(T[I]\right.$$

$$\left. + (EA\lambda_R - T_R)\{\theta\}[\theta]^T)[\partial N]\,d\xi\right) \tag{5.2.9}$$

for use in Eq. (5.2.8c).

For any selected isoparametric element, substitute the appropriate $[N]$ and $[\partial N]$ from Eqs. (5.2.3) and (5.2.4) and Table 5.2.1 into Eqs. (5.2.7) to (5.2.9). Note that if a linear element is selected

$$[N] = [(1 - \xi)I \; \xi I]$$

$$[\partial N] = [-I \; I]$$

and the results presented in Sec. 3.3.2 for a straight nonlinear element are obtained. Further, note that Eqs. (5.2.7) and (5.2.9) require integrals over the range 0 to 1 of functions of ζ *and* of terms multiplied by $\{r\}$, the cartesian locations of element nodes [see Eqs. (5.2.5)]. Thus, numerical integration will be required for each element other than a linear element. Gaussian quadrature is recommended in the numerical integration of isoparametric elements. The need to integrate numerically each element at each time step is the reason isoparametric elements are more expensive to use than straight elements.

As an example of the use of isoparametric elements [90], consider the problem of determining the natural frequencies of a stretched cable as shown in Fig. 5.2.2. Three finite element models were used [90]: 12 linear elements, 6 quadratic elements, and 4 cubic elements. Thus each model had an equal number of degrees of freedom. The first 10 frequencies and mode shapes were calculated with each model, and the frequencies are compared with the exact solutions in Table 5.2.2.

It can be seen that all three models provide reasonably good approximations in the lower modes, but that accuracy deteriorates in the higher modes. Higher-order elements perform better than lower-order elements in the lower modes while the situation is reversed in the higher modes. This can be traced to the fact that in the lower modes, higher-order elements are better able to interpolate the mode shapes. In the higher modes, excessive deformation of elements is needed to duplicate the mode shapes and extra constraints are encountered in the higher-order elements. This can be seen in the mode shape for the ninth mode shown in Fig. 5.2.3. Although it has not been proved conclusively for nonlinear structures, the dynamic response of a structure is usually dominated by a few lower modes. Hence, one would choose higher-order elements if accuracies in the lower modes are needed. The computational times used in the three

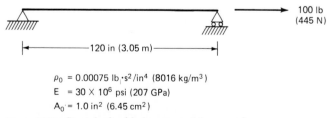

ρ_0 = 0.00075 lb $\cdot$ s^2/in^4 (8016 kg/m^3)
E   = 30 $\times$ 10^6 psi (207 GPa)
A$_0$ = 1.0 in^2 (6.45 cm^2)

Figure 5.2.2 Stretched cable isoparametric example.

TABLE 5.2.2 Natural Frequencies of a Stretched Cable

Natural frequencies, rad/s

Model	Mode									
	1	2	3	4	5	6	7	8	9	10
Exact solution	9.560	19.119	28.679	38.238	47.798	57.357	66.917	76.476	86.040	95.596
4 cubic elements	9.560	19.120	28.699	38.490	48.319	58.867	70.270	78.881	106.295	127.570
6 quadratic elements	9.560	19.134	28.786	38.655	48.914	57.735	73.735	87.457	103.570	120.690
12 linear elements	9.587	19.338	29.420	40.000	51.235	63.245	76.051	89.443	102.776	114.736

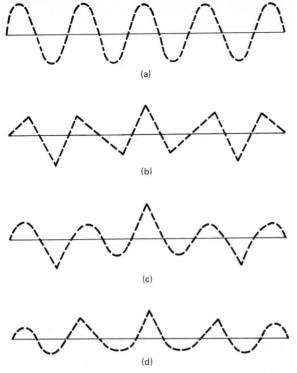

Figure 5.2.3 Ninth mode of stretched cable [90]. (*a*) Exact solution; (*b*) 12 linear elements; (*c*) 6 quadratic elements; (*d*) 4 cubic elements.

models were comparable, with lower-order models requiring slightly less time though the differences were negligible.

5.3. Hydrodynamic Forces

We have seen that the discrete equation of motion for a multiple-degree-of-freedom model of a structure is

$$[M]\{\ddot{y}\} + [C]\{\dot{y}\} + [K]\{y\} = \{F\} \qquad (5.3.1)$$

where $[M]$, $[C]$, and $[K]$ are structural mass, damping, and stiffness matrices. For structures vibrating in a fluid medium the force vector consists of [96]

$$\{F\} = \{F_s\} + \{F_f\} \qquad (5.3.2a)$$

where $\{F_s\}$ = vector of time-dependent nonfluid forces
$\{F_f\}$ = vector of hydrodynamic coupling forces due to dynamic

pressures induced by the relative motions between the structure and the fluid.

The term $\{F_f\}$ is given by [96]

$$\{F_f\} = \{F_{f0}\} - ([M_f]\{\ddot{y}\} + [C_f]\{\dot{y}\} + [K_f]\{y\})   (5.3.2b)$$

where $[M_f]$ = added mass matrix
   $[C_f]$ = viscous, or radiation, damping matrix (both corresponding to motions of the structure in an otherwise still fluid, e.g., wave radiation)
   $[K_f]$ = buoyant stiffness matrix due to fluid displaced during the structural motion
   $\{F_{f0}\}$ = incident fluid force on the structure due to fluid velocities $\{v\}$ and accelerations $\{\dot{v}\}$, e.g., wave particle motion.

The term $\{F_{f0}\}$ is given by

$$\{F_{f0}\} = ([M_{f0}] + [M_f])\{\dot{v}\} + [C_f]\{v\}   (5.3.2c)$$

where $[M_{f0}]$ = mass of fluid displaced by the structure.

For structural components and bodies with large dimensions relative to the lengths of waves incident on and radiated by the structure, diffraction-radiation theory is needed to quantify $[M_f]$ and $[C_f]$ [94]. They are functions of the wave frequencies. For small-diameter structures, a semi-empirical approach called the Morison equation approach [21, 83, 94, 102], is used to define $[M_f]$ and $[C_f]$. In this instance $[C_f]$ is due to real fluid effects such as fluid viscosity, Reynolds number, vortex shedding, wake formation, and various other parameters rather than to wave radiation or diffraction. Reference 167 discusses aspects of the determination of wave forces on structures and the various parameters which affect $[M_f]$ and $[C_f]$.

For cables we can adopt the Morison equation approach [168] in which the fluid flow around the cable is assumed to be undisturbed by the presence of the cable. In this instance $[M_{f0}]$ and $[K_{f0}]$ are negligible and we can combine Eqs. (5.3.1) and (5.3.2a to c) to obtain

$$([M] + [M_f])\{\ddot{y}\} + ([C] + [C_f])\{\dot{y}\} + [K]\{y\}$$

$$= \{F_s\} + [M_f]\{v\} + [C_f]\{v\}   (5.3.3)$$

where $[M + M_f]$ is called the virtual mass of the cable and $([C] + [C_f])$ can be called the apparent damping. Unfortunately, $[C_f]$ has been shown to be a nonlinear function of the relative velocity ($\{\dot{y}\} - \{v\}$) [94]. One approach to this nonlinearity in a direct time integration solution is to use the Newmark method with iterations and evaluate $[C_f]$ at the latest iterate $\{y^p\}$ to the velocity. $[C_f]$ shows a strong dependency on the Reynolds number for the relative motion normal to the cable axis. This and other aspects are discussed in Ref. 21.

5.3.3 Fluid effects in finite element models

A finite element model of hydrodynamic forces on cables can be generated using isoparametric elements. Aspects of such a derivation for a straight element are discussed in Ref. 168. The following is an example of the use of the isoparametric formulation presented in Sec. 5.2 when hydrodynamic loads are applied to strongly curved cable segments and large dynamic deflections occur. These results were obtained by A. Lo [90].

Small-scale experiments on the relaxation of a cable-buoy system have been conducted [169] for the purpose of validating cable computer programs. The experimental model consisted of a submerged spherical buoy anchored to the bottom by a single cable. The buoy was then statically displaced to a specified position and then released. The dynamic responses of the system were recorded. The problem has been successfully simulated by various finite element models. The displaced configuration of the cable-buoy system is shown in Fig. 5.3.1, and the properties of the cable and buoy are [169]

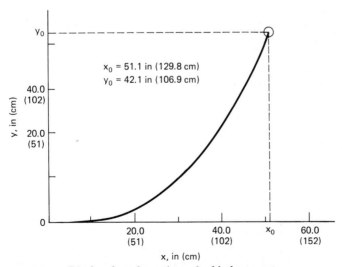

Figure 5.3.1 Displaced configurations of cable-buoy system.

Cable:

$$AE \qquad\qquad = 4.8 \text{ lb}$$

Weight in air $\quad = 11.443 \times 10^{-3}$ lb/ft

Weight in water $\quad = 2.381 \times 10^{-3}$ lb/ft

Unstretched length $= 71.56$ in

Diameter $\qquad\quad = 0.163$ in

Buoy:

Diameter $\qquad = 2.0$ in

Weight in air $\quad = 0.025$ lb

Weight in water $= -0.121$ lb

The finite element model used in the present study consisted of four quadratic cable elements with a concentrated mass element to model the buoy. The displaced configuration was obtained by the viscous relaxation technique with the cable laying in a horizontal position initially. The release of the buoy was simulated by applying to the buoy the buoyancy of the buoy and forces required to cancel the static restraints at the buoy. The drag and inertia forces due to the presence of water and the motion of the cable-buoy system were represented by a discrete hydrodynamic model based on the Morison equation approach. Constant drag and added mass coefficients were used; they are

Cable normal drag coefficient $\quad = 1.20$

Cable tangential drag coefficient $= 0.02$

Cable added mass coefficient $\quad = 1.00$

Buoy drag coefficient $\qquad\qquad = 0.50$

Buoy added mass coefficient $\qquad = 0.50$

The response of the system during the first 3.0 s after release was simulated. The size of the time step used was 0.005 s, and modified Newton-Raphson iteration was used at every step to ensure a convergence tolerance of less than 0.1. The global stiffness matrix was updated after every two steps. The results of the simulation are shown in Figs. 5.3.2 to 5.3.5.

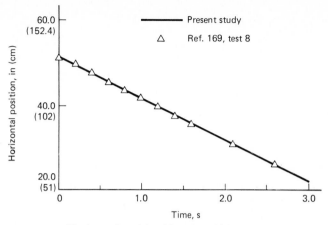

Figure 5.3.2 Horizontal position histories of buoy.

It can be seen in Figs. 5.3.2 and 5.3.3 that the finite element model provided excellent estimates of the horizontal and vertical positions of the buoy in comparison to experimental results [169]. The total velocity of the buoy is shown in Fig. 5.3.4 and the agreement is quite acceptable. The fact that the model seems to have overpredicted the velocity and the ver-

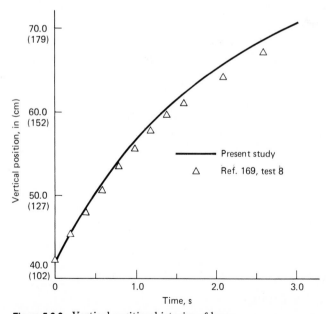

Figure 5.3.3 Vertical position histories of buoy.

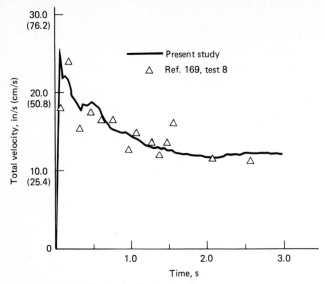

Figure 5.3.4   Total velocity histories of buoy.

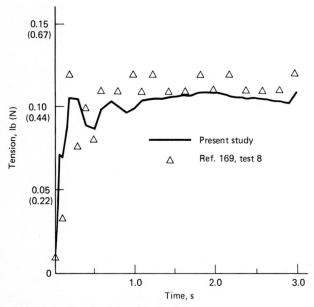

Figure 5.3.5   Tension histories of cable.

tical displacement of the buoy is likely due to the inadequacy of the constant drag coefficients used. It should be noted that the velocity history from the experiment was obtained by numerically differentiating the displacement data. The tension histories are shown in Fig. 5.3.5. The model has underestimated the tension in general, especially the peak tension. This is probably due to the fact that the tension history was measured at the anchor in the experiment while it was computed at an interior integration point in the finite element model.

Statics and Dynamics of Membrane Structures

6

Linearized Behavior of Membranes

6.1 Introduction

In this part, we consider the static and dynamic behavior of membrane structures. Membrane structures are comprised of extremely thin sheets formed into flat or curved surfaces. They transmit loads to the supporting medium by means of planar direct and shear stresses. Their bending and transverse shear rigidities are negligible.

Membrane structures depend on two-dimensional planar prestressing to maintain their shape and to support loads. As shown in Fig. 6.1.1, the prestressing can be provided by either pressurization, dead weight, or edge tractions. The prestressing requires a reactive medium to maintain the desired shape. The medium may be either foundations, compression members, pressurized gases, fluids, or counterstressing members.

If pressurized air is used as the supporting medium, this class of tension structure can be subdivided into four categories [9]: (1) air-restrained structures wherein the internal pressure is less than the ambient air pressure; (2) air-supported roofs or enclosures wherein an enclosed space is slightly pressurized; (3) air-inflated members wherein considerably pressurized and sealed tubes or mats are the load-carrying members used as an enclosure; and (4) hybrid systems of combinations of the other classes.

Membranes can be considered as two-dimensional spaces. However, loadings occur in a three-dimensional world and, therefore, if components of load transverse to the membrane are to be supported, the two-dimen-

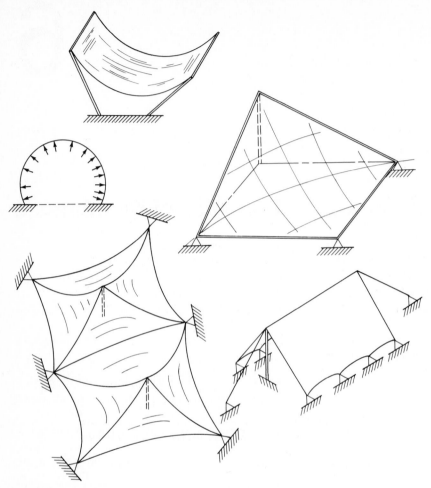

Figure 6.1.1 Prestressed membranes.

sional membrane must be curved. In Appendix B the mathematics of the differential geometry of curved surfaces is summarized.

The behavior of membrane structures is inherently nonlinear, and we can, therefore, expect a sparsity of closed-form solutions to predict the behavior. Approximate and numerical methods of analysis are the main subjects of this chapter. We will here focus on linearized models and on approximate solutions for the stresses during both the prestressing and in-service phases. Such solutions would serve as the bases for initial design studies. In Chap. 7 we will examine the effect of prestress on subsequent in-service behavior of a class of membrane structures. We will develop approximate equations to include prestress in a simplified fashion and illustrate the importance of including prestress. In Chap. 8 we will focus on nonlinear techniques for numerically modeling membrane behavior under both static and dynamic loads. A semianalytic technique to predict the prestressing behavior of pressurized membranes of revolution will be presented. We will also develop in Chap. 8 a superparametric finite element model for nonlinear static problems and a nonlinear isoparametric element for predicting membrane response under dynamic loads.

6.1.1 Mechanics of membranes

Membranes are fabricated from thin sheets of filmlike material, rubber-like or polymer materials, woven fabrics, or composite fabrics in which the coating provides additional in-plane shear rigidity to the base fabric. Characterizations of such materials was the subject of Chap. 1. In this chapter, we will either disregard the constitutive properties of the membrane or assume an isotropic linearly elastic material. Hyperelastic and other nonlinear models of membrane materials can be included using the techniques described subsequently in Chap. 8. See Refs. 19, 51, and 112 for extensive discussions of material properties for membranes. In Chap. 5 of Ref. 112 fabrication techniques for cutting, patterning, joining, attaching, and anchoring component parts of sheets for membrane structures are described. Some discussion of long-term creep and viscoelastic behavior of membranes is included in Ref. 178.

Because nonmetallic materials used commonly for membranes have lower tensile strengths, they are often reinforced by cable nets in order to increase unsupported span lengths. The techniques described in this part can be used for approximate continuous models of cables interconnected in a single flat or curved surface and loaded transversely [77, 78]. The approximate techniques described in the present chapter can be used not only for global representations of cable-reinforced membranes but also for macroscopic models of membrane segments spanning between cables.

The continuous approximation for networks is most applicable for dense networks of interlaced cables. We could, for example, develop equations for a continuous equivalent membrane (orthotropic with negligible in-plane shear rigidity) for a dense network and then proceed to discretize those equations into a set equivalent to another, less dense, collection of equivalent cables.

Shown in Fig. 6.1.2 is a differential volume (dx, dy, dz) of a membrane of thickness h. The origin of a curvilinear coordinate system x, y, z lies on the middle surface of the membrane with z directed toward the center of curvature. The averaged resultants, per unit length on the middle surface, of the direct stresses σ_x, σ_y on an element are defined as

$$N_x = \int_{-h/2}^{h/2} \sigma_x \left(1 - \frac{z}{R_x} \right) dz \tag{6.1.1a}$$

$$N_y = \int_{-h/2}^{h/2} \sigma_y \left(1 - \frac{z}{R_y} \right) dz \tag{6.1.1b}$$

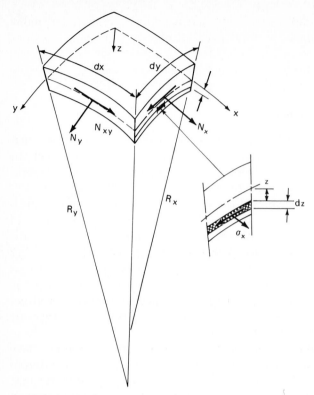

Figure 6.1.2 Membrane stress resultants.

where R_x, R_y are radii of curvatures along x, y coordinate lines. Equations (6.1.1) can be approximated [79] as

$$N_x = \int_{-h/2}^{h/2} \sigma_x dz \qquad N_y = \int_{-h/2}^{h/2} \sigma_y dz$$

if h/R_x and $h/R_y \ll 1$, which is the case for membranes. The averaged in-plane shear stress per unit length of the shear stress τ_{xy} is, likewise, approximated as

$$N_{xy} = \int_{-h/2}^{h/2} \tau_{xy} dz$$

If we succeed in calculating the planar stress resultants N_x, N_y, N_{xy} in one planar coordinate system (x, y), for a specified load distribution, we can find the principal resultants (maximum and minimum values in alternate coordinate directions oriented at an angle θ with the original axes).

The principal resultants are represented graphically in Fig. 6.1.3 on a Mohr's circle [28] for planar stress and are given by

$$N_{\min} = \frac{N_x - N_y}{2} - \left[\left(\frac{N_y - N_x}{2} \right)^2 + N_{xy}^2 \right]^{1/2} \cos 2\theta \quad (6.1.2a)$$

$$N_{\max} = \frac{N_x - N_y}{2} + \left[\left(\frac{N_y - N_x}{2} \right)^2 + N_{xy}^2 \right]^{1/2} \cos 2\theta \quad (6.1.2b)$$

where

$$\tan 2\theta = \frac{2N_{xy}}{N_y - N_x} \quad (6.1.2c)$$

Note that even if N_x and N_y are both positive, it is possible, if N_{xy} were large enough, for the $N_{\min}$ in Eq. (6.1.2a) to be negative, as shown in Fig. 6.1.3.

It is important to consider the potential occurrence of negative stresses in membranes. Because they have negligible bending resistance, membranes have negligible buckling resistance. Instead of realizing a negative stress, as predicted by a methematical model, the membrane will wrinkle (buckle), with the wrinkles running in the direction of predicted negative stress. The two-dimensional stress state will be replaced by a uniaxial stress state in the direction of the predicted maximum principal stress. Tension field theory [173–175] has been used to predict extents of wrinkled regions and magnitudes of tension stresses for static problems.

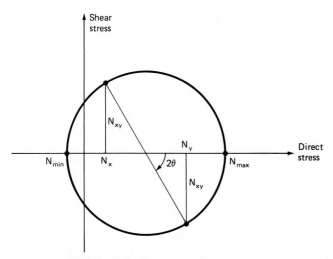

Figure 6.1.3 Mohr's circle of stress resultants.

6.1.2   Scope of chapter

In the following sections we will consider the approximate behavior of linearized models for flat and curved membranes. The focus of this chapter is on the equilibrium of forces and stress resultants on preestablished geometries for membranes. We emphasize the calculation of stresses rather than displacements, except when required for a slightly nonlinear analysis or for predicting dynamic responses.

In Sec. 6.2 flat sheets are considered. Analytic techniques for calculating prestress in select geometries are developed and a simple finite element model is presented for problems involving prestressing of more complex geometries. Closed-form and numerical techniques for predicting the slight transverse static and dynamic deflections of initially flat taut sheets are also developed.

In Sec. 6.3 we consider singly curved (cylindrical) and doubly curved (translational) surfaces for membranes. Techniques from the momentless theory of shells are adopted to calculate stresses in slightly curved (shallow) membranes and for complete cylindrical membranes with various cross-sectional curves.

Membranes of revolution are the subject of Sec. 6.4. Axisymmetric solutions for stress resultants in doubly curved revolutes with arbitrary meridianal contours are presented. Symmetric and nonsymmetric stresses due to additional in-service loads are calculated assuming an undisturbed geometry and no effect of prestress on the additional deformation.

6.2   Flat Membranes and Networks

In this section we examine membranes which are assumed to be flat in their prestressed configurations. This implies that dead weight of the membrane is negligible compared to the in-plane loadings. When we consider the in-service static and dynamic behavior in Secs. 6.2.3 and 6.2.4, we will assume that the excursions of the membrane normal to the initial plane are small enough that the geometry of the initial plane can be used as the reference and that initial stresses remain unchanged in magnitude and direction. In the succeeding section, we will consider slightly curved membranes to arrive at a more realistic model of membrane behavior.

One approach to the analysis of cable networks is to assume the discrete network to be replaced by an equivalent continuous membrane with modified material properties. In the limit, a continuous differentiable surface through the nodes in a network is governed by the same differential equations as the equivalent orthotropic membrane with negligible shear rigidity. This model is only accurate in cases involving large numbers of cables with nearly equal tension and mass distribution. When we consider the vibration of continuous membranes, we will extend the discussion of solution techniques to include vibration of discrete cable networks.

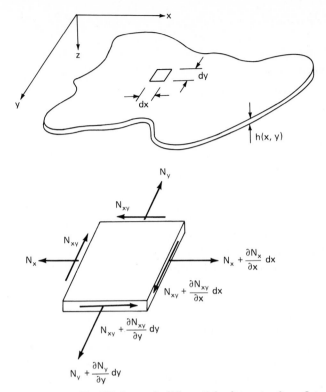

Figure 6.2.1 Equilibrium of differential element of a flat membrane.

We will first consider analytic solutions to flat rectangular and circular membranes. Solutions to more general problems are then developed by a simple finite element modeling technique. The transverse motion and superposed stresses due to in-service transverse static loads are considered in Sec. 6.2.3. The dynamic in-service behavior is the subject of Sec. 6.2.4.

6.2.1 Analytic prestressing solutions

Equilibrium of stress resultants on a differential element of the flat membrane shown in Fig. 6.2.1. requires that

$$\frac{\partial N_x}{\partial x} + \frac{\partial N_{xy}}{\partial y} + f_x = 0 \qquad (6.2.1a)$$

$$\frac{\partial N_{xy}}{\partial x} + \frac{\partial N_y}{\partial y} + f_y = 0 \qquad (6.2.1b)$$

where f_x, f_y are the distributed forces per unit area in the x, y directions, respectively.

Assuming negligible motion in the z direction during the prestressing phase, we find the linear strain-displacement relation to be [28]

$$\varepsilon_x = \frac{\partial u}{\partial x} \tag{6.2.2a}$$

$$\varepsilon_y = \frac{\partial v}{\partial y} \tag{6.2.2b}$$

$$\gamma_{xy} = \frac{\partial u}{\partial y} + \frac{\partial v}{\partial x} \tag{6.2.2c}$$

where u, v are the displacements in the x, y directions, respectively, from the unstrained position. For an elastic isotropic material we have the following constitutive relations between the strains and stress resultants:

$$\varepsilon_x = \frac{1}{Eh} (N_x - \nu N_y) \tag{6.2.3a}$$

$$\varepsilon_y = \frac{1}{Eh} (N_y - \nu N_x) \tag{6.2.3b}$$

$$\gamma_{xy} = \frac{1}{Gh} N_{xy} = 2 \left(\frac{1 + \nu}{Eh} \right) N_{xy} \tag{6.2.3c}$$

We cannot arbitrarily specify u, v and expect to obtain strains by Eqs. (6.2.2), which are physically realizable and are compatible with one other. The displacements u, v must be such that the strains satisfy the following compatibility relation:

$$\frac{\partial^2 \varepsilon_x}{\partial y^2} + \frac{\partial^2 \varepsilon_y}{\partial x^2} - \frac{\partial^2 \gamma_{xy}}{\partial x\, \partial y} = 0 \tag{6.2.4a}$$

If we substitute Eqs. (6.2.3) into Eq. (6.2.4a) and use derivatives of Eqs. (6.2.1) to eliminate N_{xy}, we obtain

$$\nabla^2 (N_x + N_y) + (1 + \nu)(f_{x,x} + f_{y,y}) = 0 \tag{6.2.4b}$$

where $\nabla^2 = $ the Laplacian operator $= (\partial^2/\partial x^2 + \partial^2/\partial y^2)$.

Equations (6.2.1) are solvable in terms of a stress function ϕ which is defined such that

$$N_x = Eh \frac{\partial^2 \phi}{\partial y^2} - V_x \tag{6.2.5a}$$

$$N_y = Eh \frac{\partial^2 \phi}{\partial x^2} - V_y \qquad (6.2.5b)$$

$$N_{xy} = -Eh \frac{\partial^2 \phi}{\partial x\, \partial y} \qquad (6.2.5c)$$

where

$$V_x = \int_x f_x\, dx$$

$$V_y = \int_y f_y\, dy$$

If we substitute Eqs. (6.2.5) into Eqs. (6.2.1) we see that Eqs. (6.2.1) are satisfied identically. A governing equation for the new variable ϕ is obtained by substituting Eqs. (6.2.5) into Eq. (6.2.4b):

$$En\, \nabla^2(\nabla^2 \phi) = \nabla^2(V_x + V_y) - (1 + \nu)(f_{x,x} + f_{y,y}) \qquad (6.2.6)$$

Thus, the problem of determining stress resultants in a flat sheet due to in-plane loads reduces to the boundary-value problem posed by the biharmonic equation, Eq. (6.2.6), where for cartesian coordinates

$$\nabla^4 = \frac{\partial^4}{\partial x^4} + 2\frac{\partial^4}{\partial x^2\, \partial y^2} + \frac{\partial^4}{\partial y^4}$$

subject to boundary conditions on N_x, N_y, N_{xy}.

There are two popular approaches to the analytic solution of Eq. (6.2.6): solutions by polynomials and by Fourier series. Solutions by polynomials is essentially an inverse method in which some of the coefficients a_{ij} of

$$\phi = \sum_i \sum_j a_{ij} x^i y^j$$

are determined so as to satisfy the governing differential equation, Eq. (6.2.6), and the remainder of a_{ij} so as to construct elementary solutions with prescribed variations in tractions on the boundaries. Solutions to several practical cases can be synthesized by combinations of these elemental solutions. In some cases, actual tractions can be approximated by static equivalents pertaining to elemental solutions, and St. Venant's principle [28] can be used to justify a solution away from the edges with approximated tractions.

Let us consider an example in which no body forces are applied, i.e., $V_x = V_y = 0$. A third-degree polynomial

$$\phi = a_{02} y^2 + a_{03} y^3 + a_{11} xy + a_{12} xy^2$$

$$+ a_{20} x^2 + a_{21} x^2 y + a_{30} x^3 \qquad (6.2.7)$$

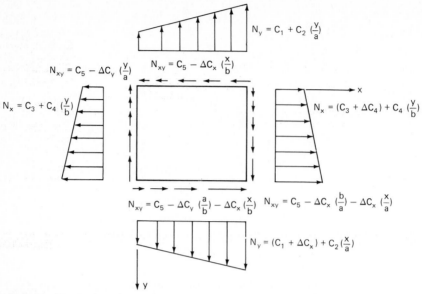

Figure 6.2.2 Linearly varying edge tractions on a rectangular membrane.

satisfies Eq. (6.2.6) identically. Substituting Eq. (6.2.7) into Eqs. (6.2.5), we obtain

$$N_x = Eh(2a_{02} + 6a_{03}y + 2a_{12}x) \qquad (6.2.8a)$$

$$N_y = Eh(2a_{20} + 6a_{30}x + 2a_{21}y) \qquad (6.2.8b)$$

$$N_{xy} = -Eh(a_{11} + 2a_{12}y + 2a_{21}x) \qquad (6.2.8c)$$

Shown in Fig. 6.2.2 is a rectangular sheet subject to a linearly varying edge traction for which Eqs. (6.2.8) can provide solutions.

We see that only seven independent coefficients C_1 to C_5, ΔC_x, ΔC_y, in Fig. 6.2.2 can be specified in the elemental solution, Eqs. (6.2.7) and (6.2.8), for a 1:1 match with the seven coefficients in the cubic equation, Eq. (6.2.7). Evaluation of Eqs. (6.2.8) on $x = 0$, a, and on $y = 0$, b leads to expressions for a_{ij} in terms of C_1 to C_5, ΔC_x, ΔC_y and, therefore

$$N_x = C_3 + C_4\left(\frac{y}{b}\right) + \Delta C_y\left(\frac{x}{a}\right) \qquad (6.2.9a)$$

$$N_y = C_1 + C_2\left(\frac{x}{a}\right) + \Delta C_x\left(\frac{y}{b}\right) \qquad (6.2.9b)$$

$$N_{xy} = C_5 - \Delta C_y \left(\frac{y}{a}\right) - \Delta C_x \left(\frac{x}{b}\right) \qquad (6.2.9c)$$

By specifying various combinations of C_1 to C_5, ΔC_x, ΔC_y, several elemental solutions can be obtained.

Note that in the example considered above, only stress conditions on the edges were considered. To include displacement conditions, we need to integrate Eqs. (6.2.2) for u, v after ε_x, ε_y, γ_{xy} have been expressed in terms of N_x, N_y, N_{xy} by way of Eqs. (6.2.3). Often the real displacement conditions will lead to discrepancies at corners, and St. Venant's principle will have to be invoked. The finite element technique described in a succeeding subsection provides a numerical procedure for determining solutions to problems involving displacement conditions. Higher-order polynomials can be considered as solutions to Eq. (6.2.6), but relationships between the coefficients will have to be specified so as to identically satisfy Eq. (6.2.6).

The second solution method for Eq. (6.2.6) is that of Fourier expansions. This method is useful in cases for which the loads and edge tractions are not expressible in terms of simple polynomials and in cases involving discontinuities in load. The homogeneous form of Eq. (6.2.1) is satisfied by

$$\phi = \sum_m Y_m(y)\sin\frac{m\pi x}{a} \qquad (6.2.10)$$

where a = span in x direction, and $Y_m(y)$ = undetermined. This equation will satisfy diaphragm-like conditions at $x = 0$, a. That is, on $y = 0$, a, $N_y = 0$, and N_{xy} is nonzero and able to assume any values necessary to maintain global equilibrium. Upon substitution of Eq. (6.2.10) into the homogeneous form of Eq. (6.2.6), we obtain

$$\frac{d^4 Y_m}{dy^4} - 2\left(\frac{m\pi}{a}\right)^2 \frac{d^2 Y_m}{dy^2} + \left(\frac{m\pi}{a}\right)^4 Y_m = 0 \qquad (6.2.11)$$

the solution to which is [28]

$$Y_m + C_{1m}\cosh\frac{m\pi y}{a} + C_{2m}\sinh\frac{m\pi y}{a}$$

$$+ C_{3m}\left(\frac{m\pi y}{a}\right)\cosh\frac{m\pi y}{a} + C_{4m}\left(\frac{m\pi y}{1}\right)\sinh\frac{m\pi y}{a} \qquad (6.2.12)$$

The homogeneous stress resultants are then

$$N_x = Eh \sum_m \left(\frac{m\pi}{a}\right)^2 \sin \frac{m\pi x}{a} \left\{ C_{1m} \cosh \frac{m\pi y}{a} + C_{2m} \sinh \frac{m\pi y}{a} \right.$$

$$+ C_{3m} \left[2 \cosh \frac{m\pi y}{a} + \left(\frac{m\pi y}{a}\right) \sinh \frac{m\pi y}{a} \right]$$

$$\left. + C_{4m} \left[2 \sinh \frac{m\pi y}{a} + \left(\frac{m\pi y}{a}\right) \cosh \frac{m\pi y}{a} \right] \right\}$$

$$(6.2.13a)$$

$$N_y = - Eh \sum_m \left(\frac{m\pi}{a}\right)^2 \sin \frac{m\pi x}{a} \left\{ C_{1m} \cosh \frac{m\pi y}{a} + C_{2m} \sinh \frac{m\pi y}{a} \right.$$

$$\left. + C_{3m} \left(\frac{m\pi y}{a}\right) \cosh \frac{m\pi y}{a} + C_{4m} \left(\frac{m\pi y}{a}\right) \sinh \frac{m\pi y}{a} \right\}$$

$$(6.2.13b)$$

$$N_{xy} = - Eh \sum_m \left(\frac{m\pi}{a}\right)^2 \cos \frac{m\pi x}{a} \left\{ C_{1m} \sinh \frac{m\pi y}{a} + C_{2m} \cosh \frac{m\pi y}{a} \right.$$

$$+ C_{3m} \left[\cosh \frac{m\pi y}{a} + \left(\frac{m\pi y}{a}\right) \sinh \frac{m\pi y}{a} \right]$$

$$\left. + C_{4m} \left[\sinh \frac{m\pi y}{a} + \left(\frac{m\pi y}{a}\right) \cosh \frac{m\pi y}{a} \right] \right\}$$

$$(6.2.13c)$$

The coefficients C_{1m} to C_{4m} remain to be determined from traction conditions on edges with $y = $ constant. For general traction conditions, Fourier series representations can be used. For example, consider the case of rectangular sheet loaded as shown in Fig. 6.2.3.
The Fourier expansion of the traction is

$$\bar{N} = \sum_{m \text{ odd}} \bar{N}_m \sin \frac{m\pi x}{a} \qquad (6.2.14a)$$

where

$$\bar{N}_m = \frac{2}{a} \int_0^a \bar{N} \sin \frac{m\pi x}{a} \, dx$$

$$= \frac{4\bar{N}}{m\pi} \sin \frac{m\pi(v - d)}{a} \sin \frac{m\pi v}{a} \qquad (6.2.14b)$$

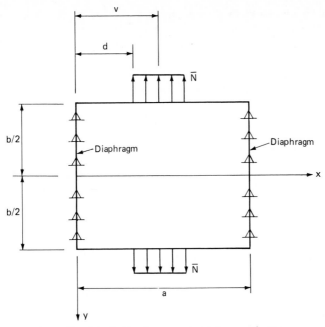

Figure 6.2.3 Partial edge load on a rectangular membrane.

In the limit as $(v - d) \to 0$ and $2\overline{N}(v - d) \to P$, we have

$$\overline{N}_m = \frac{2P}{a} \sin \frac{m\pi v}{a} \qquad (6.2.14c)$$

for a concentrated load. Evaluating the stress resultants N_y, N_{xy} on $y = \pm b/2$ using Eqs. (6.2.3b, c), we obtain ($C_{2m} = C_{3m} = 0$ by symmetry)

$$C_{1m} \cosh \frac{m\pi b}{2a} + C_{4m} \left(\frac{m\pi b}{2a} \right) \sinh \frac{m\pi b}{2a} = -\frac{\overline{N}_m}{Eh} \left(\frac{a}{m\pi} \right)^2$$

$$C_{1m} \sin \frac{m\pi b}{2a} + C_{4m} \left[\sin \frac{m\pi b}{2a} + \left(\frac{m\pi b}{2a} \right) \cosh \frac{m\pi b}{2a} \right] = 0$$

and therefore

$$C_{4m} = \frac{(\overline{N}_m/Eh)(a/m\pi)^2 \sinh(m\pi b/2a)}{[m\pi b/2a + \sinh(m\pi b/2a) \cosh(m\pi b/2a)]}$$

$$C_{1m} = -\frac{(\overline{N}_m/Eh)(a/m\pi)^2 [\sinh(m\pi b/2a) + (m\pi b/2a) \cosh(m\pi b/2a)]}{[(m\pi b/2a) + \sinh(m\pi b/2a) \cosh(m\pi b/2a)]}$$

for use in Eqs. (6.2.13).

Thus far we have considered only problems in rectangular coordinates. The biharmonic equation, Eq. (6.2.6), can also be cast into cylindrical coordinates with the polar form of the Laplacian operator:

$$\nabla^2 = \left(\frac{\partial}{\partial r^2} + \frac{1}{r}\frac{\partial}{\partial r} + \frac{1}{r^2}\frac{\partial^2}{\partial \theta^2} \right)$$

where r = radial coordinate, and θ = circumferential angle. If we now consider axisymmetric problems, we have as our governing equation

$$\frac{d^4\phi}{dr^4} + \frac{2}{r^2}\frac{d^3\phi}{dr^3} - \frac{1}{r^2}\frac{d^2\phi}{dr^2} + \frac{1}{r^3}\frac{d\phi}{dr} = 0 \tag{6.2.15}$$

which has as a general homogeneous solution

$$\phi = C_1 \log r + C_2 r^2 \log r + C_3 r^2 + C_4 \tag{6.2.16}$$

with corresponding stress resultants

$$N_r = Eh \left[\frac{C_1}{r^2} + C_2 (1 + 2 \log r) + 2C_3 \right] \tag{6.2.17a}$$

$$N_\theta = Eh \left[-\frac{C_1}{r^2} + C_2(3 + 2 \log r) + 2C_3 \right] \tag{6.2.17b}$$

$$N_{r\theta} = 0 \tag{6.2.17c}$$

If the membrane is continuous at the origin, $C_1 = C_2 = 0$ to guarantee finite stresses at $r = 0$. Therefore, for a continuous membrane with no hole at the origin

$$N_r = N_\theta = 2EhC_3$$

and we see that only constant symmetric stress may exist.

For annular membranes with openings at the origin, various elemental solutions can be constructed with C_1 and C_2 nonzero. For instance, consider the annular membrane stressed in Fig. 6.2.4 by a traction N_i on the inner edge $r = a$, and by a traction N_0 on the outer edge $r = b$. With $C_2 = 0$, we have

$$N_r = Eh \left(\frac{C_1}{a^2} + 2C_3 \right) = N_i \ @ \ r = a$$

$$N_r = Eh \left(\frac{C_1}{b^2} + 2C_3 \right) = N_0 \ @ \ r = b$$

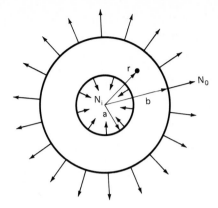

Figure 6.2.4 Annular membrane with edge tractions.

Therefore, after solving for C_1 and C_3,

$$N_r = \frac{1}{b^2 - a^2}\left\{ N_0 b^2 \left[1 - \left(\frac{a}{r}\right)^2 \right] - N_i a^2 \left[1 - \left(\frac{b}{r}\right)^2 \right] \right\} \qquad (6.2.18a)$$

$$N_\theta = \frac{1}{b^2 - a^2}\left\{ N_0 b^2 \left[1 + \left(\frac{a}{r}\right)^2 \right] - N_i a^2 \left[1 + \left(\frac{b}{r}\right)^2 \right] \right\} \qquad (6.2.18b)$$

If, for example, $N_0 = 0$

$$N_r = \frac{a^2 N_i}{a^2 - b^w}\left[\left(\frac{b}{r}\right)^2 - 1 \right]$$

$$N_\theta = -\frac{a^2 N_i}{a^2 - b^2}\left[\left(\frac{b}{r}\right)^2 + 1 \right]$$

and we see that instead of a compressive circumferential stress, we would have wrinkles running radially.

Several interesting cases of nonsymmetric distribution of stress can be constructed from solutions to Eq. (6.2.6) tabulated in Chap. 4 of Ref. 28 for equivalent two-dimensional elasticity problems in cylindrical coordinates, e.g., stress distributions around holes in uniformly stretched membranes (Fig. 6.2.5a), and stresses in wedge-shaped membranes due to tip loads (Fig. 6.2.5b).

In the wedge-shaped membrane, we have

$$N_r = \frac{P \cos \theta}{r(\alpha + \frac{1}{2} \sin 2\alpha)}$$

where P = tip load, and α = opening angle.

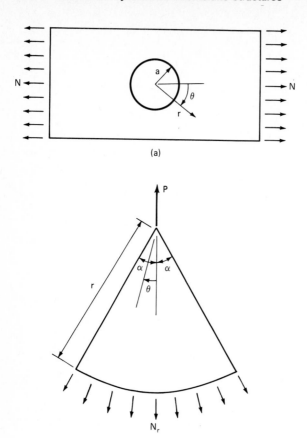

(a)

(b)

Figure 6.2.5 Example nonsymmetric flat membrane problems. (a) Hole in uniformly loaded membrane; (b) wedge-shaped membrane with trip load.

For the uniformly loaded membrane with a hole

$$N_r = \frac{N}{2}\left\{\left[1 - \left(\frac{a}{r}\right)^2\right] + \left[1 + 3\left(\frac{a}{r}\right)^4 - 4\left(\frac{a}{r}\right)^2\right]\cos 2\theta\right\}$$

$$N_\theta = \frac{N}{2}\left\{\left[1 + \left(\frac{a}{r}\right)^2\right] - \left[1 + 3\left(\frac{a}{r}\right)^4\right]\cos 2\theta\right\}$$

$$N_{r\theta} = -\frac{N}{2}\left\{1 - 3\left(\frac{a}{r}\right)^4 + 2\left(\frac{a}{r}\right)^2\right\}\sin 2\theta$$

At $r = a$, the maximum value is $N_\theta = 3N$ at $\theta = \pm\pi/2$. At $\theta = 0$ and π, $N_\theta = -N$ is predicted and we see that wrinkles occur if $\cos 2\theta > 0.5$.

6.2.2   Finite element solutions [165]

In cases involving either irregular geometries of the membrane, distributed loads on the membrane, irregular edge traction, or edge displacement conditions, the analytic methods of the previous subsection are of little help. Instead a numerical procedure is required. The finite element method provides a means to develop matrix equations describing the behavior of flat membranes in terms of a discrete number of displacements at nodal points. The fundamentals of finite element modeling theory were discussed in Sec. 3.1 where a representative problem in linear elasticity was considered.

In Sec. 3.1 we wrote the governing matrix equations as

$$[K]\{D\} = \{Q\} \tag{6.2.19a}$$

where $[K] = \Sigma_e [K_e]$ = assembled stiffness matrix
$\{D\}$ = nodal displacements
$\{Q\} = \Sigma_e \{Q_e\}$ = assembled nodal loads

The elemental stiffness matrix is

$$[K_e] = \int_{V_e} [A]^T [\overline{E}][A] dv_e \tag{6.2.19b}$$

where $[A]$ is developed from the strain equation

$$\{\varepsilon\} = [A]\{d\} \tag{6.2.19c}$$

and $[\overline{E}]$ is the elasticity matrix. We will present in this section equations for $[A]$ and $[\overline{E}]$ appropriate for flat membranes modeled as constant strain triangles as shown in Fig. 6.2.6. The elemental load vector is

$$\{Q_e\} = \int_{V_e} [N]^T [N] dv_e \{p\} \tag{6.2.19d}$$

where $[N]$ is the matrix of approximating functions used to interpolate the displacements

$$\{u\} = [N]\{d\} \tag{6.2.19e}$$

and $\{p\}$ = nodal values of distributed loads. We will present expressions for $[N]$ and $\{Q_e\}$ for constant strain triangles. In Chap. 8, more refined models are considered and isoparametric forms for $[N]$ and, hence, $[A]$, $\{Q_e\}$, and $[K_e]$ are developed.

Assume displacements u, v at a point, x, y in a triangular element to be expressible as

$$u = [\psi]\{a\} \qquad\qquad (6.2.20a)$$

$$v = [\psi]\{b\} \qquad\qquad (6.2.20b)$$

where $\{a\} = \{a_1 a_2 a_3\}^T$ and $\{b\} = \{b_1 b_2 b_3\}^T$ are undetermined parameters and $\{\psi\} = [1, x, y] = $ linear shape functions for which

$$\begin{Bmatrix} \varepsilon_x \\ \varepsilon_y \\ \gamma_{xy} \end{Bmatrix} = \begin{Bmatrix} \dfrac{\partial u}{\partial x} \\[2mm] \dfrac{\partial v}{\partial y} \\[2mm] \dfrac{\partial u}{\partial y} + \dfrac{\partial v}{\partial x} \end{Bmatrix} = \begin{Bmatrix} a_3 \\ b_3 \\ a_3 + b_2 \end{Bmatrix} = \begin{Bmatrix} \text{constants} \end{Bmatrix}$$

and, hence, a constant strain field prevails throughout the element. Now, we convert from parameters $\{a\}$, $\{b\}$ to physically meaningful parameters u, v evaluated at three node points i, j, k on the triangular element shown

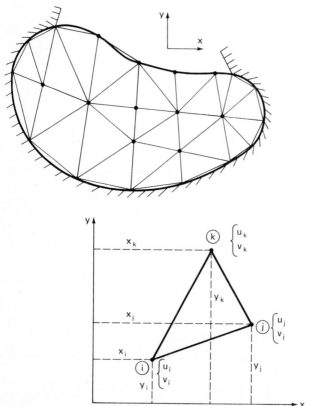

Figure 6.2.6 Constant-strain triangular element.

in Fig. 6.2.6. For u we evaluate Eq. (6.2.20a) at x_i, y_i; x_j, y_j; and x_k, y_k to obtain

$$\{\bar{u}\} = \begin{Bmatrix} u_i \\ u_j \\ u_k \end{Bmatrix} = \begin{bmatrix} \psi_i \\ \psi_j \\ \psi_k \end{bmatrix} \{a\} = \begin{bmatrix} 1 & x_i & y_i \\ 1 & x_j & y_j \\ 1 & x_k & y_k \end{bmatrix} \{a\}$$

which can be inverted so that

$$\{a\} = [C]\{\bar{u}\}$$

with

$$[C] = \frac{1}{2\ \text{area}} \begin{bmatrix} (x_j y_k - x_k y_j) & (x_k y_i - x_i y_k) & (x_i y_j - x_j y_i) \\ (y_j - y_k) & (y_k - y_i) & (y_i - y_j) \\ (x_k - x_j) & (x_i - x_k) & (x_i - x_j) \end{bmatrix}$$

Hence, we find our matrix $[N]$ of approximating functions in Eq. (6.2.19e) to be

$$u = [\psi]\{a\} = [N_i N_j N_k] \begin{Bmatrix} u_i \\ u_j \\ u_k \end{Bmatrix} \qquad (6.2.21a)$$

$$N_i = \frac{1}{2\ \text{area}} [(x_j y_k - x_k y_j) + x(y_j - y_k) + y(x_k - x_j)] \quad (6.2.21b)$$

with N_j and N_k obtained by permuting the indices i, j, k, in N_i of Eq. (6.2.21b). Similarly, for v we have

$$v = [\psi]\{b\} = [N_i N_j N_k] \begin{Bmatrix} u_i \\ u_j \\ u_k \end{Bmatrix} \qquad (6.2.21c)$$

Substituting Eqs. (6.2.21) into Eq. (6.2.19c), we obtain

$$\{\varepsilon\} = \begin{Bmatrix} \varepsilon_x \\ \varepsilon_y \\ \varepsilon_{xy} \end{Bmatrix} = [A_i A_j A_k] \begin{Bmatrix} d_i \\ d_j \\ d_k \end{Bmatrix} \qquad (6.2.22a)$$

where

$$[A_i] = \frac{1}{2 \cdot \text{area}} \begin{bmatrix} (y_j - y_k) & 0 \\ 0 & (x_j - x_k) \\ (x_j - x_k) & (y_j - y_k) \end{bmatrix} \qquad (6.2.22b)$$

$$\{d_i\} = \begin{Bmatrix} u_i \\ v_i \end{Bmatrix} \tag{6.2.22c}$$

and $[A_j]$, $[A_k]$ are permutations of i, j, k in $[A_i]$. Then, for a plane stress condition,

$$\{N\} = \begin{Bmatrix} N_x \\ N_y \\ N_{xy} \end{Bmatrix} = [\overline{E}]\{\varepsilon\} \tag{6.2.23a}$$

$$[\overline{E}] = \frac{Eh}{(1 - \nu^2)} \begin{bmatrix} 1 & \nu & 0 \\ \nu & 1 & 0 \\ 0 & 0 & \dfrac{1 - \nu}{2} \end{bmatrix} \tag{6.2.23b}$$

Thus, Eq. (6.2.19b) gives us the elemental stiffness matrix as

$$[K_e] = \begin{bmatrix} K_{ii} & K_{ij} & K_{ik} \\ K_{ji} & K_{jj} & K_{jk} \\ K_{ki} & K_{kj} & K_{kk} \end{bmatrix}$$

$$K_{ii} = [A_i]^T[\overline{E}][A_i]$$

$$K_{ij} = [A_i]^T[\overline{E}][A_j]$$

and so forth with A_i given by Eq. (6.2.22b) and $[\overline{E}]$ by Eq. (6.2.23b). If we assume the load intensities f_x, f_y to be constant over each element, Eqs. (6.2.19d) and (6.2.21) give the elemental load vector as

$$\{Q_e\} = \frac{h \cdot \text{area}}{3} \begin{Bmatrix} f_i \\ f_j \\ f_k \end{Bmatrix} \qquad f_i = f_j = f_k = \begin{Bmatrix} f_x \\ f_y \end{Bmatrix}$$

Once Eq. (6.2.19a) has been solved for the nodal displacements $\{D\}$, the stress resultants are determined from Eqs. (6.2.22) and (6.2.23). Note that the stresses will be constant throughout each element, and, hence, discontinuities in stress will occur at edges of elements. More sophisticated elements [87, 88, 95, 165] can be developed to alleviate this problem and to provide more accurate predictions for problems involving curved boundaries. Superparametric and isoparametric two-dimensional membrane elements with curvature in three dimensions are described in Chap. 8.

6.2.3   In-service phase

No membrane or network will remain flat. Dead weight alone will cause transverse curvature in order to provide transverse components of the in-plane stress resultants to equilibrate the load. However, we can often assume that the additional curvatures due to in-service transverse loads are small. We will first derive nonlinear equations reflecting the coupling of the in-plane stresses with the transverse deflection w of the initially flat membrane. Then we will make the decoupling assumption that the stress resultants are predetermined for the prestressing phase (as calculated in Sec. 6.2.1 or 6.2.2) and remain unchanged in magnitude as the membrane undergoes additional transverse displacement.

As we saw in Sec. 6.2.1., the stress resultants satisfy equilibrium equations, Eqs. (6.2.1), where we now interpret N_x, N_y, N_{xy} as projections onto the plane of the initially flat membrane. (See Sec. 6.3.1 for elaborations on projections for shallow curvature membranes.) Thus, we can assume, as in Sec. 6.2.1, a stress function solution to Eqs. (6.2.1), i.e., Eq. (6.2.5).

To arrive at an equivalent equation to Eq. (6.2.6), but now including the effects of transverse deflection w, we add to the strain-displacement equations, Eq. (6.2.2), nonlinear terms reflecting large rotations $\partial w/\partial x$ and $\partial w/\partial y$:

$$\varepsilon_x = \frac{\partial u}{\partial x} + \tfrac{1}{2}\left(\frac{\partial w}{\partial x}\right)^2$$

$$\varepsilon_y = \frac{\partial v}{\partial y} + \tfrac{1}{2}\left(\frac{\partial w}{\partial y}\right)^2$$

$$\gamma_{xy} = \frac{\partial u}{\partial x} + \frac{\partial v}{\partial y} + \frac{\partial w}{\partial x}\frac{\partial w}{\partial y}$$

The compatibility equation, Eq. (6.2.4), becomes

$$\frac{\partial^2 \varepsilon_x}{\partial y^2} + \frac{\partial^2 \varepsilon_y}{\partial x^2} - \frac{\partial^2 \gamma_{xy}}{\partial x\,\partial y} = \left(\frac{\partial^2 w}{\partial x\,\partial y}\right)^2 - \left(\frac{\partial^2 w}{\partial x^2}\right)\left(\frac{\partial^2 w}{\partial y^2}\right) \qquad (6.2.24)$$

Now, substituting Eqs. (6.2.3) and (6.2.5) into Eq. (6.2.24), we obtain the equation

$$\nabla^4 \phi = \left(\frac{\partial^2 w}{\partial x\,\partial y}\right)^2 - \left(\frac{\partial^2 w}{\partial x^2}\right)\left(\frac{\partial^2 w}{\partial y^2}\right) \qquad (6.2.25)$$

Another equation relating ϕ and w is obtained from equilibrium in the transverse (z) direction of transverse projections of forces on the differential element shown in Fig. 6.2.7.

Summing transverse components, we have

$$0 = q + N_x \frac{\partial^2 w}{\partial x^2} + N_y \frac{\partial^2 w}{\partial y^2} + 2N_{xy} \frac{\partial^2 w}{\partial x\,\partial y} \qquad (6.2.26)$$

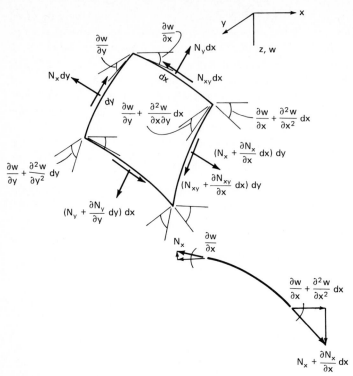

Figure 6.2.7   Vertical equilibrium of slightly curved membrane.

where Eqs. (6.2.1) have been used to eliminate terms and q = intensity of transverse load. Substituting Eqs. (6.2.5) into Eq. (6.2.26), we obtain the second governing equation for ϕ and w as

$$\frac{q}{Eh} + \frac{\partial^2 \phi}{\partial y^2}\frac{\partial^2 w}{\partial x^2} + \frac{\partial^2 \phi}{\partial x^2}\frac{\partial^2 w}{\partial y^2} - 2\frac{\partial^2 \phi}{\partial x\,\partial y}\frac{\partial^2 w}{\partial x\,\partial y} = 0 \qquad (6.2.27a)$$

In cylindrical coordinates, Eq. (6.2.27a) is replaced by

$$\frac{q}{Eh} + \frac{\partial^2 w}{\partial r^2}\left(\frac{1}{r}\frac{\partial \phi}{\partial r} + \frac{1}{r^2}\frac{\partial^2 \phi}{\partial \theta^2}\right) + \left(\frac{1}{r}\frac{\partial w}{\partial r} + \frac{1}{r^2}\frac{\partial^2 w}{\partial \theta^2}\right)\frac{\partial^2 \phi}{\partial r^2}$$

$$- 2\frac{\partial}{\partial r}\left(\frac{1}{r}\frac{\partial w}{\partial \theta}\right)\frac{\partial}{\partial r}\left(\frac{1}{r}\frac{\partial \phi}{\partial \theta}\right) = 0 \qquad (6.2.27b)$$

Equations (6.2.25) and (6.2.27) are two coupled nonlinear equations for ϕ and w. If we assume that the initial stresses in a flat sheet stiffen the membrane sufficiently that transverse displacements due to added load are small, we can assume N_x, N_y, N_{xy} (and, hence, ϕ) to be predetermined and unvarying in magnitude under added loads q. That is, we first assume $w = 0$ in Eq. (6.2.25) and solve for ϕ (or N_x, N_y, N_{xy}) as in the previous subsections. Second, we assume ϕ known in Eq. (6.2.27a or b) and solve for w. This suggests an iterative solution that can be used for the coupled nonlinear problem in which the Gauss-Seidel method (see Sec. 3.3.1) is used to solve Eqs. (6.2.25) and (6.2.27) recursively for approximate ϕ and w. Solutions to Eqs. (6.2.27) are discussed in Sec. 6.3 where shallow translational membranes are considered.

Approximate solutions for nearly flat membranes can be developed from the Rayleigh-Ritz method in which the total potential energy of the membrane is minimized with respect to the parameters of approximate compatible functions assumed for the displacements, u, v, w. The total potential energy of a slightly deformed membrane is

$$V = -\iint qw \, dx \, dy + \iint \tfrac{1}{2}(N_x\varepsilon_x + N_y\varepsilon_y + 2N_{xy}\varepsilon_{xy})dx \, dy \qquad (6.2.28a)$$

$$= -\iint qw \, dx \, dy + \frac{Eh}{2(1 - \nu^2)} \iint [\varepsilon_x^2 + \varepsilon_y^2 + 2\nu\varepsilon_x\varepsilon_y$$

$$+ \tfrac{1}{2}(1 - \nu)\gamma_{xy}^2] \, dx \, dy \qquad (6.2.28b)$$

Substituting for ε_x, ε_y, γ_{xy}, we have

$$V = -\iint qw \, dx \, dy + \frac{Eh}{2(1 - \nu^2)} \iint \left\{ \left(\frac{\partial u}{\partial x} \right)^2 + \left(\frac{\partial w}{\partial x} \right)^2 + \left(\frac{\partial v}{\partial y} \right)^2 \right.$$

$$+ \left(\frac{\partial v}{\partial y} \right) \left(\frac{\partial w}{\partial y} \right)^2 + \frac{1}{4} \left[\left(\frac{\partial w}{\partial x} \right)^2 + \left(\frac{\partial w}{\partial y} \right)^2 \right]^2$$

$$+ 2\nu \left[\frac{\partial u}{\partial x} \frac{\partial v}{\partial y} + \frac{1}{2} \left(\frac{\partial v}{\partial y} \right) \left(\frac{\partial w}{\partial x} \right)^2 + \frac{1}{2} \left(\frac{\partial u}{\partial x} \right) \left(\frac{\partial w}{\partial y} \right)^2 \right]$$

$$+ \frac{1 + \nu}{2} \left[\left(\frac{\partial u}{\partial x} \right)^2 + 2 \left(\frac{\partial u}{\partial y} \right) \left(\frac{\partial v}{\partial x} \right) + \left(\frac{\partial v}{\partial x} \right)^2 \right.$$

$$+ 2 \left(\frac{\partial u}{\partial y} \right) \left(\frac{\partial w}{\partial x} \right) \left(\frac{\partial w}{\partial y} \right)$$

$$\left. \left. + 2 \left(\frac{\partial v}{\partial x} \right) \left(\frac{\partial w}{\partial x} \right) \left(\frac{\partial w}{\partial y} \right) \right] \right\} dx \, dy$$

$$(6.2.29)$$

Now, assume functions (e.g., polynomials) of u, v, w which satisfy displacement boundary conditions and minimize Eq. (6.2.29) with respect to the coefficients multiplying each of the functions. This will lead to a set of nonlinear algebraic equations for the coefficients.

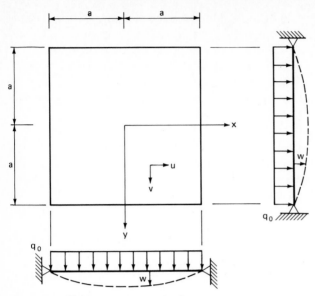

Figure 6.2.8   Uniform transverse load on a square membrane.

In Ref. 184 is considered the example shown in Fig. 6.2.8 of a square membrane subject to a uniform transverse load q. Assume

$$u = u_0 \sin \frac{\pi x}{a} \cos \frac{\pi y}{2a} \qquad (6.2.30a)$$

$$v = v_0 \sin \frac{\pi y}{a} \cos \frac{\pi x}{2a} \qquad (6.2.30b)$$

$$w = w_0 \cos \frac{\pi x}{2a} \cos \frac{\pi y}{2a} \qquad (6.2.30c)$$

which satisfies the conditions of $u = v = w = 0$ on $x = \pm a$ and $y = \pm a$, and reproduces the symmetry properties needed for u, v, w if we set $v_0 = u_0$. Upon substitution of Eqs. (6.2.30) into Eq. (6.2.29) and integration, we obtain ($v = 0.25$, $v_0 = u_0$)

$$V = -\frac{16a^2 q_0 w_0}{\pi^2} + \frac{Eh}{7.5} \left[\frac{5\pi^4}{64} \frac{w_0^4}{a^2} - \frac{17\pi^2}{6} \frac{u_0 w_0^2}{a} + u_0^2 \left(\frac{35\pi^2}{4} + \frac{80}{9} \right) \right]$$

Now, minimizing with respect to u_0 and w_0, we obtain

$$\frac{\partial V}{\partial w_0} = 0 = -\frac{16a^2 q_0}{\pi^2} + \frac{Eh}{7.5} \left(\frac{5\pi^4}{16a^2} w_0^3 - \frac{17\pi^2}{6} \frac{u_0 w_0}{a} \right) \qquad (6.2.31a)$$

$$\frac{\partial V}{\partial u_0} = 0 = 0 + \frac{Eh}{7.5}\left[-\frac{17\pi^2}{6}\frac{w_0^2}{a} + u_0\left(\frac{35\pi^2}{2} + \frac{160}{9}\right)\right] \quad (6.2.31b)$$

Solving Eq. (6.2.31b) for u_0 in terms of w_0^2 and then Eq. (6.2.31a) for w_0, we find

$$u_0 = \frac{17\pi^2 w_0^2}{(105\pi^2 + 320/3)\,a} = 0.147\,\frac{w_0^2}{a}$$

$$w_0 = 0.802\left(\frac{q_0 a^4}{Eh}\right)^{1/3}$$

6.2.4 Dynamic in-service behavior

If we assume that the in-plane stress resultants remain constant, we can develop equations of motion for the transverse vibrations of flat sheets from Eq. (6.2.26). With an inertial load $q = -\rho h \ddot{w}$ we can write Eq. (6.2.26) as

$$\rho h \ddot{w} = N_x \frac{\partial^2 w}{\partial x^2} + N_y \frac{\partial^2 w}{\partial y^2} - 2N_{xy}\frac{\partial^2 w}{\partial x\,\partial y} \quad (6.2.32)$$

By separation of variables [$w = W(x,\, y)\cdot T(t)$], we obtain from Eq. (6.2.32)

$$\frac{\ddot{T}}{T} = \frac{1}{W\rho h}\left(N_x\frac{\partial^2 W}{\partial x^2} + N_y\frac{\partial^2 W}{\partial y^2} - 2N_{xy}\frac{\partial^2 W}{\partial x\,\partial y}\right) = -\omega_n^2$$

Therefore, we have a harmonic solution

$$T = e^{i\omega t}$$

and we must solve

$$N_x\frac{\partial^2 W}{\partial x^2} + N_y\frac{\partial^2 W}{\partial y^2} - 2N_{xy}\frac{\partial^2 W}{\partial x\,\partial y} + \rho h \omega^2 W = 0 \quad (6.2.33)$$

for the natural frequencies ω and mode shapes $W(x,\, y)$ corresponding to each frequency.

For example, if $N_{xy} = 0$ and $N_x = N_y = N$, Eq. (6.2.33) becomes

$$\nabla^2(W) + \frac{\omega^2 \rho h}{N}\,W = 0 \quad (6.2.34)$$

For an $a \times b$ rectangular sheet restrained on all edges, we have the mode shapes

$$W = \sum_m \sum_n W_{mn} \sin \frac{m\pi x}{a} \sin \frac{n\pi y}{b} \qquad (6.2.35)$$

with natural frequencies

$$\omega_{mn} = \frac{\pi}{a} \sqrt{\frac{N}{\rho h} \left[(m)^2 + \left(n\frac{a}{b} \right)^2 \right]}$$

See Chap. 3 of Ref. 185 for further solutions to the free vibrations of flat membranes.

Solutions for the vibration of flat cable networks and cable-reinforced membranes have been developed [105] using a membrane analogy based on Eq. (6.2.27). It was assumed that $N_{xy} = 0$ and that N_x and N_y are given by

$$N_x = \overline{N}_x + \sum_i T_{xi}\, \delta(y - y_i) \qquad (6.2.36a)$$

$$N_y = \overline{N}_y + \sum_j T_{yj}\, \delta(x - x_j) \qquad (6.2.36b)$$

where $\overline{N}_x$, $\overline{N}_y$ are membrane pretensions; T_{xi}, T_{yi} are cable pretensions in the ith, jth cables in x, y directions; and $\delta(\ldots)$ denotes the Dirac delta function. Galerkin's method [18, 104, 166] for irregular domains was used to specify functions W as in Eqs. (6.2.33) to (6.2.35):

$$W_{mn} = g(x, y)x^{m-1}y^{n-1}$$

where $g(x, y) = 0$ is the equation of the bounding curve of the membrane. See Ref. 105 for numerical results of frequencies in several case studies. The same authors extended the work described above to treat the case of cable networks with small initial transverse deformations [186].

6.3   Translational and Cylindrical Membranes

As a result of prestressing, membranes assume a doubly curved shape. In this section, equilibrium of stress resultants and forces on a membrane deformed into either a translational or cylindrical surface is considered. A translational surface is one which can be generated by translating a plane curve (generator) parallel to itself over another curve (directrix) which is perpendicular to the generator. See Fig. 6.3.1. If we adopt as surface coordinates the projected plane coordinates x, y of the embedding three-dimensional coordinate system, x, y, z, then the equation of the surface can be written as

$$z = f_1(x) + f_2(y)$$

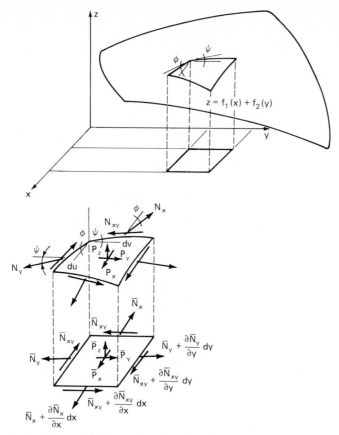

Figure 6.3.1 Projection and equilibrium of a translational surface element.

If we denote

$$z_x = \frac{\partial z}{\partial x} = \tan \phi \qquad z_{xx} = \frac{\partial^2 z}{\partial x^2}$$

$$z_y = \frac{\partial z}{\partial y} = \tan \psi \qquad z_{yy} = \frac{\partial^2 z}{\partial y^2}$$

then we can relate the geometry of the curved element to that of an element du, dv projected onto the horizontal plane by

$$dx = du \cos \phi \tag{6.3.1a}$$

$$dy = dv \cos \psi \tag{6.3.1b}$$

where

$$\cos \phi = \frac{1}{\sqrt{1 + z_x^2}} \qquad (6.3.1c)$$

$$\cos \psi = \frac{1}{\sqrt{1 + z_y^2}} \qquad (6.3.1d)$$

The differential area of the curved element is

$$d(\text{area}) = \sqrt{1 + z_x^2 + z_y^2} \, dx \, dy \qquad (6.3.1e)$$

We will assume that the rise to span ratio of the membrane is small enough that we can consider the xy coordinates on the projection of the membrane onto the horizontal plane to be single-valued coordinates.

The curvature of the surface will require expressions for the second derivatives z_{xx}, z_{yy}, z_{xy}. The surface can be classified as synclastic (positive gaussian) if

$$z_{xx} z_{yy} - z_{xy}^2 > 0$$

or as anticlastic (negative gaussian) if

$$z_{xx} z_{yy} - z_{xy}^2 < 0$$

or as developable (zero gaussian) if

$$z_{xx} z_{yy} - z_{xy}^2 = 0$$

One special class of developable translational shell is that of cylinders which are described in a following subsection.

6.3.1 Translational shallow shells

To arrive at equations of equilibrium we consider pseudo-stress resultants $\overline{N}_x$, $\overline{N}_y$, $\overline{N}_{xy}$ which are assumed to act on the projected differential element, as shown in Fig. 6.3.1. They are defined such that they provide the same total horizontal component of force on the horizontally projected element as do the real stress resultants acting on the deformed surface of the membrane, i.e.,

$$\overline{N}_x \, dy = N_x \, dv \cos \phi$$

$$\overline{N}_y \, dx = N_y \, du \cos \psi$$

$$\overline{N}_{xy} \, dx = \overline{N}_{xy} \, dy = N_{xy} \, du \cos \phi$$

Therefore, we find our pseudo-stress resultants to be

$$\overline{N}_x = N_x \frac{\cos \phi}{\cos \psi} = N_x \sqrt{\frac{1 + z_y^2}{1 + z_x^2}} \qquad (6.3.2a)$$

$$\overline{N}_y = N_y \frac{\cos \psi}{\cos \phi} = N_y \sqrt{\frac{1 + z_x^2}{1 + z_y^2}} \qquad (6.3.2b)$$

$$\overline{N}_{xy} = N_{xy} \qquad (6.3.2c)$$

Now, if we enforce equilibrium of horizontal components of force on either the deformed or projected element, we arrive at two first-order partial differential equations for $\overline{N}_x$, $\overline{N}_y$, $\overline{N}_{xy}$ which resemble those of Sec. 6.2.

$$\frac{d\overline{N}_x}{\partial x} + \frac{\partial \overline{N}_{xy}}{\partial y} + \overline{P}_x = 0 \qquad (6.3.3a)$$

$$\frac{\partial \overline{N}_y}{\partial y} + \frac{\partial \overline{N}_{xy}}{\partial x} + \overline{P}_y = 0 \qquad (6.3.3b)$$

where $\overline{P}_x$, $\overline{P}_y$ are the equivalent forces on the horizontal projection of the element

$$\overline{P}_x = P_x \frac{d(\text{area})}{dx \, dy} = P_x \sqrt{1 + z_x^2 + z_y^2} \qquad (6.3.4a)$$

$$\overline{P}_y = P_y \frac{d(\text{area})}{dx \, dz} = P_y \sqrt{1 + z_x^2 + z_y^2} \qquad (6.3.4b)$$

Also

$$\overline{P}_z = P_z \frac{d(\text{area})}{dx \, dy} = P_z \sqrt{1 + z_x^2 + z_y^2} \qquad (6.3.4c)$$

If we now sum vertical components of force on the deformed element and express equilibrium in terms of pseudo-stress resultants we find

$$\frac{\partial}{\partial x} (\overline{N}_x \tan \phi) + \frac{\partial}{\partial y} (\overline{N}_y \tan \psi) + \frac{\partial}{\partial x} (\overline{N}_{xy} \tan \psi)$$

$$+ \frac{\partial}{\partial y} (\overline{N}_{xy} \tan \phi) \, \overline{P}_z = 0 \qquad (6.3.5a)$$

Then, if we expand Eq. (6.3.5a) and use Eqs. (6.3.3) to eliminate terms, we obtain an algebraic equation for $\overline{N}_x$, $\overline{N}_y$, $\overline{N}_{xy}$:

$$z_{xx}\overline{N}_x + 2z_{xy}\overline{N}_{xy} + z_{yy}\overline{N}_y = \overline{P}_z + z_x\overline{P}_x + z_y\overline{P}_y \qquad (6.3.5b)$$

Noting that Eqs. (6.3.3) are similar to those of Sec. 6.2, we see that we may reduce the three equations of equilibrium (Eqs. 6.3.3 and 6.3.5b) further by the introduction of a stress function (Pucher's stress function [79, 120]):

$$\overline{N}_x = \frac{\partial^2 H}{\partial y^2} - \int \overline{P}_x \, dx \qquad (6.3.6a)$$

$$\overline{N}_y = \frac{\partial^2 H}{\partial x^2} - \int \overline{P}_y \, dy \qquad (6.3.6b)$$

$$\overline{N}_{xy} = -\frac{\partial^2 H}{\partial x \, \partial y} \qquad (6.3.6c)$$

If we substitute Eqs. (6.3.6) into Eqs. (6.3.3) we see that Eqs. (6.3.3) are identically satisfied. Rather than considering compatibility of strains to arrive at a governing equation for H, as was done in Sec. 6.2, in this instance we substitute Eqs. (6.3.6) into the vertical equilibrium equation, Eq. (6.3.5b), to obtain a second-order partial differential equation for the stress function:

$$z_{xx}H_{yy} - 2z_{xy}H_{xy} + z_{yy}H_{xx}$$

$$= \overline{P}_x z_x + \overline{P}_y z_y + \overline{P}_z + z_{xx} \int \overline{P}_x \, dx + z_{yy} \int \overline{P}_y \, dy \quad (6.3.7a)$$

Therefore, the solution to the stress state in a slightly curved membrane reduces to the determination of H which satisfies Eq. (6.3.7a) and prescribed boundary conditions on $\overline{N}_x$, $\overline{N}_y$, $\overline{N}_z$. Once H is found, pseudo-stresses $\overline{N}_x$, $\overline{N}_y$, $\overline{N}_{xy}$ are determined by Eqs. (6.3.6) and real stress resultants N_x, N_y, N_{xy} by Eqs. (6.3.2).

In most instances of shallow membranes, $\overline{P}_x$, $\overline{P}_y$ can be neglected. Further, for translational shells we have that $z_{xy} = 0$. Thus we can rewrite Eq. (6.3.7a) as

$$z_{xx}H_{yy} + z_{yy}H_{xx} = P_z \sqrt{1 + z_x^2 + z_y^2} \qquad (6.3.7b)$$

which in general represents a partial differential equation with variable coefficients. This equation would serve to predict additional stresses due to dead weight superposed during the in-service phase of a membrane. Alternatively, if we seek approximate solutions to the pressurization phase of an air-supported membrane [with an assumption that $z = f_1(x) + f_2(y)$], we need not neglect $\overline{P}_x$, $\overline{P}_y$ and find our governing equations to be

$$z_{xx}H_{yy} + z_{yy}H_{xx} = p \sqrt{1 + z_x^2 + z_y^2} \qquad (6.3.7c)$$

where p = pressure. In either of Eqs. (6.3.7b) or (6.3.7c) it is often convenient to assume $\sqrt{1 + z_x^2 + z_y^2} = 1$, which is consistent with the assumption of shallowness.

For complex, architecturally interesting shapes of membranes and networks of cables, we can make use of shell analysis procedures and programs for the numerical analysis of shallow shells. For a segment of a membrane structure, perhaps spanning within a network of reinforcing cables, it is reasonable to assume that its rise-to-span ratio is relatively flat. Then certain simplifications in the expressions for the curvatures z_{xx} and z_{yy} can be effected and equivalent loads on the planar projection obtained with simplifications such as z_x^2 and $z_y^2 \ll 1$. For cable-reinforced segments, the edge tractions would be the loads transmitted to the cables. Other discussions of shallow shell analysis are given in Refs. 119 to 125.

Elliptic and hyperbolic paraboloids.   In particular cases, analytic solutions to Eqs. (6.3.7) can be generated. For example, an elliptic or hyperbolic paraboloid can be described by

$$z = C_1 \left(\frac{x}{a}\right)^2 + C_2 \left(\frac{y}{b}\right)^2 \tag{6.3.8}$$

and

$$z_{xx} = \frac{2C_1}{a^2}$$

$$z_{yy} = \frac{2C_2}{b^2}$$

If $C_1 C_2 > 0$, we have an elliptic paraboloid which can be visualized as in Fig. 6.3.2 as a concave parabola moving over another set of concave par-

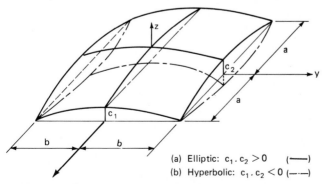

(a) Elliptic: $c_1 . c_2 > 0$ (——)
(b) Hyperbolic: $c_1 . c_2 < 0$ (—·—)

Figure 6.3.2 Elliptic and hyperbolic paraboloids.

abolas. If $C_1 C_2 < 0$, we have a hyperbolic paraboloid which is visualized in Fig. 6.3.2 as a convex parabola moving over a concave parabola [5] (or the converse).

We will assume shallow membranes and assume the right-hand sides of Eq. (6.3.7b or c) to be constant, e.g., superposed snow load or uniform pressure. In that instance, analytic solutions can be developed by separation of variables in series form [176, 177] for boundary conditions of $\overline{N}_x = 0$ at $x = \pm a$ and $\overline{N}_y = 0$ at $y = \pm b$.

These conditions will serve well for cases of superposed stresses due to superposed loads. However, the solutions for prestressing resultants need to satisfy nonzero conditions on N_x, N_y, and will therefore lead to restrictions on either the shape or the magnitude of load to maintain a desired shape. For example, if we impose $\overline{N}_x = N_1$ on $x = \pm a$ and $\overline{N}_y = N_2$ on $y = \pm b$, we can satisfy Eq. (6.3.6) by

$$H = \frac{N_1 y^2}{2} + \frac{N_2 x^2}{2}$$

and obtain constant $\overline{N}_x$, $\overline{N}_y$ throughout if

$$\frac{2C_1}{a^2} N_1 + \frac{2C_2}{b^2} N_2 = P_z \tag{6.3.9}$$

Now, if $P_z = p = $ internal pressure, Eq. (6.3.9) determines one of either C_1, C_2, or p for specified N_1, N_2. With specified p this constitutes what can be called a "shape-finding" problem [111, 112, 137–141, 178]. If, on the other hand, $P_z = 0$, then Eq. (6.3.9) requires that one of C_1 or C_2 be negative in order to maintain tension throughout. The latter case requires a hyperbolic paraboloid as in Fig. 6.3.2b and is an example of cases where, in the absence of pressure, opposing curvatures of the surface are required to provide prestressing of membranes.

In the case of superposed in-service loads on a hyperbolic paraboloid, it is often convenient to rotate the planar coordinate system so that the x, y axes coincide with the asymptotes. Then the equation for the surface can be written as

$$z = \frac{xy}{C_3}$$

and Eq. (6.3.7a) becomes

$$\frac{2\overline{N}_{xy}}{C_3} = \Omega = \overline{P}_z + \overline{P}_x \frac{y}{C_3} + \overline{P}_y \frac{x}{C_3}$$

from which a direct solution to $\overline{N}_{xy}$ is obtained. Then, $\overline{N}_x$, $\overline{N}_y$ are determined by Eqs. (6.3.6) as

$$\overline{N}_x = -\frac{C_3}{2} \int \left(\frac{\partial \Omega}{\partial y} + \overline{P}_x \right) dx + g_1(y)$$

$$\overline{N}_y = -\frac{C_3}{2} \int \left(\frac{\partial \Omega}{\partial x} + \overline{P}_y \right) dy + g_2(x)$$

with arbitrary functions g_i to be set from boundary conditions on $\overline{N}_x$, $\overline{N}_y$.

Suppose $\overline{N}_x = N_1$ on $x = \pm a$ and $\overline{N}_y = N_2$ on $y = \pm b$, $\overline{P}_x = \overline{P}_y = 0$ throughout, and $\overline{P}_z = $ constant. The solution is

$$\overline{N}_{xy} = \frac{P_z C_3}{2}$$

$$\overline{N}_x = N_1 \qquad \overline{N}_y = N_2$$

Although $\overline{N}_x$ and $\overline{N}_y$ are both positive, we must still beware of compressive wrinkles because the shear $\overline{N}_{xy}$ is nonzero. Considering principal stresses, we see that a zone of compressive principal stress will be predicted if

$$\overline{N}_{xy} > \sqrt{\overline{N}_x \cdot \overline{N}_y}$$

Instead, wherever that condition would prevail, there will be a zone of wrinkles aligned with the direction of the positive principal stress.

Catenoidal shape. As another example of translational membranes, consider the problem of determining a shape for a membrane (or a dense network of cables) required to maintain constant tension and zero shear under dead-weight loading, i.e., with $P_z = -\gamma$ and

$$\overline{N}_x = \overline{N}_y = C$$

In this case, Eq. (6.3.7b) gives

$$\left(\frac{\partial^2 z}{\partial x^2} + \frac{\partial^2 z}{\partial y^2} \right) C = -\gamma \sqrt{1 + z_x^2 + z_y^2}$$

This nonlinear equation could be solved numerically to determine z as a function of x and y. If we assume a shallow surface such that $\sqrt{1 + z_x^2 + z_y^2} = 1$, then the equation reduces to Poisson's equation, solutions to which have been tabulated for many problems in structural mechanics, e.g., torsion of bars [28]. If we consider a rectangular planform, as shown in Fig. 6.3.3a, we can write the solution as [176]

$$z = \frac{-16\gamma a^2}{C\pi^3} \sum_{m \text{ odd}} \frac{1}{m} (-1)^{m-1/2} \left(1 - \frac{\cosh m\pi y/2a}{\cosh m\pi b/2a} \right) \cos \frac{m\pi x}{2a}$$

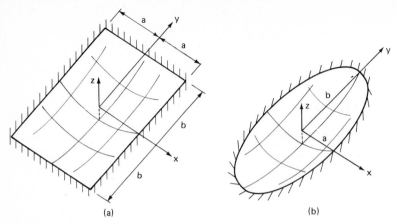

Figure 6.3.3 Suspended catenoidal shapes. (*a*) Rectangular planform; (b) elliptical planform. ′

For an elliptic planform as shown in Fig. 6.3.3*b*, we have

$$z = \frac{-\gamma}{2C} \frac{a^2 b^2}{a^2 + b^2} \left[\left(\frac{x}{a} \right)^2 + \left(\frac{y}{b} \right)^2 - 1 \right]$$

These represent shapes, that will be assumed, for example, by hanging roofs supporting snow loads or suspended panels.

6.3.2 Cylindrical membranes

Cylinders constitute a special class of translational shapes which has particular significance in membrane analysis. In this class, the generator is a straight line which is translated over a closed curve as a directrix. See Fig. 6.3.4 for three example cylinders.

This class of shapes is significant in that the gaussian curvature is zero and therefore membranes with these shapes are fabricable from flat sheets without straining. The gaussian curvature is zero because the generator is normal to the directrix, lies along a direction of principal curvature, and is straight, implying an infinite radius of curvature. Note that if the closed curves acting as directrices are circles, the cylinder is right circular and could be classified as a membrane of revolution, the analysis of which class is the subject of Sec. 6.4.

The cylinder shown in Fig. 6.3.5 is embedded in a three-dimensional space in which we take the x axis parallel to the generator. We need not restrict our attention to shallow cylindrical surfaces, and instead of using projected planform coordinates, we will adopt as our surface coordinates the axial coordinate x and the angle θ a line normal to the surface makes in the xy plane (see Fig. 6.3.5)

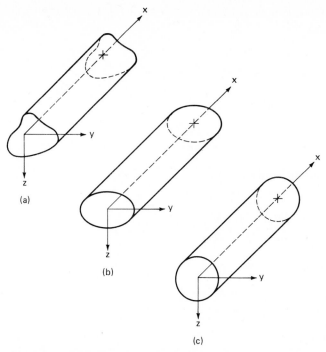

Figure 6.3.4 Representative cylindrical membranes. (*a*) General directrix; (*b*) elliptic directrix; (*c*) circular directrix.

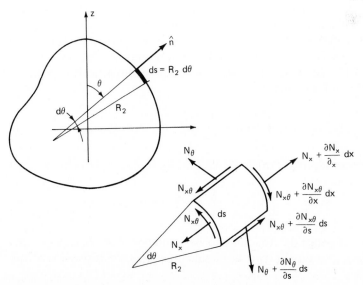

Figure 6.3.5 Coordinates and equilibrium of a cylinder.

If we write the equations of a directrix as $z = f(y)$, then the radius of curvature R_2 is

$$\frac{1}{R_2} = \frac{-d^2f/dy^2}{[1 + (df/dy)^2]^{1/2}} = -\frac{d^2f}{dy^2}\cos^3\theta$$

and the angle θ is

$$\theta = \tan^{-1}\left(\frac{df}{dy}\right)$$

Equilibrium of the differential segment shown in Fig. 6.3.5 gives in the axial direction

$$\frac{\partial N_x}{\partial x} + \frac{1}{R_2}\frac{\partial N_{x\theta}}{\partial\theta} + P_x = 0 \qquad (6.3.10a)$$

in the circumferential direction

$$\frac{\partial N_{x\theta}}{\partial x} + \frac{1}{R_2}\frac{\partial N_\theta}{\partial\theta} + P_\theta = 0 \qquad (6.3.10b)$$

and in the normal direction

$$N_\theta - pR_z = 0 \qquad (6.3.10c)$$

where P_x, P_θ = tangential loads and p = normal pressure.

Note that the circumferential stress resultant N_θ is determined directly by Eq. (6.3.10c) independently of any boundary conditions. If the cylinder is not closed, tractions on θ = constant edges must be provided to maintain the stress state predicted in Eq. (6.3.10c) if R_2 is preset, or the radius R_2 must be changed to allow matching of the prescribed traction.

Equations (6.3.10) have an analytic solution. Substitute Eq. (6.3.10c) into (6.3.10b) and integrate to obtain

$$N_{x\theta} = -\int\left(P_\theta + \frac{1}{R_2}\frac{\partial N_\theta}{\partial\theta}\right)dx + g_1(\theta) \qquad (6.3.11a)$$

Then substitute Eq. (6.3.11a) into (6.3.10a) and integrate to obtain

$$N_x = -\int\left(P_x + \frac{1}{R_2}\frac{\partial N_{x\theta}}{\partial\theta}\right)dx + g_2(\theta) \qquad (6.3.11b)$$

where $g_1(\theta)$ are arbitrary functions to be evaluated from boundary conditions on N_x, $N_{x\theta}$ on the two terminal directrices, or traverses, of the cylinder.

Note that only two of the four possible combinations of N_x, $N_{x\theta}$ on the two ends can be used to determine g_1 and g_2. In other words, we must specify both of N_x, N_{xy} (free condition) if the other end has no traction condition (fixed condition), or we must specify N_x values on both ends if N_{xy} is unspecified. This latter condition is called a diaphragm condition. For example, in air-supported cylinders, a diaphragm-like condition is often realized in calottes (cuplike caps) on the ends.

If the loading is not a function of x, we have from Eqs. (6.3.10c) and (6.3.11)

$$N_\theta = pR_2 \tag{6.3.12a}$$

$$N_{x\theta} = g_1(\theta) - g_0(\theta) \tag{6.3.12b}$$

$$N_\theta = g_2(\theta) + \frac{1}{2R_2}\frac{dg_0}{d\theta}x^2 - P_x x - \frac{dg_1}{d\theta}x \tag{6.3.12c}$$

where

$$g_0(\theta) = P_\theta + \frac{1}{R_2}\frac{dN_\theta}{d\theta} = P_\theta + \frac{p}{R_2}\frac{dR_2}{d\theta} + \frac{dp}{d\theta} \tag{6.3.12d}$$

Note that the form of the x dependence in Eqs. (6.3.12) is the same as those used to calculate the shear and moment resultants for uniformly loaded statically determinate beams. This is the basis of the beam analogy described in the sequel.

Elliptic cylinder.   As an example, consider an elliptic cylinder cantilevered as shown in Fig. 6.3.6 and subjected to uniform internal pressure p_0. We

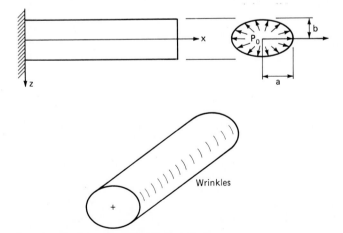

Figure 6.3.6 Cantilevered elliptic cylinder.

reemphasize that the geometry we prescribed is based on the shape of the end profiles and the pressure p_0: the initial shape needed to obtain the prestressed shape remains to be determined from a nonlinear analysis.

Since for an ellipse with semiaxis dimensions a, b (Fig. 6.3.6)

$$R_2 = \frac{a^2 b^2}{(a^2 \sin^2 \theta + b^2 \cos^2 \theta)^{3/2}} = \frac{a\rho^2}{(\sin^2 \theta + \rho^2 \cos^2 \theta)^{3/2}}$$

where $\rho = b/a$ = ratio of semiaxis dimensions. We have from Eq. (6.3.12)

$$N_\theta = \frac{\rho^2}{(\sin^2 \theta + \rho^2 \cos^2 \theta)^{3/2}} (p_0 a)$$

Since $N_x = N_{x\theta} = 0$ at $x = 0$, and thus $g_1(\theta) = g_2(\theta) = 0$, we have

$$N_{x\theta} = f_1(\theta)(p_0 x) \tag{6.3.13a}$$

$$f_1(\theta) = \frac{3(1 - \rho^2) \sin \theta \cos \theta}{(\cos^2 \theta + \rho^2 \cos^2 \theta)} \tag{6.3.13b}$$

$$N_x = f_2(\theta) \left(\frac{p_0}{a} \frac{x^2}{2} \right) \tag{6.3.13c}$$

$$f_2(\theta) = \frac{3(1 - \rho^2)}{\rho^2} \frac{\sin^4 \theta - \rho^2 \cos^4 \theta}{(\sin^2 \theta + \rho^2 \cos^2 \theta)^{1/2}} \tag{6.3.13d}$$

Note that if $\rho = 1$, i.e., a right circular cylinder,

$$N_\theta = p_0 a \qquad N_{x\theta} = N_x = 0$$

We see that if $\rho < 1$, then N_x is negative near $\theta = 0$, π. If $\rho > 1$, then N_x is negative near $\theta = \pi/2$, $3\pi/2$. Thus, uniformly pressurized elliptic cylinders will always possess a region with wrinkles running in the circumferential direction (see Fig. 6.3.6) unless they are subject to an additional axial prestressing.

Approximate nonlinear analysis.   At the beginning of the previous subsection we noted that the elliptic shape was assumed a priori as the deformed shape. In fact, the cylinder will deform under changes in load distribution as illustrated in Fig. 6.3.7 for an air-supported cylinder subjected to a crosswind. The original air-pressurized shape is a right circular cylinder with

$$N_\theta = p_0 a \qquad N_{x\theta} = 0 \qquad N_x = g_2(\theta)$$

Under wind, the geometry and stress will change. If changes in geometry are neglected, the load could be approximated as

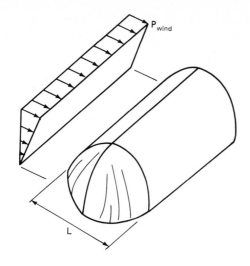

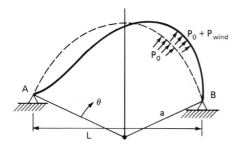

Figure 6.3.7 Crosswind on an air-supported cylinder.

$$p = p_0 + p_w \sin \theta \qquad (6.3.14)$$

and Eqs. (6.3.12) could be used. The representation of wind load given by Eq. (6.3.14) is a gross simplification, even if changes in geometry are neglected. Shown in Fig. 6.3.8 is a representative pressure profile for a fully closed circular cross section [111, 178, 179]. In Ref. 178, the following formulas for the wind pressure and shape parameter are suggested:

$$p_w = \tfrac{1}{2} C_\theta \rho V_0^2$$

$$C = \cos \frac{\theta \pi}{1.2} \qquad \text{if } 0 \leqslant \theta \leqslant 0.6 \text{ rad}$$

$$= 4.23 - 9.02\theta + 3.20\theta^2 \qquad \text{if } 0.6 < \theta \leqslant 2.07$$

$$= -0.4 \qquad \text{if } 2.07 < \theta \leqslant \pi$$

where ρ = air density
V_0 = wind velocity

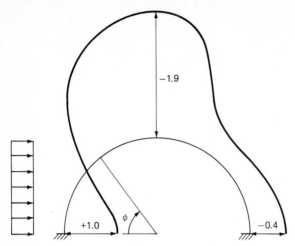

Figure 6.3.8 Pressure profile on wind-loaded circular cylinder.

It is possible to also include the effects of changes in geometry in an approximate fashion if we assume the cylinder is long relative to its diameter. In the region far from the end calottes the cylinder cross section will deform similarly to a cable spanning between points A and B in Fig. 6.3.7. Iteration techniques for predicting the cable shape (the cylindrical profile) and tension (N_θ) have been described previously in Sec. 2.2, and also in Refs. 112 and 180. For example, the profile can be approximated as a series of short, uniformly loaded straight sections and the technique described in Sec. 2.2 can be used to develop an iterative solution to the deformed profile. Tests have shown that these approximations give reasonable results even for shorter cylinders [161].

As an example in which an analytic solution to a nonlinear profile finding problem is possible, let us consider the shape and stresses in a load-adaptive cylinder containing a fluid. See Fig. 6.3.9 for a definition sketch.

If we allow the flexible membrane cylinder to adapt its geometry to maintain a constant circumferential stress, we will have by Eq. (6.3.12a)

$$N_\theta = R_2 p = R_2 \gamma z = N_0 = \text{constant} \tag{6.3.15}$$

where γ = weight density of the fluid, z = coordinate located such that the cylinder sustains a pressure head H at the apex, and $R_2(\theta)$ = instantaneous radius of curvature of the cylinder cross section. If no other loads but fluid pressure were considered and the cylinder were assumed to be supported on diaphragms at its ends, the other stress resultants would be zero.

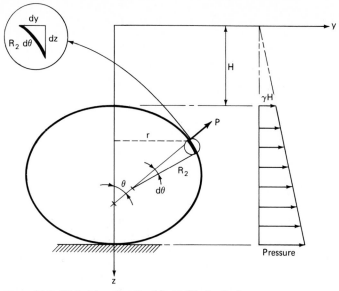

Figure 6.3.9 Definition sketch of fluid-filled cylinder.

We wish to determine the geometry y, z, R_2 to maintain $N_\theta = N_0$. By consideration of the differential element in Fig. 6.3.9, we see

$$\frac{1}{R_2} = \frac{d\theta}{ds} = \frac{d\theta}{dy}\frac{dy}{ds} = \frac{d(\sin\theta)}{dy}$$

and therefore Eq. (6.3.15) can be rewritten as

$$\frac{d\xi}{dt} = a\eta \qquad (6.3.16a)$$

where we defined nondimensional variables

$$\xi = \sin\theta \qquad t = \frac{y}{H} \qquad \eta = \frac{z}{H}$$

and the parameter $a = \gamma H^2/N_0$. In addition, we see from the differential segment in Fig. 6.3.9 that

$$\frac{dz}{dy} = \tan\theta$$

and therefore in terms of our nondimensional variables

$$\frac{d\eta}{dt} = \frac{\xi}{\sqrt{1-\xi^2}} \qquad (6.3.16b)$$

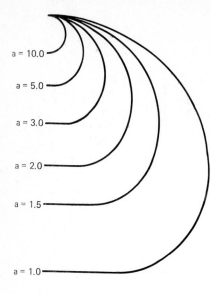

$a = 10.0$

$a = 5.0$

$a = 3.0$

$a = 2.0$

$a = 1.5$

$a = 1.0$

Figure 6.3.10 Shapes of fluid-filled cylinders: $a = \gamma h^2 / N$.

Equations (6.3.16) constitute two simultaneous first-order equations for ξ and η in terms of t with initial conditions $\xi = 0$, $\eta = 1$ at $t = 0$ (apex). Combining these two equations, we obtain a single second-order equation:

$$\frac{d^2\xi}{dt^2} = \frac{a\xi}{\sqrt{a - \xi^2}} \tag{6.3.17}$$

which is a Newton equation with a solution:

$$\frac{d\xi}{dt} = a\eta = a\left[1 + \frac{2}{a}\left(1 - \sqrt{1 - \xi^2}\right)\right]^{1/2} \tag{6.3.18a}$$

and thus

$$t = \int_0^\xi \frac{d\xi}{a\left[1 + \frac{2}{a}\left(1 - \sqrt{1 - \xi^2}\right)\right]^{1/2}} \tag{6.3.18b}$$

Solutions to Eqs. (6.3.18) are displayed in Fig. 6.3.10 for various values of the parameter a. It can be seen that as a become larger, i.e., H very large, the cross-sectional geometry approaches that of a circular cylinder subject to a uniform internal pressure.

Beam analogies. One class of air-supported structure is highly pressurized tubes used as beam or column segments to support external loads. Applications of, and stress analyses of, the nonlinear response of such tubes are the subjects of Refs. 130 to 133. Particular attention is devoted there to the calculation of limit loads and to failure mechanisms.

Upon examination of Eqs. (6.3.12) we note that in determining $N_{x\theta}$ and N_x characteristics, the cross-sectional geometry and the θ dependence of the stresses are embedded in the functions $g_0(\theta)$, $g_1(\theta)$, $g_2(\theta)$. If $P_x = 0$ and if P_θ and $p = p(\theta)$ only, then the x dependence of N_x, $N_{x\theta}$, is explicit and resembles the equations of shear and moment in a statically determinate beam under uniform lateral load. Thus, within the linearized membrane theory as applied to air-inflated cylindrical tubes or segments, the beam analogy developed for momentless cylindrical shells could be adopted.

In that approach, equivalent moment and shear resultants on the total cross sections are defined in terms of the shear and direct axial stresses on the cylinder. Then, beam analysis can be used to calculate those equivalent forces and hence the stresses to be superposed on the stresses due to internal pressure. This would give the axial variation in the stress resultants. The circumferential variations follow from Eqs. (6.3.12) for different cylindrical geometries and loads, e.g., Eqs. (6.3.13) for a cantilevered elliptical cylinder.

This technique could be used even for noncircular cross sections and partially closed cross sections. Analysis procedures developed for barrel roof shells could be used. The cable reinforcement on the edges of adjoining segments would be in a sense equivalent to edge beams between barrel roof segments. See Refs. 126 to 129 for alternative techniques for analysis of momentless cylindrical shells.

Vibrations of cylindrical tubes. For long cylindrical tubes, we can use beam flexural modes to approximate the lower natural frequencies of the tubes. These approximations can also be used for shorter cylinders if we include in the beam analysis the effects of rotatory inertia and shear deformations. However, for the higher modes of vibration in which significant circumferential distortions of the cylinder occur, techniques of shell analysis must be adopted.

It is common in considering the dynamic response of highly prestressed cylindrical tubes to assume the oscillations to be small and to neglect changes in cross-sectional geometry. For noncircular cross sections, numerical techniques, such as described in Chap. 8, should be used. For fully closed circular cross sections, predictions of modal frequencies and modal shapes, both circumferential and axial, can be developed using the procedures described for prestressed membranes of revolution in Sec. 7.4.

In Ref. 178 are considered those solutions in which axial variations can be neglected. In that instance, solutions to the governing differential equations of free radial deflection w

$$\frac{\mu a}{p_0} \frac{\partial^2 w}{\partial t^2} + \frac{\partial^2}{\partial \theta^2}\left(w - \frac{\mu a}{p_0}\frac{\partial^2 w}{\partial t^2}\right) + \frac{\partial^4 w}{\partial \theta^4} = 0 \qquad (6.3.19)$$

are derived from long cylinders with partially closed cross sections. In Eq. (6.3.19) μ = density, a = radius, p_0 = internal pressure. Forced vibrations are determined using modal superposition techniques as described in Chap. 4 for discrete systems and in Sec. 7.4.4 for continuous systems.

In Ref. 181 are given formulas for natural frequencies of cylinders with various supporting conditions. The natural frequencies of a cylinder with finite length L, radius a, and thickness h are functions of prestress levels and are given by

$$\omega_{nm}^2 = (\overline{\omega}_{nm})^2 + \frac{N_x}{\mu h}\left(\frac{m\pi}{L}\right)^2 + \frac{N_\theta}{\mu h}\left(\frac{n}{a}\right)^2 \qquad (6.3.20a)$$

where N_x, N_θ = constant prestress

n = number of circumferential waves in mode shape

m = number of axial waves

$\overline{\omega}_{nm}$ = natural frequency of equivalent cylinder with no prestressing

For example, for a circular cylinder supported on diaphragms

$$\overline{\omega}_{nm} = \frac{\lambda_{nm}}{a}\sqrt{\frac{E}{\mu(1-\nu^2)}} \qquad (6.3.20b)$$

$$\lambda_{nm} = \frac{\{(1-\nu^2)(m\pi a/L)^4 + (h^2/12a^2)[n^2 + (m\pi a/L)^2]^4\}^{1/2}}{\left[n^2 + \left(\frac{m\pi a}{L}\right)^2\right]} \qquad (6.3.20c)$$

The effects of prestress on both frequency and mode shape will be discussed further in Sec. 7.4.

6.4 Approximate Analysis of Axisymmetric Membranes

One common class of prestressed membrane is that of axisymmetric shapes, such as spheres, ellipsoids, and so forth, in which the desired configuration of the membrane subsequent to prestressing is assumed to be generated by revolving a curve about a central axis. The differential geometry of this class of surface is described in Sec. 6.4.1.

The prestressing phase of behavior of an axisymmetric membrane is nonlinear and, hence, the undeformed and deformed configuration are distinct from one another. In Sec. 6.4.2 an approximate analysis technique for determining the stress resultants on the deformed configuration is presented. No consideration is given to determining the nonlinear displacements to that configuration. That subject is covered in Chap. 8 wherein nonlinear methods are developed.

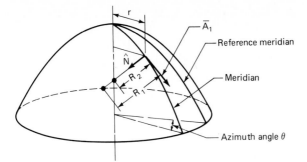

Figure 6.4.1 Geometry of membrane of revolution.

It is not illogical to assume the behavior of an axisymmetric membrane during the in-service phase to be linear in that the membrane has been considerably stiffened during prestressing. In Sec. 6.4.3 techniques for determining the superposed stress resultants due to static nonsymmetric in-service loads are described. Those techniques are approximate in that the effect of prestressing on the subsequent behavior is neglected in order to use classical techniques of momentless theory for shells of revolution. The effect of prestressing on the in-service behavior of both stresses and displacements is described in the following chapter (Secs. 7.2 and 7.3). Dynamic effects are also described in a subsequent section (Sec. 7.4).

6.4.1 Differential geometry of revolutes

A surface of revolution is generated by rotating a plane curve about an axis in its plane (see Fig. 6.4.1). The generating curve is called a meridian and lies in the so-called meridional plane. The cross-sectional curve cut by a plane perpendicular to the axis of revolution is a circle of radius r. The two surface coordinates most commonly selected are $x^1 = x =$ coordinate measure in the meridional plane and $x^2 = \theta =$ azimuth angle which the meridional plane makes with a reference plane. These constitute an orthogonal coordinate system. There are several choices for x: (1) angle ϕ between the principal normal $\hat{N}$ to the meridian and the axis of revolution (common for spheres); (2) arc length s_m along the meridian (common for cones); and (3) distance z along the axis of revolution to a cross-sectional plane (common for cylinders).

The surface coordinates x, θ are orthogonal, i.e., the length of a differential segment is given in terms of the metric tensor $A_{\alpha\beta}$ by

$$ds^2 = A_{\alpha\beta}\, dx^\alpha\, dx^\beta \tag{6.4.1a}$$

$$= A^2\, (dx)^2 + r^2\, (d\theta)^2 \tag{6.4.1b}$$

where $\sqrt{A_{22}} = r =$ radius of azimuth circle and $\sqrt{A_{11}} = A(x) =$ function of selected coordinate measure x. See Appendix B for a review of differential geometry of surfaces using tensor calculus. Orthogonality is evident from $A_{12} = A_{21} = 0$.

The unit normal $\hat{N}$ to the surface is assumed positive when directed inward. The surface is embedded in a three-dimensional space, and we must therefore be concerned with the curvature of the surface through calculation of rates of change of the unit normal with respect to x and θ. The curvature of the curve on the normal plane section of the surface is

$$k_n = -\frac{d\hat{N}}{ds} \cdot \hat{t} \tag{6.4.2a}$$

where $\hat{t}$ = tangent vector. In terms of surface coordinates and the metric tensor, we have

$$k_n = \frac{B_{\alpha\gamma}dx^\alpha dx^\gamma}{A_{\beta\rho}dx^\beta dx^\rho} \tag{6.4.2b}$$

where $B_{\alpha\gamma}$ is called the curvature tensor.

At every point on a surface there exists two perpendicular directions for which k_n is an extremum. Those directions are directions of principal curvature. For a surface of revolution, they coincide with the coordinate lines, and thus

$$B_{12} = B_{21} = 0 \tag{6.4.3a}$$

$$B_{11} = \frac{A^2}{R_1} \tag{6.4.3b}$$

$$B_{22} = \frac{r^2}{R_2} \tag{6.4.3c}$$

where $R_i = 1/k_{ni}$ = principal radii of curvature. The geometric interpretations of R_1 and R_2 are given in Fig. 6.4.2, where it is seen that $R_1 =$

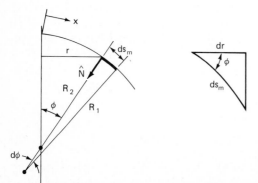

Figure 6.4.2 Radii of curvature for membrane of revolution.

radius of curvature of the meridional line s_m and R_2 = distance to the axis of revolution along a line normal to the surface. From Fig. 6.4.2 some convenient geometrical relationships can be deduced (ds_m = differential length along the meridian):

$$r = R_2 \sin \phi \tag{6.4.4a}$$

$$ds_m = A \, dx = R_1 \, d\phi \tag{6.4.4b}$$

$$\frac{dr}{ds_m} = \frac{(1/A) \, dr}{dx} = \cos \phi \tag{6.4.4c}$$

$$\frac{dz}{ds_m} = \frac{(1/A) \, dz}{dx} = \sin \phi \tag{6.4.4d}$$

$$\frac{dr}{d\phi} = R_1 \cos \phi \tag{6.4.4e}$$

$$\frac{dz}{d\phi} = R_1 \sin \phi \tag{6.4.4f}$$

$$\left(\frac{1}{r}\right)\left(\frac{dr}{d\phi}\right) = \frac{R_1}{R_2} \cot \phi \tag{6.4.4g}$$

Note that if $x = \phi$, then $A = R_1$, or if $x = s_m$, then $A = 1$.

The gaussian curvature (see Appendix B) is an invariant of the surface. For a surface of revolution.

$$(K) = \left(\frac{1}{R_1}\right)\left(\frac{1}{R_2}\right)$$

and thus we see that K is positive if the centers of curvature lie on the same side of the surface, negative if they lie on opposite sides, or zero if one of R_1, R_2 is infinite. The only surfaces realizable from an initially flat unstrained surface (radii of infinite curvature) are therefore those for which $K = 0$: cylinders, cones, or combinations of conical frustums. Other surfaces fabricated from flat sheets can only be approximated as surfaces of revolution.

6.4.2 Symmetric analysis for prestressing behavior

In the initial analysis of both the prestressing and in-service phases of membranes, it is common practice to use linear momentless theory for shells in which it is assumed that moments in the shells are negligible and that the prestressed shape is the reference geometry for all calculations.

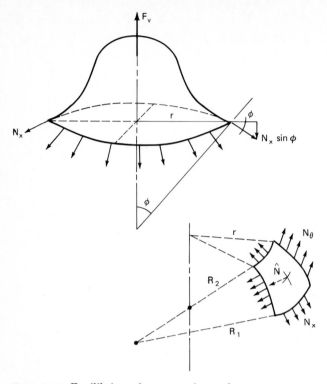

Figure 6.4.3 Equilibrium of segment of a revolute.

In linear momentless theory the stress calculations can be decoupled from the displacement calculations. For certain boundary conditions and geometries the equilibrium equations for the prestressing phase are statically determinate and often can be solved in closed form without recourse to numerical models.

One broad class of shells for which such closed-form prestressing solutions can be derived are shells of revolution which are loaded axisymmetrically. In that case the shear stress resultants $N_{x\theta} = 0$ and equilibrium of vertical components of forces on a segment of the membrane above an azimuth circle (see Fig. 6.4.3) can be used to write

$$(2\pi r)\, N_x \sin \phi = F_v \tag{6.4.5a}$$

where F_v = total vertical component of external axisymmetric force on the membrane segment above the azimuth circle. If we can express r, ϕ, and F_v in terms of x, the meridional stress resultant N_x is then given by

$$N_x = \frac{F_v}{2\pi r \sin \phi} \tag{6.4.5b}$$

Then, from consideration of equilibrium in the normal $\hat{N}$ direction of forces on the differential segment shown in Fig. 6.4.3

$$\frac{N_\theta}{R_2} + \frac{N_x}{R_1} - p = 0 \qquad (6.4.5c)$$

where p = outward-directed pressure and the hoop stress resultant N_θ can be determined as

$$N_\theta = pR_2 - \frac{R_2}{R_1} N_x \qquad (6.4.5d)$$

For example, if we have a sphere, $R_1 = R_2 = R$, subjected to uniform internal pressure p

$$F_v = p\pi r^2$$

and therefore, since $r = R \sin \phi$, we have from Eqs. (6.4.5)

$$N_x = N_\theta = \frac{pR}{2}$$

Ellipsoids.   Consider the ellipsoid of revolution generated by revolving an ellipse about the vertical axis. In this instance

$$R_1 = \frac{a^2 b^2}{(a^2 \sin^2 \phi + b^2 \cos^2 \phi)^{3/2}} \qquad (6.4.6a)$$

$$R_2 = \frac{a^2}{(a^2 \sin^2 \phi + b^2 \cos^2 \phi)^{1/2}} \qquad (6.4.6b)$$

where a, b are the lengths of the major axes of the ellipse. Due to internal pressure, we have from Eqs. (3.6.5)

$$F_v = p\pi r^2 = p\pi R_2^2 \sin^2 \phi \qquad (6.4.6c)$$

$$N_x = \frac{pR_2}{2} \qquad (6.4.6d)$$

$$N_\theta = \frac{pR_2}{2} \left(2 - \frac{R_2}{R_1} \right) \qquad (6.4.6e)$$

and we see than when $R_2 > 2R_1$, the hoop stress resultant N_θ is predicted as negative. For the ellipsoid this occurs at values of (a/b) and ϕ for which

$$2 - \left(\frac{a}{b} \right)^2 \sin^2 \phi - \cos^2 \phi < 0 \qquad (6.4.7a)$$

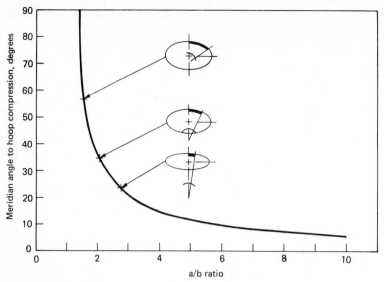

Figure 6.4.4 Transition point from tension to compression in an ellipsoid.

The roots of Eq. (6.4.7a) satisfy

$$\sin \phi = 1 / \sqrt{\left(\frac{a}{b}\right)^2 - 1} \qquad (6.4.7b)$$

and are displayed graphically in Fig. 6.4.4 for various values of the ratio (a/b).

Since the membrane can only transmit tensile stresses, wrinkles will occur in the meridional direction in the regions near the equator where compressive stresses were predicted. See Refs. 137, 174, and 175 for further discussions of the wrinkling phenomenon.

Approximate pole-supported tent.   As a final example of the axisymmetrically loaded membranes of revolution, consider the simulation of an initially flat membrane supported over a planform by poles. As shown in Fig. 6.4.5, this problem can be represented by that of a single pole-supported membrane in which the meridional curve is approximated as a catenary.

The geometric quantities can then be obtained by adopting the radius r as the meridional coordinate. Thus, we assume the catenary solution

$$z = \frac{\cosh u - 1}{\beta} \qquad (6.4.8a)$$

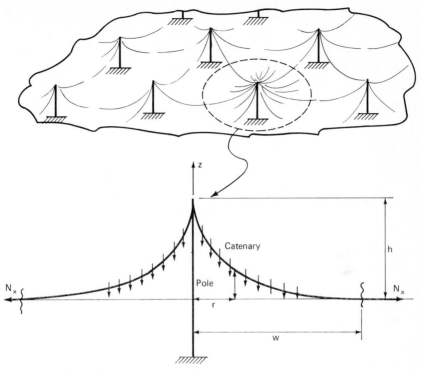

Figure 6.4.5 Approximate pole-supported tent.

where $u = \beta (w - r)$ and β = geometric parameter related to the sag ratio (h/w) and evaluated from the transcendental relation between h, w, and β obtained by evaluation of Eq. (6.4.8a) at $r = 0$ where $z = h$, i.e.,

$$(\beta w) \left(\frac{h}{w} \right) = \cosh \beta w - 1 \qquad (6.4.8b)$$

Solutions to Eq. (6.4.8b) are displayed in Fig. 6.4.6 as functions of βw for various sag ratios h/w. The parameter βw is determined from points of zero crossing in Fig. 6.4.6. Equations (6.4.4) then lead to

$$\sin \phi = -\tanh u \qquad (6.4.9a)$$

$$\cos \phi = \operatorname{sech} u \qquad (6.4.9b)$$

$$A = \cosh u \qquad (6.4.9c)$$

$$R_2 = \frac{-r}{\tanh u} \qquad (6.4.9d)$$

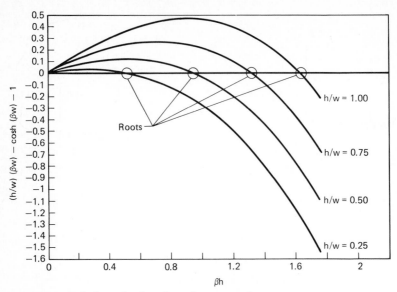

Figure 6.4.6 Relations of ρ, h, w for pole-supported tent.

$$R_1 = \frac{+\cosh^2 u}{\beta} \qquad (6.4.9e)$$

$$\frac{ds}{dr} = \cosh u \qquad (6.4.9f)$$

$$\frac{dz}{dr} = -\sinh u \qquad (6.4.9g)$$

Equilibrium of a symmetric segment above an azimuth circle gives by Eq. (6.4.5b)

$$N_x = \frac{F_v}{2\pi r \tanh u} \qquad (6.4.10a)$$

where we use Eqs. (6.4.9) to evaluate the vertical force resultant F_v as

$$F_v = Q - q \int_0^r 2\pi\xi \left(\frac{ds}{d\xi}\right) d\xi$$

$$= Q + \frac{2Hq}{\beta}\left[r \sinh u + (z - h)\right] \qquad (6.4.10b)$$

in which q = dead weight per unit area of membrane and Q = concentrated force due to the supporting pole. The term Q is determined from Eq. (6.4.10) evaluated at $r = w$ ($u = 0$). At that point

$$\lim_{r \to w} (N_x) = \frac{q}{\beta}$$

and therefore

$$Q = \frac{2\pi qh}{\beta}$$

which leads via Eqs. (6.4.10) to

$$N_x = \frac{q}{\beta} \left(\cosh u + \frac{z}{r} \tanh u \right) \qquad (6.4.11a)$$

The hoop stress resultant N_θ is evaluated using Eq. (6.4.5d), in which the normal component of the dead weight is

$$p = -q \cos \phi = \frac{-q}{\cosh u}$$

and therefore, by Eqs. (6.4.5d), (6.4.9e), and (6.4.11a)

$$N_\theta = \frac{qz}{\sinh^2 u}$$

In Fig. 6.4.7 is displayed the meridional profile of pole-supported membranes with various sag ratios. In Fig. 6.4.8 are displayed the corresponding distributions of meridional and hoop stress resultants.

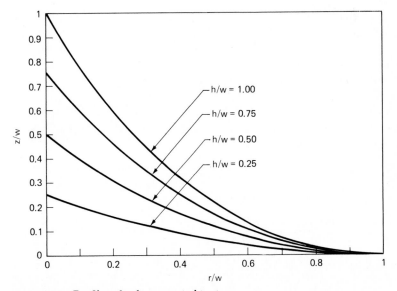

Figure 6.4.7 Profiles of pole-supported tent.

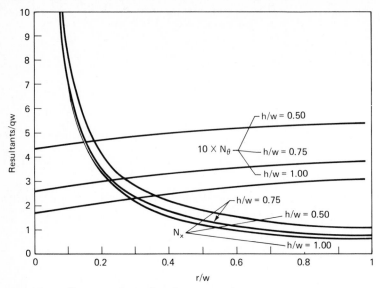

Figure 6.4.8 Stress resultants in pole-supported tent.

6.4.3 Linear analysis for in-service behavior

The in-service behavior of prestressed membranes is affected by the level of prestress. In this section we will neglect that effect in order to arrive at a simplified analysis. In Chap. 7, the prestressing effect on in-service behavior is described in detail and estimates of its importance for both static and dynamic response are discussed there.

The solution method described in Sec. 6.4.2 for axisymmetric loads on membranes of revolution would serve for the approximate stress analysis of the prestressing phase. If the in-service loads were also symmetric, the additional stresses could be calculated in the same manner.

The in-service loads are more likely to be nonsymmetric. In this instance approximate solutions for membranes of revolution could be obtained using Fourier analysis. Equilibrium of forces acting on a differential segment of the membrane leads to [79, 119, 122]

$$\frac{\partial (rN_x)}{\partial x} + A\frac{\partial N_{x\theta}}{\partial \theta} - \frac{dr}{dx}N_\theta + ArP_x = 0 \qquad (6.4.12a)$$

$$\frac{\partial (rN_{x\theta})}{\partial x} + A\frac{\partial N_\theta}{\partial \theta} + \frac{dr}{dx}N_{x\theta} + ArP_\theta = 0 \qquad (6.4.12b)$$

$$\frac{N_x}{R_1} + \frac{N_\theta}{R_2} - p = 0 \qquad (6.4.12c)$$

where P_x, P_θ are meridional and circumferential components of external load.

For nonsymmetric loads, expand the loads and stress resultants as Fourier series with respect to the circumferential angle θ:

$$P_x = \sum_{n=0}^{\infty} P_{xn} \cos n\theta \qquad (6.4.13a)$$

$$P_\theta = \Sigma \, P_{\theta n} \sin n\theta \qquad (6.4.13b)$$

$$p = \Sigma \, p_n \cos n\theta \qquad (6.4.13c)$$

$$N_x = \Sigma \, N_{xn} \cos n\theta \qquad (6.4.13d)$$

$$N_\theta = \Sigma \, N_{\theta n} \cos n\theta \qquad (6.4.13e)$$

$$N_{x\theta} = \Sigma \, N_{x\theta n} \sin n\theta \qquad (6.4.13f)$$

where the coefficients P_{xn}, N_{xn}, etc., are functions of x only. The load coefficients are determined in the interval $-\pi \leqslant \theta \leqslant \pi$ by

$$\begin{Bmatrix} P_{x0} \\ p_0 \end{Bmatrix} = \frac{1}{\pi} \int_0^\pi \begin{Bmatrix} p_x \\ p \end{Bmatrix} d\theta \qquad (6.4.14a,b)$$

$$\begin{Bmatrix} P_{xn} \\ p_n \end{Bmatrix} = \frac{2}{\pi} \int_0^\pi \begin{Bmatrix} P_x \\ p \end{Bmatrix} \cos n\theta \, d\theta \qquad (6.4.14c,d)$$

$$P_{\theta n} = \frac{2}{\pi} \int_0^\pi P_\theta \sin n\theta \, d\theta \qquad (6.4.14e)$$

To determine N_{xn}, $N_{\theta n}$, $N_{x\theta n}$, first substitute Eqs. (6.4.13) into Eqs. (6.4.12) and eliminate $N_{\theta n}$ using Eq. (6.4.12c) to obtain two first-order differential equations for each value of n:

$$\frac{dN_{xn}}{dx} = \frac{A}{r} \left[R_2(p_n \cos \phi - P_{xn} \sin \phi) \right.$$

$$\left. - nN_{x\theta n} - \left(a + \frac{R_2}{R_1} \right) \cos \phi N_{xn} \right] \qquad (6.4.15a)$$

$$\frac{dN_{x\theta n}}{dx} = \frac{A}{r} \left[R_2(np_n - P_{\theta n} \sin \phi) \right.$$

$$\left. - z_n \cos \phi N_{x\theta n} - n \frac{R_2}{R_1} N_{xn} \right] \qquad (6.4.15b)$$

There is no general solution to Eqs. (6.4.15). Numerical solutions can be generated by numerical integrations along a meridional line [170]. Two boundary conditions on either N_{xn} or $N_{x\theta n}$ must be specified for each integer n at either end of the meridional line. Techniques of superposing initial-value solutions to satisfy two-point boundary-value problems can be used [170]. One such technique is described in Sec. 7.3. Care must be taken if the apex is closed ($r = 0$) in that it is a regular singular point of Eqs. (6.4.15): Taylor expansions of N_{xn}, $N_{x\theta n}$ must be taken and Eqs. (6.4.15) used to develop recursive relations for the Taylor coefficients. See Sec. 7.3 for further details of this technique.

If the membrane is a spherical shape, it is possible to develop a closed-form solution to Eqs. (6.4.15). In that case, take $x = \phi$, $A = R_1$, and rewrite Eqs. (6.4.15) as

$$\frac{dN_{xn}}{d\phi} + 2N_{xn} \cot \phi + nN_{x\theta n} \cos \phi = -R(P_{xn} - p_n \cos \phi) \qquad (6.4.16a)$$

$$\frac{dN_{x\theta n}}{d\phi} + 2N_{x\theta n} \cot \phi + nN_{xn} \cos \phi = -R(P_{\theta n} - np_n \cos \phi) \qquad (6.4.16b)$$

Then, let

$$U_1 = N_{xn} + N_{x\theta n} \qquad (6.4.17a)$$

$$U_2 = N_{xn} - N_{x\theta n} \qquad (6.4.17b)$$

and take the sum and difference of Eqs. (6.4.16) to obtain

$$\frac{dU_i}{d\phi} + Q_i(\phi)\, U_i + \Lambda_i(\phi) = 0 \qquad i = 1, 2 \qquad (6.4.18a,b)$$

where

$$Q_1 = 2 \cot \phi + n \cos \phi \qquad (6.4.18c)$$

$$Q_2 = 2 \cot \phi - n \cos \phi \qquad (6.4.18d)$$

$$\Lambda_1 = -R \left[P_{xn} + P_{\theta n} - p_n \left(\cot \phi + n \cos \phi \right) \right] \qquad (6.4.18e)$$

$$\Lambda_2 = -R \left[P_{xn} - P_{\theta n} - p_n \left(\cot \phi - n \cos \phi \right) \right] \qquad (6.4.18f)$$

Equations (6.4.18) have an integrating factor

$$I_i = e \exp \left(\int Q_i \, d\phi \right) \qquad (6.4.19a)$$

and, therefore,

$$U_i = \frac{1}{I_i} [C_i - \int \Lambda_i I_i \, d\phi] \qquad (6.4.19b,c)$$

where C_i = arbitrary constants to be determined from boundary conditions, and

$$I_1 = \sin^2 \phi \tan^n \left(\frac{\phi}{2}\right) \qquad (6.4.19d)$$

$$I_2 = \sin^2 \phi \cot^n \left(\frac{\phi}{2}\right) \qquad (6.4.19e)$$

Once Eqs. (6.4.19b,c) have been integrated, the stress resultants are recovered from Eqs. (6.4.17) as

$$N_{xn} = \frac{U_1 + U_2}{2} \qquad (6.4.20a)$$

$$N_{x\theta n} = \frac{U_1 - U_2}{2} \qquad (6.4.20b)$$

As an example, consider a wind load superposed on a spherical membrane. Adopting a grossly simplified representation of the wind load

$$P_x = P_\theta = 0$$

$$p = - \bar{p} \sin\phi \cos \theta$$

we have $U_1 = U_2 = 0$ for all $n \neq 1$. For $n = 1$, Eqs. (6.4.19) give

$$U_1 = \frac{1 + \cos \phi}{\sin^3 \phi} [C_1 + \bar{p}R \cos \phi \, (1 - \tfrac{1}{3} \cos^2 \phi)]$$

$$U_2 = \frac{1 - \cos \phi}{\sin^3 \phi} [C_1 - \bar{p}R \cos \phi \, (1 - \tfrac{1}{3} \cos^2 \phi)]$$

and therefore

$$N_x = \frac{\cos \theta}{\sin^3 \phi} \left[\frac{C_1 + C_2}{2} + \frac{C_2 - C_1}{1} \cos \phi + \bar{p}R \cos^2 \phi \, (1 - \tfrac{1}{3} \cos^2 \phi) \right]$$

$$N_{x\theta} = \frac{\sin \theta}{\sin^3 \phi} \left[\frac{C_1 - C_2}{2} + \frac{C_1 + C_2}{2} \cos \phi + \bar{p}R \cos \phi \, (1 - \tfrac{1}{3} \cos^2 \phi) \right]$$

For N_x and $N_{x\theta}$ to be finite at $\phi = 0$, $C_2 = -C_1 = \frac{2}{3}\,\overline{p}R$ and finally

$$N_x = -\frac{\overline{p}R \cos \theta \cos \phi}{3 \sin^3 \phi} (2 - 3 \cos \phi + \cos^3 \phi)$$

$$N_{x\theta} = -\frac{\overline{p}R \sin \theta}{3 \sin^3 \phi} (2 - 3 \cos \phi + \cos^3 \phi)$$

$$N_\theta = -\overline{p}R \cos \theta - N_x$$

Note that these are superposed stress resultants and therefore the final stress N_x will be decreased from the prestressed level on the windward side and will be increased on the leeward side. We would need to check if the combined prestress and superposed stress on the windward side were negative. If so, local wrinkling would occur.

Prestress Effects on Behavior of Membranes of Revolution

In Chap. 6 the effect of prestressing on the subsequent behavior of membranes during in-service loading was not considered. That is, other than serving the role of maintaining an equilibrium geometry for further calculations, the prestress was assumed not to change the additional displacements or stresses due to additional load. In this chapter a linearized theory is developed in Sec. 7.1 for prestressed membranes which does account for at least the first-order effects of the prestress. Solutions based on numerical integration along a typical meridian line of a membrane of revolution are developed in Secs. 7.2 and 7.3 for symmetrical [38] and nonsymmetrical [170] static loadings. A method for calculating natural frequencies and modes which include prestress effects [42] is described in Sec. 7.4.

It is assumed in the subsequent sections that (1) the thickness of the membrane is extremely small compared to the radii of curvature of the membrane; (2) the moment resultants and transverse shears are negligible compared to the planar force resultants; (3) the material is elastic, homogeneous, initially unstrained, and isotropic; (4) the additional displacements are infinitesimal; and (5) the additional strains are linearly elastic.

Tensor calculus is used as the basis for derivation of the field equations: a review of the conventions and rules for algebra and calculus of tensors is given in Appendix B, with special attention given to the differential geometry of curvilinear surfaces. In the subsequent sections the following conventions for notation are adopted: a lowercase Latin italic subscript or superscript is understood to have the range 1, 2, 3 unless specified otherwise; Greek indices denote a range of 1, 2 only on the surface; when an index is repeated—once as a subscript and once as a superscript—summation with respect to that index is implied over the range of the index; indices enclosed in parentheses denote physical components of tensors and the summation convention is suppressed; and vectors are denoted by superimposed bars (unit vectors by superimposed carets).

7.1   Field Equations for General Membranes [38]

Once the configuration of a prestressed membrane has been determined, that configuration can be used as the reference surface $*C$ for any additional deformation. The determination of the nonlinear displacements to obtain $*C$ is the subject of Chap. 8. The additionally deformed membrane is considered as distinct from the previous configurations, especially when equilibrium is considered. A special notation is used for quantities on the additionally deformed middle surface $'C$, e.g., the base vectors of $'C$ are denoted by $(*\overline{A}_\alpha + \varepsilon \overline{A}'_\alpha)$. This implies that $(*\overline{A}_\alpha + \varepsilon \overline{A}'_\alpha)$ differs from the base vectors $*\overline{A}_\alpha$ to $*C$ by an amount $\varepsilon \overline{A}'_\alpha$, where ε is a small parameter which is a function of the added loads.

The in-service phase is the superposition of an additional load vector $\varepsilon \overline{F}'$ on the prestressed membrane. For convenience, the total load vector $(*\overline{F} + \varepsilon \overline{F}')$ is expressed in terms of its components in the directions of $*\overline{A}_\alpha$ and $*\hat{N}$ where $*\hat{N}$ is the unit normal to $*C$. Because of $\varepsilon \overline{F}'$, a displacement vector $\varepsilon \overline{V}'$ is superimposed on $*\overline{V}$, the displacement from the unstressed state to the prestressed state.

Let us assume: (1) the additional displacements are infinitesimal; and (2) the additional deformation is linearly elastic. These assumptions are reasonable since the membrane is stiffened in its desired shape due to the prestressing. Only terms of order ε are retained in the sequel. The succeeding equations include terms of order $(\varepsilon)^0$, which cancel exactly with the equivalent terms in the prestressing equations, and terms of order $(\varepsilon)^1$.

7.1.1   Geometry and strain

The vector $\varepsilon \overline{V}'$ causes $*C$ to deform into $'C$, whose position vector $(*\overline{R} + \varepsilon \overline{R}')$ differs from $*\overline{R}$ by an infinitesimal amount. The vector $\varepsilon \overline{V}'$ is expressible in terms of the $*\overline{A}_\alpha$ and $*\hat{N}$ as

$$\overline{V}' = V'^{\gamma *}\overline{A}_\gamma + V'^{3*}\hat{N} \tag{7.1.1}$$

and

$$\frac{\partial \overline{V}'}{\partial x^\alpha} = U_\alpha'^\gamma * \overline{A}_\gamma + U_\alpha'^3 * \hat{N} \tag{7.1.2}$$

in which V'^γ and V'^3 = the tensor components of $\overline{V}'$, and

$$U_\alpha'^\gamma = V_0'^\gamma|_\alpha - V'^3 B_\alpha^\gamma \tag{7.1.3a}$$

$$U_\alpha'^3 = V'^3|_\alpha + V_0'^\gamma B_{\gamma\alpha} \tag{7.1.3b}$$

where the vertical bar denotes the covariant derivative on $*C$. The base vectors of $'C$ are

$$(*\overline{A}_\alpha + \varepsilon \overline{A}_\alpha') = \frac{\partial(*\overline{R} + \varepsilon \overline{R}')}{\partial x^\alpha} = *\overline{A}_\alpha + \varepsilon \frac{\partial \overline{V}'}{\partial x^\alpha} \tag{7.1.4}$$

Therefore, it can be shown that the terms to be added to the metric tensors of $*C$ to form those of $'C$ are

$$A_{\alpha\gamma}' = *A_{\alpha\rho} U_\gamma'^\rho + *A_{\gamma\rho} U_\alpha'^\rho \tag{7.1.5a}$$

$$A'^{\alpha\gamma} = -*A^{\alpha\rho} U_\rho'^\gamma - *A^{\gamma\rho} U_\rho'^\alpha \tag{7.1.5b}$$

Let $(*A + \varepsilon A')$ denote the determinant of $(*A_{\alpha\gamma} + \varepsilon A_{\alpha\gamma}')$:

$$\frac{A'}{*A} = *A^{\alpha\gamma} A_{\alpha\gamma}' = 2U_\gamma'^\gamma \tag{7.1.5c}$$

The unit normal of $'C$ is $(*\hat{N} + \varepsilon \overline{N}')$; it can be shown that

$$\overline{N}' = -*A^{\gamma\rho} U_\gamma'^3 * \overline{A}_\rho \tag{7.1.6a}$$

Also, from the Gauss-Weingarten relations [see Eqs. (B.36 to 38) of Appendix B]

$$\frac{\partial \overline{N}'}{\partial x^\alpha} = -(*B_\alpha^\gamma U_\gamma'^3 * \hat{N} + A^{\rho\gamma} U_\gamma'^3|_\alpha * \overline{A}_\rho) \tag{7.1.6b}$$

To derive Eqs. (7.1.6), it is necessary to use

$$(\varepsilon_{\alpha\gamma} + \varepsilon \varepsilon_{\alpha\gamma}') = \varepsilon_{\alpha\gamma}(1 + \varepsilon U_\rho'^\rho) \tag{7.1.7a}$$

$$(\varepsilon^{\alpha\gamma} + \varepsilon \varepsilon'^{\alpha\gamma}) = \varepsilon^{\alpha\gamma}(1 - \varepsilon U_\rho'^\rho) \tag{7.1.7b}$$

for the permutation tensors of $'C$. The curvature tensor of $'C$ is $(*B_{\alpha\gamma} + \varepsilon B_{\alpha\gamma}')$, in which

$$B'_{\alpha\gamma} = \tfrac{1}{2}(U'^3_\alpha|_\gamma + U'^3_\gamma|_\alpha + {}^*B_{\alpha\sigma}U'^\sigma_\gamma + {}^*B_{\gamma\sigma}U'^\sigma_\alpha) \tag{7.1.8a}$$

$$B'^\rho_\alpha = {}^*A^{\rho\gamma}U'^3_\gamma|_\alpha - {}^*B^\gamma_\alpha U'^\rho_\gamma$$

Covariant differentiation on $'C$ is expressible in terms of the coordinates on $^*C$. The Christoffel symbols of $'C$ are $({}^*\Gamma^\rho_{\alpha\gamma} + \varepsilon\Gamma'^\rho_{\alpha\gamma})$, in which

$$\Gamma'^\rho_{\alpha\gamma} = \frac{1}{2}{}^*A^{\sigma\rho}\left(\frac{\partial A'_{\sigma\alpha}}{\partial x^\gamma} + \frac{\partial A'_{\sigma\gamma}}{\partial x^\alpha} - \frac{\partial A'_{\alpha\gamma}}{\partial x^\sigma}\right) - {}^*A^{\sigma\rho}A'_{\sigma\beta}{}^*\Gamma^\beta_{\alpha\gamma} \tag{7.1.9a}$$

Two vertical lines denote covariant differentiation on C:

$$({}^*T_\alpha + \varepsilon T'_\alpha)\|_\gamma = {}^*T_\alpha|_\gamma + \varepsilon T'_\alpha\|_\gamma \tag{7.1.9b}$$

in which $T'_\alpha\|_\gamma$ is defined as

$$T'_\alpha\|_\gamma = T'_\alpha|_\gamma - \Gamma'^\rho_{\alpha\gamma}T_\rho \tag{7.1.9c}$$

To obtain the strain tensor of the additionally deformed membrane, $({}^*A_{ij} + \varepsilon A'_{ij})$ is compared to the metric tensor $^0A_{ij}$ of the initially unstrained membrane as

$$2({}^*\gamma_{\alpha\beta} + \varepsilon\gamma'_{\alpha\beta}) = ({}^*A_{\alpha\beta} + \varepsilon A'_{\alpha\beta}) - {}^0A_{\alpha\beta} = 2{}^*\gamma_{\alpha\beta} + \varepsilon A'_{\alpha\beta} \tag{7.1.10}$$

and therefore, by Eq. (7.1.5a)

$$\gamma'_{\alpha\gamma} = {}^*A_{\rho\alpha}U'^\rho_\gamma + {}^*A_{\rho\gamma}U'^\rho_\alpha \tag{7.1.11}$$

7.1.2 Equilibrium

It is not assumed that the stress resultants and the loads of the prestressing phase are simply "carried over" to the new configuration. The components of the stress tensor $(\sigma^{\alpha\beta} + \varepsilon\sigma'^{\alpha\beta})$ are replaced by an equivalent system of force resultant vectors acting on $'C$, namely

$$(\bar{n}^\alpha + \varepsilon\bar{n}'^\alpha) = ({}^*n^{\alpha\gamma} + \varepsilon n'^{\alpha\gamma})\frac{{}^*\bar{A}_\gamma + \varepsilon\bar{A}'_\gamma}{\sqrt{{}^*A^{\alpha\alpha} + \varepsilon A'^{\alpha\alpha}}} \text{ (no sum on } \alpha) \tag{7.1.12}$$

in which $({}^*n^{\alpha\gamma} + \varepsilon n'^{\alpha\gamma}) = $ the force resultant tensors in the $({}^*\bar{A}_\alpha + \varepsilon\bar{A}'_\alpha)$ directions.

The load vector $({}^*\bar{F} + \varepsilon\bar{F}')$ is the force on $'C$ per unit area of $'C$:

$$(\bar{F} + \varepsilon\bar{F}') = [({}^*F^\alpha + \varepsilon F'^\alpha)({}^*\bar{A}_\alpha + \varepsilon\bar{A}'_\alpha)$$

$$+ ({}^*F^3 + \varepsilon F'^3)({}^*\hat{N} + \varepsilon\bar{N}')] \tag{7.1.13}$$

in which $(*F^\alpha + \varepsilon F'^\alpha)$ and $(*F^3 + \varepsilon F'^3)$ = the tensor components of $(\overline{F} + \varepsilon \overline{F'})$.

Considering a differential element of the membrane, we find that equilibrium of forces in terms of Taylor expansions of the force resultants

$$\sqrt{*A} \, *n^{\alpha\gamma} *\overline{A}_\gamma \qquad (\text{no sum, } \alpha \neq \gamma)$$

acting on the differential sides of the element, and of the force $(*\overline{F} + \varepsilon \overline{F'}) \sqrt{*A + \varepsilon A'} \, dx^1 \, dx^2$ acting on the surface area of the element will yield equilibrium equations of the form $(\varepsilon)^0 \phi_1 + (\varepsilon)^1 \phi_2 = 0$. The $(\varepsilon)^0$ terms cancel exactly with the equilibrium equations for the prestressing phase (considered in Chap. 8). Thus, the in-service equilibrium equations are

$$(n'|_\alpha^{\alpha\gamma} + F'^\gamma) + (*n^{\rho\gamma}\Gamma'^\alpha_{\rho\alpha} + *n^{\alpha\rho}\Gamma'^\gamma_{\rho\alpha}) = 0 \qquad (7.1.14)$$

$$(n'^{\alpha\gamma} *B_{\alpha\gamma} + F'^3) + (*n^{\alpha\gamma} B'_{\alpha\gamma}) = 0 \qquad (7.1.15)$$

The first group of terms in Eqs. (7.1.14) and (7.1.15) represents the effects of the changes in the stress resultants. The second group of terms shows the effects of the changes in geometry interacting with the original stress resultants.

7.1.3 Stress-strain

In forming the stress-strain relations the total stresses and strains in the additionally deformed membrane must be considered—not just their changes. Since the material is perfectly elastic, it can be seen that the stress tensor of the additionally deformed membrane is

$$(*\sigma^{\alpha\beta} + \varepsilon\sigma'^{\alpha\beta}) = \frac{1}{2}\frac{\sqrt{*A}}{(A + \varepsilon A')}\left[\frac{\partial *U}{\partial(*\gamma_{\alpha\beta} + \varepsilon\gamma'_{\alpha\beta})} + \frac{\partial *U}{\partial(*\gamma_{\alpha\beta} + \varepsilon\gamma'_{\alpha\beta})}\right] \tag{7.1.16}$$

in which $*U$ = the strain energy per unit volume of the new membrane and where $(*A + \varepsilon A')$ = the determinant of $(*A_{\alpha\beta} + \varepsilon A'_{\alpha\beta})$. If it is assumed that the additional deformation is linearly elastic and that the material is isotropic, homogeneous, and initially unstrained, then

$$\sigma'^{\alpha\gamma} = \left\{\frac{E}{2(1 + \nu)}\left[*A^{\alpha\beta} *A^{\gamma\rho} + *A^{\alpha\rho} *A^{\gamma\beta}\right.\right.$$

$$\left.\left. + \frac{2\nu}{(1 - \nu)} *A^{\alpha\gamma} *A^{\beta\rho}\right] - *A^{\beta\rho} *\sigma^{\alpha\gamma}\right\}\gamma'_{\beta\rho} \tag{7.1.17}$$

Substituting Eq. (7.1.11) into Eq. (7.1.17) and integrating through the thickness h, we obtain the relations between the additions to the force resultants and the additional displacements:

$$n'^{\alpha\gamma} = \frac{Eh}{2(1+\nu)} \left[{}^*A^{\alpha\rho} U_\rho'^\gamma + {}^*A^{\gamma\rho} U_\rho'^\alpha \right.$$

$$\left. + \frac{2\nu}{(1-\nu)} {}^*A^{\alpha\gamma} U_\rho'^\rho \right] - {}^*n^{\alpha\gamma} U_\rho'^\rho \quad (7.1.18)$$

7.1.4 Physical components

The physical components of the vectors on C are the magnitudes in the directions of their unit reference vectors and are denoted by enclosing subscripts in parentheses. From Eq. (7.1.1)

$$\bar{V}' = \sum_\gamma V_{(\gamma)}' \frac{{}^*\bar{A}_\gamma}{\sqrt{{}^*A_{\gamma\gamma}}} + V_{(3)}' {}^*\hat{N} \quad (7.1.19)$$

in which the physical components of $\bar{V}'$ are

$$V_{(\gamma)}' = V'^\gamma \sqrt{{}^*A_{\gamma\gamma}} \text{ (no sum)} \qquad V_{(3)}' = V'^3 \quad (7.1.20)$$

Similarly, from Eq. (7.1.12)

$$(\bar{n}^\alpha + \varepsilon\bar{n}^\alpha) = \sum_\gamma [{}^*n_{(\alpha\gamma)} + \varepsilon n_{(\alpha\gamma)}'] \frac{{}^*\bar{A}_\gamma + \varepsilon\bar{A}_\gamma'}{\sqrt{{}^*A_\gamma + \varepsilon A_{\gamma\gamma}'}} \quad (7.1.21)$$

in which the physical components $n_{(\alpha\gamma)}'$ are

$$n_{(\alpha\gamma)}' = \left[n'^{\alpha\gamma} + \frac{1}{2} \left(\frac{A_{\gamma\gamma}'}{{}^*A_{\gamma\gamma}} - \frac{A'^{\alpha\alpha}}{{}^*A^{\alpha\alpha}} \right) {}^*n^{\alpha\gamma} \right] \sqrt{\frac{{}^*A_{\gamma\gamma}}{{}^*A^{\alpha\alpha}}} \text{ (no sum)} \quad (7.1.22)$$

Finally, the physical components of the load vector are found from

$$(\bar{F} + \varepsilon\bar{F}') = \left[\sum_\gamma ({}^*F_{(\gamma)} + \varepsilon F_{(\gamma)}') \frac{{}^*\bar{A}_\gamma + \varepsilon\bar{A}_\gamma'}{\sqrt{{}^*A_{\gamma\gamma} + \varepsilon A_{\gamma\gamma}'}} \right.$$

$$\left. + ({}^*F_{(3)} + \varepsilon F_{(3)}')({}^*\hat{N} + \varepsilon\bar{N}') \right] \quad (7.1.23)$$

From Eq. (7.1.13),

$$F_{(\gamma)}' = \sqrt{{}^*A_{\gamma\gamma}} \left(F'^\gamma + \frac{1}{2} \frac{A_{\gamma\gamma}'}{{}^*A_{\gamma\gamma}} {}^*F^\gamma \right) \text{ (no sum)}, \ F_{(3)}' = F'^3 \quad (7.1.24)$$

The field equations have been developed for the in-service phase of a membrane of arbitrary shape. Only two assumptions have been made regarding the additional displacements: (1) the additional displacements are infinitesimal; and (2) the behavior during the additional deformation is linearly elastic. These restrictions could be removed. A finite additional displacement could be approximated by piecing together successive small displacements.

7.2 Axisymmetrically Loaded Membranes of Revolution

In this section the field equations are specialized for membranes of revolution with axisymmetric loads, the equations are reduced by means of substitutions into the equilibrium equations, and a method of solution is presented. Details of the definition of geometric quantities on a surface of revolution are given in Sec. 6.4.1. Let $x = x^1$ be some regular coordinate along the meridian (Fig. 6.4.1) and let x^2 be the azimuth angle θ. Therefore, using results from Sec. 6.4.1 and 7.1.1, we obtain the nonzero geometric quantities as

$$U_1'^1 = \frac{dV'^1}{dx} + \frac{1}{2} {}^*A^{11} \frac{dA_{11}}{dx} V'^1 - \frac{1}{R_1} V'^3 \tag{7.2.1a}$$

$$U_2'^2 = \frac{1}{r} V'^1 \frac{dr}{dx} - \frac{1}{R_2} V'^3 \tag{7.2.1b}$$

$$U_1'^3 = \frac{d}{dx}(V'^3) + {}^*A_{11} \frac{V'^1}{R_1} \tag{7.2.1c}$$

$$A_{11}' = 2 {}^*A_{11} U_1'^1 \tag{7.2.2a}$$

$$A_{22}' = 2r^2 U_2'^2 \tag{7.2.2b}$$

$$B_1'^1 = {}^*A^{11} U_1'^3|_1 - \frac{U_1'^1}{R_1} \tag{7.2.3a}$$

$$B_2'^2 = \frac{U_2'^3|_2}{r^2} - \frac{U_2'^2}{R_2} \tag{7.2.3b}$$

$$\Gamma_{11}'^1 = \frac{d}{dx}(U_1'^1) \qquad \Gamma_{12}'^2 = \frac{d}{dx}(U_2'^2) \tag{7.2.4a}$$

$$\Gamma_{22}'^1 = 2 {}^*A^{11}(U_1'^1 - U_2'^2) \frac{r\,dr}{dx} - {}^*A^{11}r^2 \frac{d}{dx}(U_2'^2) \tag{7.2.4b}$$

in which $U_1^{'3}|_1$ and $U_2^{'3}|_2$ can be evaluated by taking the covariant derivatives of Eq. (7.1.3b).

If Eqs. (7.1.17) and (7.2.1) to (7.2.4) are substituted into Eqs. (7.1.14) and (7.1.15), the equilibrium equations can be obtained in terms of the displacements. These will be two ordinary second-order differential equations. For convenience in satisfying boundary conditions the equations should be in terms of physical components. Substituting Eqs. (7.2.1) to (7.3.5) into Eq. (7.1.17) and subsequently into Eq. (7.1.22), we obtain

$$
n'_{(11)} = \left\{ V_{(1)} \frac{1}{r\sqrt{*A_{11}}} \frac{dr}{dx} \left[\frac{Eh}{(1-\nu^2)} \nu - n_{(11)} \right] - V_{(3)} \left[\frac{Eh}{1-\nu^2} \frac{R_2 + \nu R_1}{R_1 R_2} \right. \right.
$$

$$
\left. + n_{(11)} \left(\frac{1}{R_1} - \frac{1}{R_2} \right) \right] + \frac{dV_{(1)}}{dx} \frac{1}{\sqrt{*A_{11}}} \left(\frac{Eh}{1-\nu^2} + n_{(11)} \right) \right\}
$$

$$\text{(7.2.5)}$$

$$
n'_{(22)} = \left\{ V_{(1)} \frac{1}{r\sqrt{*A_{11}}} \frac{dr}{dx} \left(\frac{Eh}{1-\nu^2} + n_{(22)} \right) - V_{(3)} \left[\frac{Eh}{1-\nu^2} \frac{R_1 + \nu R_2}{R_1 R_2} \right. \right.
$$

$$
\left. - n_{(22)} \left(\frac{1}{R_1} - \frac{1}{R_2} \right) \right] + \frac{dV_{(1)}}{dx} \frac{1}{\sqrt{*A_{11}}} \left(\frac{Eh}{1-\nu^2} \nu + n_{(22)} \right) \right\}
$$

$$\text{(7.2.6)}$$

With all of the above substitutions, the physical forms of the equilibrium equations (7.1.14) and (7.1.15) (the equilibrium equations of the prestressing phase are used to eliminate and condense terms) become

$$
V'_{(1)} f1(x) - V'_{(3)} f2(x) + \frac{dV'_{(1)}}{dx} f3(x) - \frac{dV'_{(3)}}{dx} f4(x)
$$

$$
+ \frac{d^2 V'_{(1)}}{dx^2} f5(x) + F'_{(1)} = 0 \quad \text{(7.2.7a)}
$$

$$
V'_{(1)} h1(x) - V'_{(3)} h2(x) + \frac{dV'_{(1)}}{dx} h3(x) - \frac{dV'_{(3)}}{dx} h4(x)
$$

$$
+ \frac{d^2 V'_{(3)}}{dx^2} h5(x) + \sqrt{A_{11}} \, F'_{(3)} = 0 \quad \text{(7.2.7b)}
$$

in which $f1(x)$ to $f5(x)$ and $h1(x)$ to $h5(x)$ = the following functions of x:

$$
f1(x) = \frac{Eh *A^{11}}{1-\nu^2} \left[\frac{1}{r} \frac{dr}{dx} \left(\frac{\nu}{h} \frac{dh}{dx} - \frac{\nu}{2} *A^{11} \frac{d *A_{11}}{dx} - \frac{1}{r} \frac{dr}{dx} \right) + \frac{\nu}{r} \frac{d^2 r}{dx^2} \right]
$$

$$
+ n_{(22)} *A^{11} \left[\frac{1}{r} \frac{dr}{dx} \left(\frac{*A^{11}}{2} \frac{d *A_{11}}{dx} - \frac{1}{r} \frac{dr}{dx} \right) - \frac{1}{r} \frac{d^2 r}{dx^2} \right] + \frac{1}{\sqrt{*A_{11}}} \frac{1}{r} \frac{dr}{dx} F_{(1)}
$$

$$\text{(7.2.8a)}$$

$$
\begin{aligned}
f2(x) = \frac{Eh^*A^{11}}{1 - \nu^2} & \left[\frac{1}{h}\frac{dh}{dx}\left(\frac{1}{R_1} + \frac{\nu}{R_2} \right) + \frac{1}{r}\frac{dr}{dx}(1 - \nu)\left(\frac{1}{R_1} - \frac{1}{R_2} \right) \right. \\
& \left. - \left(\frac{1}{R_1^2}\frac{dR_1}{dx} + \frac{\nu}{R_2^2}\frac{dR_2}{dx} \right) \right] \sqrt{^*A_{11}} + \frac{n_{(22)}}{\sqrt{^*A_{11}}} \left\{ \left[\frac{2}{r}\frac{dr}{dx}\left(\frac{1}{R_1} - \frac{1}{R_2} \right) \right. \right. \\
& \left. \left. + \frac{1}{R_2^2}\frac{dR_2}{dx} \right] \right\} + \frac{n_{(11)}}{\sqrt{^*A_{11}}}\frac{1}{R_1^2}\frac{dR_1}{dx^1} + \frac{F_{(1)}}{R_2}
\end{aligned}
$$

$$\text{(7.2.8b)}$$

$$
\begin{aligned}
f3(x) = \frac{Eh^*A^{11}}{1 - \nu^2} & \left(\frac{1}{h}\frac{dh}{dx^1} + \frac{1}{r}\frac{dr}{dx} - \frac{^*A^{11}}{2}\frac{d^*A_{11}}{dx} \right) \\
& + \left(^*A^{11}n_{(22)}\frac{1}{r}\frac{dr}{dx} - ^*A^{11}n_{(11)}\frac{^*A^{11}}{2}\frac{d^*A_{11}}{dx} \right) \quad \text{(7.2.8c)}
\end{aligned}
$$

$$
f4(x) = \left(\frac{Eh^*A^{11}}{1 - \nu^2}\frac{R_2 + \nu R_1}{R_1 R_2} + \frac{n_{(11)}}{^*A_{11}}\frac{1}{R_1} - \frac{n_{(22)}}{^*A_{11}}\frac{1}{R_2} \right)\sqrt{^*A_{11}} \quad \text{(7.2.8d)}
$$

$$
f5(x) = \frac{Eh^*A^{11}}{1 - \nu^2} + ^*A^{11}n_{(11)} \quad \text{(7.2.8e)}
$$

and

$$
h1(x) = \frac{Eh}{1 - \nu^2}\frac{1}{r}\frac{dr}{dx}\frac{R_1 + \nu R_2}{R_1 R_2} - \frac{n_{(11)}}{R_1}\left(\frac{1}{R_1}\frac{dR_1}{dx} + \frac{1}{r}\frac{dr}{dx} \right) + \frac{n_{(22)}}{R_1}\frac{1}{r}\frac{dr}{dx}
$$

$$\text{(7.2.9a)}$$

$$
\begin{aligned}
h2(x) = \frac{Eh}{1 - \nu^2} & \left(\frac{R_1 + \nu R_1}{R_1} + \frac{R_1 + \nu R_2}{R_2} \right)\frac{\sqrt{^*A_{11}}}{R_1 R_2} \\
& - \frac{n_{(22)}\sqrt{^*A_{11}}}{R_1 R_2} - \frac{n_{(11)}\sqrt{^*A_{11}}}{R_1 R_2} \quad \text{(7.2.9b)}
\end{aligned}
$$

$$
h3(x) = \frac{Eh}{1 - \nu^2}\frac{R_2 + \nu R_1}{R_1 R_2} - \frac{n_{(22)}}{R_2} + \frac{n_{(11)}}{R_1} \quad \text{(7.2.9c)}
$$

$$
h4(x) = \frac{n_{(11)}}{\sqrt{^*A_{11}}}\frac{^*A^{11}}{2}\frac{d^*A_{11}}{dx} - \frac{n_{(22)}}{\sqrt{^*A_{11}}}\frac{1}{r}\frac{dr}{dx} \quad \text{(7.2.9d)}
$$

$$
h5(x) = \frac{n_{(11)}}{\sqrt{^*A_{11}}} \quad \text{(7.2.9e)}
$$

7.2.1 Solution procedure

In general, the solution to an nth-order linear boundary-value problem, e.g., Eqs. (7.2.7) to (7.2.9), can be obtained as a linear combination of the solutions to $m + 1$ initial-value problems, where $m < n$. The constants of the combination $\pi_{(j)}$, $j = 1, 2, \ldots, m$ are determined from a set of m linear algebraic equations.

The Newmark method (as described in Sec. 5.1.3) was used to solve each of the set of initial-value problems numerically at discrete points $i = 1, 2, \ldots, n$ along the meridian. The Newmark method determines the displacements and their derivatives at the end of a constant increment $H = x_{i+1} - x_i$ in terms of the displacements and their first two derivatives at the beginning of the increment, and in terms of the assumed values of the second derivatives at the end of the increment. Since initial values of the second derivatives are assumed at the end of the increment, it can be seen that within each increment the displacements are determined by successive approximations.

The algorithm [38] for the solution of the boundary-value problem posed by Eqs. (7.2.7) to (7.2.9) is as follows:

1. Values of $_1V'_{(1)}$, $_1V'_{(3)}$, $d_1V'_{(1)}/dx$, and $d_1V'_{(3)}/dx$ are selected so as to satisfy the boundary conditions at the initial point. The leading subscript i on a symbol, e.g., $_1V'_{(1)}$, denotes the value of the symbol at the ith point on the meridian. Then, $d_1^2V'_{(1)}/dx^2$ and $d_1^2V'_{(3)}/dx^2$ are found from Eqs. (7.2.7) to (7.2.9) with the loads applied. Note that m of these initial values is arbitrary, where m is the number of boundary conditions at the last point on the meridian.

2. Newmark's method is applied to this initial-value problem.

3. The residuals $\Gamma_{(j)}$, $j = 1, \ldots, m$ of the boundary conditions at $i = n$ are calculated.

4. If at least one of these boundary conditions is not satisfied, m $(= 2)$ independent sets of assumptions of $_1V'_{(1)}$, $_1V'_{(3)}$, $d_1V'_{(1)}/dx$, $d_1V'_{(3)}$ dx, $d_1^2V'_{(1)}/dx^2$, and $d_1^2V'_{(3)}/dx^2$ are assumed which satisfy the boundary conditions and homogeneous differential equations at the initial point.

5. Newmark's method is applied for each of m independent initial-value problems.

6. The residuals $\theta_{(j,k)}$, j, $k = 1, 2, \ldots, m$, in the boundary conditions at $i = n$ are calculated.

7. The following matrix equation is solved for the correction factors $\Pi_{(k)}$, $k = 1, \ldots, m$:

$$[\theta_{(j,k)}]\{\Pi_{(k)}\} + \{\Gamma_{(j)}\} = 0 \qquad (7.2.10)$$

8. Then, the correct values of $_iV'_{(1)}$ and $_iV'_{(3)}$ at each point on the meridian are

$$_i V'_{(1)} = {}^{(0)}_i V'_{(1)} + \sum_{j=1}^{m} \Pi_{(j)} {}^{(j)}_i V'_{(1)} \tag{7.2.11a}$$

$$_i V'_{(3)} = {}^{(0)}_i V'_{(3)} + \sum_{j=1}^{m} \Pi_{(j)} {}^{(j)}_i V'_{(3)} \tag{7.2.11b}$$

in which ${}^{(j)}_i V'_{(1)}$ and ${}^{(j)}_i V'_{(3)}$ = the last iterates of $_i V'_{(1)}$ and $_i V'_{(3)}$ for each of the m homogeneous initial-value problems solved in step 5. Also, $_i V'_{(1)}$ and $_i V'_{(3)}$ denote the last iterates for the particular solution as found in step 2.

7.2.2 Boundary conditions

To completely specify a particular membrane problem, it is necessary to stipulate the in-service phase boundary conditions. This is more difficult than for the prestressing phase because the in-service phase behavior is not truly that of a membrane. In the prestressing phase the desired final shape could be obtained while still satisfying the membrane assumption by introducing external ties. However, it is not possible to add extra ties during the in-service phase. Consequently, the membrane assumption introduces small errors which are usually made evident mathematically by the presence of an extra condition on the normal or horizontal equilibrium at an edge of the membrane, or physically by wrinkling of the membrane near the supports [173, 174].

In the in-service phase sample problems that have been considered, these extra conditions have been ignored and it is shown subsequently that the errors which are introduced are small and self-equilibrating. To be rigorous, moment equilibrium at the edge of the shell should be considered. Since the effects of the moments—small compared to the force resultants—are important only at the edges of the shell, moment equilibrium has not been studied.

It can be shown that Eq. (7.2.7a) has the form

$$\frac{d}{dx} [L(V'_{(1)}, V'_{(3)})] = 0 \tag{7.2.12}$$

in which $L(V'_{(1)}, V'_{(3)})$ is the first-order integrodifferential equation of vertical equilibrium. Therefore, Eqs. (7.2.7) are equivalent to a third-order system. However, it is preferred to solve Eq. (7.2.11) rather than the equation $L(V'_{(1)}, V'_{(3)}) = 0$. Therefore, in addition to the three boundary conditions needed for the third-order system, the extra condition that $L = 0$ be satisfied on an edge of the membrane is necessary. The requirement that the virtual work of the boundary forces be zero often yields a boundary condition on vertical equilibrium at a particular edge of the shell. However, it can be seen from the above discussion that if this case occurs, it is only necessary to satisfy vertical equilibrium at any convenient point.

7.2.3 Sample problems

The numerical solution method outlined above has been tested on various problems. The in-service phase behavior of a sphere is discussed here. The thickness is constant and the material properties are $E = 300,000$ psi (20.7 GPa); $\nu = 0.4$. The addition $\varepsilon \overline{F}'$ to the load vector $\overline{F}$ is here denoted by $\overline{P}$. If $\overline{P}$ is applied to a region R' of the membrane, a reasonable measure of the parameter ε is

$$\varepsilon' = \frac{\int_{R'} \overline{P} \, d\xi}{\int_{R'} \overline{F} \, d\xi} \tag{7.2.13a}$$

Therefore,

$$\overline{F} = \frac{\overline{P}}{\varepsilon} \tag{7.2.13b}$$

The x coordinate for the spherical segment of radius R shown in Fig. 7.2.1 is chosen as the angle ϕ between the normal and the axis of revolution. Therefore,

$$R_1 = R_2 = R \qquad *A_{11} = R^2 \qquad r = R \sin \phi \tag{7.2.14}$$

The segment is restrained at the edge $\phi = \phi_0$ such that no additional vertical displacement $\varepsilon \Delta_V'$ is allowed. The boundary conditions are

$$\Delta_V'|_{\phi=\phi_0} = -(V'_{(1)} \sin \phi_0 + V'_{(3)} \cos \phi_0) = 0 \tag{7.2.15a}$$

$$V'_{(1)}|_{\phi=0} = 0 \tag{7.2.15b}$$

$$\frac{dV'_{(3)}}{dx}\bigg|_{\phi=0} = 0 \tag{7.2.15c}$$

$$\frac{dV'_{(1)}}{dx}\bigg|_{\phi=0} = 0 \tag{7.2.15d}$$

Equation (7.2.15b) is equivalent to the condition that $n'_{(11)}$ and $n'_{(22)}$ be finite at $\phi = 0$. Equation (7.2.15c) is used to guarantee slope continuity at $\phi = 0$. Vertical equilibrium at $\phi = 0$ yields Eq. (7.2.15d). It can be seen that an auxiliary condition on horizontal equilibrium at $\phi = \phi_0$ should be satisfied. This nonmembrane condition is ignored, and the small error that is introduced is discussed later.

Equations (7.2.7), (7.2.9), and (7.2.15) are used to derive the "initial conditions." It can be seen that one of the initial values is arbitrary. For the particular solution (loads applied), let the arbitrary condition be ${}^{(0)}_1 V_{(3)} = 0$. For the homogeneous case (no applied loads), let the arbitrary condition be ${}^{(1)}_1 V_{(3)} = 1$. Newmark's method can be used to obtain the solutions to each of the initial-value problems. Equation (7.2.15a) yields the residuals.

In Table 7.2.1 the results are presented for a sample problem of an over-pressure superimposed on a prestressed sphere (Fig. 7.2.1a) in which $h = 0.002$ ft (0.61 mm), $p = 400$ psf (19.2 kPa), the sphere radius $R = 10$ ft (3.05 m), and $\phi_0 = \pi/4$. The value of ε was taken as 0.1. Therefore, $\varepsilon P_{(3)i} = p/10$ and $P_{(1)i}' = 0$ for $i = 1, 2, \ldots, n$. In Table 7.2.2 are the results of a problem of a strip load of $p/10$ superimposed (Fig. 7.2.1b) on the same pressurized sphere at $\phi = \phi_0$, i.e., $\varepsilon P_{(3)n}' = p/10$.

From the results in Table 7.2.1, the magnitude of the error introduced by neglecting the horizontal equilibrium condition at $\phi = \phi_0$ can be esti-mated. Membrane behavior during the prestressing phase is guaranteed by introducing "spokelike" ties at $\phi = \phi_0$. The stiffness of these ties is chosen such that the elongation of the ties is exactly equal to the horizon-tal displacement Δ_H at $\phi = \phi_0$ during the prestressing phase. The stiffness is therefore

$$A_t E_t = \frac{F_t}{\Delta_H} \tag{7.2.16}$$

where E_t = modulus of elasticity of the ties
A_t = area of ties per unit length of circumference
F_t = force in ties during prestressing phase

Once chosen, the stiffness of the ties cannot be changed during the in-service phase. The in-service load on the ties is denoted by $(F_t' + \varepsilon F_t')$, in which

$$F_t' = \Delta_H' A_t E_t \tag{7.2.17}$$

From Fig. 7.2.1a, the horizontal equilibrium equation at $\phi = \phi_0$ is

$$(n_{(11)} + \varepsilon n_{(11)}') \cos (\phi_0 + \varepsilon \phi_0') - (F_t + \varepsilon F_t') = 0 \tag{7.2.18}$$

in which ϕ_0' can be shown to be

$$\phi_0' = \frac{V_{(1)}' + dV_{(3)}'/dx}{R} \tag{7.2.19}$$

TABLE 7.2.1 Sphere with an Additional Overpressure

ϕ, rad (1)	$\varepsilon V'_{0(1)} \times 10^2$ (2)	$\varepsilon V'_{0(3)} \times 10^2$ (3)	$\varepsilon \dfrac{d}{d\phi}(V'_{0(3)}) \times 10^2$ (4)	$n'_{(11)}$, lb/ft (5)	$n_{(11)} + \varepsilon n'_{(11)}$, lb/ft (6)	$n'_{(22)}$, lb/ft (7)	$n_{(22)} + \varepsilon n'_{(22)}$, lb/ft (8)
0.000	0.000	−1.397	0.000	2011.680	2201.168	2011.680	2201.168
0.039	0.000	−1.396	0.040	2011.190	2201.119	2010.837	2201.084
0.079	0.000	−1.394	0.066	2010.812	2201.081	2009.569	2200.957
0.118	0.000	−1.391	0.100	2010.400	2201.040	2007.731	2200.773
0.157	0.001	−1.387	0.140	2010.311	2201.031	2005.497	2200.550
0.196	0.001	−1.380	0.187	2008.015	2200.801	2000.169	2200.017
0.236	0.002	−1.372	0.245	2006.814	2200.681	1994.850	2199.485
0.275	0.003	−1.361	0.315	2004.423	2200.442	1986.946	2198.695
0.314	0.005	−1.347	0.404	2001.625	2200.162	1976.838	2197.684
0.353	0.008	−1.329	0.515	1997.981	2199.798	1963.573	2196.357
0.393	0.011	−1.306	0.655	1993.372	2199.337	1946.318	2194.632
0.432	0.015	−1.277	0.833	1988.063	2198.806	1924.481	2192.448
0.471	0.021	−1.240	1.060	1981.476	2198.148	1896.198	2189.620
0.511	0.029	−1.193	1.349	1973.517	2197.352	1859.888	2185.298
0.550	0.039	−1.133	1.720	1963.768	2196.377	1812.981	2181.298
0.589	0.052	−1.056	2.195	1951.022	2195.102	1751.762	2175.176
0.628	0.069	−0.959	2.804	1937.543	2193.754	1674.797	2167.480
0.668	0.091	−0.834	3.586	1919.719	2191.972	1574.106	2157.411
0.707	0.120	−0.674	4.592	1898.177	2189.818	1444.115	2144.411
0.746	0.157	−0.470	5.888	1871.179	2187.118	1276.059	2127.606
0.785	0.206	−0.207	7.556	1840.345	2184.034	1059.527	2105.953

NOTE: 1 lb/ft = 14.6 N/m.

From Table 7.2.1

$$(n_{(11)} + \varepsilon n'_{(11)}) \simeq 2184 \text{ lb/ft } (31{,}886 \text{ N/m}) \qquad (7.2.20a)$$

$$\frac{dV'_{(3)n}}{dx} + \frac{dV'_{(1)n}}{dx} \simeq 0.076 \qquad (7.2.20b)$$

$$\Delta'_{Hn} = -V'_{(3)n} \sin \phi_0 + V'_{(1)n} \cos \phi_0 \simeq 0.029 \qquad (7.2.20c)$$

If the representative values $A_t = 0.01 \text{ in}^2/\text{ft}$ and $E_t = 30 \times 10^5$ psi (2.07 $\times 10^4$ MPa) are used, it can be shown that the horizontal equilibrium condition is in error by approximately 100 lb/ft (1460 N/m) of circumference. The dependency of the magnitude of error on the magnitude of $\varepsilon \overline{F}'$ is small, i.e., $n'_{(11)} \ll n_{(11)}$, and the residual in the horizontal equilibrium equation should be compared to the actual forces acting on the edge. Since the residual of 100 lb/ft (1460 N/m) is small compared to the smallest

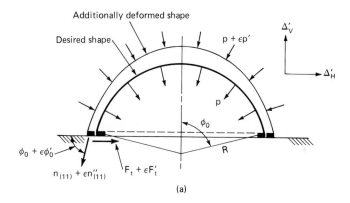

(a)

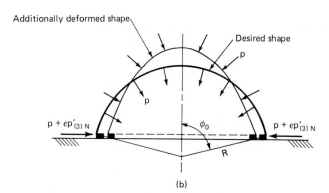

(b)

Figure 7.2.1 In-service loading of sphere with ring support. (a) Overpressure superimposed; (b) ring load superimposed.

TABLE 7.2.2 Sphere with an Additional Strip Load

ϕ, rad (1)	$\varepsilon V'_{0(1)} \times 10^2$ (2)	$\varepsilon V'_{0(3)} \times 10^2$ (3)	$\varepsilon \dfrac{d}{d\phi}(V_{0(3)}) \times 10^2$ (4)	$n'_{(11)}$, lb/ft (5)	$n_{(11)} + \varepsilon n'_{(11)}$, lb/ft (6)	$n'_{(22)}$, lb/ft (7)	$n_{(22)} + \varepsilon n'_{(22)}$, lb/ft (8)
0.000	0.000	-0.000	0.000	0.004	2000.000	0.004	2000.000
0.079	-0.000	-0.000	-0.000	0.004	2000.000	0.005	2000.000
0.157	-0.000	-0.000	-0.000	0.005	2000.000	0.006	2000.001
0.236	-0.000	-0.000	-0.000	0.006	2000.001	0.009	2000.001
0.314	-0.000	-0.000	-0.000	0.007	2000.001	0.014	2000.001
0.393	-0.000	-0.000	-0.000	0.009	2000.001	0.023	2000.002
0.471	-0.000	-0.000	-0.000	0.013	2000.001	0.037	2000.004
0.550	-0.000	-0.000	-0.000	0.017	2000.002	0.060	2000.006
0.628	-0.000	-0.000	-0.001	0.024	2000.002	0.099	2000.010
0.707	-0.000	-0.000	-0.001	0.033	2000.003	0.163	2000.016
0.785	-0.000	-0.000	-0.002	0.047	2000.005	0.270	2000.027
0.864	-0.000	-0.001	-0.003	0.067	2000.007	0.447	2000.045
0.942	-0.000	-0.001	-0.006	0.096	2000.010	0.745	2000.074
1.021	-0.000	-0.001	-0.010	0.134	2000.013	1.242	2000.124
1.100	-0.000	-0.002	-0.016	0.188	2000.019	2.080	2000.208
1.178	-0.001	-0.004	-0.026	0.255	2000.025	3.487	2000.349
1.257	-0.001	-0.007	-0.044	0.338	2000.034	5.866	2000.587
1.335	-0.002	-0.011	-0.074	0.425	2000.042	9.886	2000.989
1.414	-0.004	-0.019	-0.124	0.466	2000.047	15.597	2001.670
1.492	-0.006	-0.032	-0.209	0.421	2000.042	28.300	2002.830
1.571	0.000	0.151	7.450	147.450	2014.715	-18.340	1991.866

NOTE: 1 lb/ft = 14.6 N/m.

force acting on the edge, since the error is self-equilibrating, and since the effects of this error die out rapidly as the distance from the edge increases, the nonmembrane boundary condition on horizontal equilibrium can be neglected.

7.3 Nonsymmetrically Loaded Membranes of Revolution [170]

In this section the possibility of nonsymmetric loadings on the symmetrical geometry of the membrane of revolution is added to the results of the previous section. Also, a simplified form of the governing partial differential equations is examined in which the effects of prestressing are approximated.

Although the geometry of the prestressed shell is symmetric, the geometric quantities after additional loading are functions of the azimuth angle θ as well as meridional coordinate x. Thus, Eqs. (7.2.1) through (7.2.4) of Sec. 7.2 need to be rewritten as

$$U_1'^1 = \frac{\partial V^1}{\partial x} + \frac{1}{2} {}^*A^{11} \frac{\partial {}^*A_{11}}{\partial x} V^1 - \frac{V^3}{R_1} \qquad (7.3.1a)$$

$$U_2'^2 = \frac{\partial V^2}{\partial \theta} + \frac{1}{r} \frac{\partial r}{\partial x} V^1 - \frac{V^3}{R_2} \qquad (7.3.1b)$$

$$U_2'^1 = \frac{\partial V^1}{\partial \theta} - {}^*A^{11} r \frac{\partial r}{\partial x} V^2 \qquad (7.3.1c)$$

$$U_1'^2 = \frac{\partial V^2}{\partial x} + \frac{1}{r} \frac{\partial r}{\partial x} V^2 \qquad (7.3.1d)$$

$$A_{11}' = 2 {}^*A_{11} U_1'^1 \qquad (7.3.2a)$$

$$A_{22}' = 2r^2 U_2'^2 \qquad (7.3.2b)$$

$$A_{12}' = A_{21}' = {}^*A_{11} U_2'^1 + r^2 U_1'^2 \qquad (7.3.2c)$$

$$B_1'^1 = A^{11} U_1'^3|_1 - \frac{1}{R_1} U_1'^1 \qquad (7.3.3a)$$

$$B_2'^2 = \frac{1}{r^2} = U_2'^3|_2 - \frac{1}{R_2} U_2'^2 \qquad (7.3.3b)$$

$$B_2'^1 = {}^*A^{11} U_1'^3|_2 - \frac{1}{R_2} U_2'^1 \qquad (7.3.3c)$$

$$B_1'^2 = \frac{1}{r^2} U_2'^3|_1 = \frac{1}{R_2} U_1'^2 \tag{7.3.3d}$$

$$\Gamma_1' = \frac{\partial U_1'^1}{\partial x} \tag{7.3.4a}$$

$$\Gamma_{12}'^2 = \Gamma_{21}'^2 = \frac{\partial U_2'^2}{\partial x} \tag{7.3.4b}$$

$$\Gamma_{22}'^1 = 2*A^{11}r\frac{\partial r}{\partial x}(U_1'^1 - U_2'^2) - *A^{11}r^2\frac{\partial U_2'^2}{\partial x} \tag{7.3.4c}$$

$$\Gamma_{12}'^1 = \Gamma_{21}'^1 = -\frac{1}{r}\frac{\partial r}{\partial x}(U_2'^1 + *A^{11}r^2 U_1'^2) \tag{7.3.4d}$$

$$\Gamma_{11}'^2 = \frac{1}{2}\frac{\partial *A_{11}}{\partial x}\left(\frac{1}{r^2}U_2'^1 - *A^{11}U_1'^2\right) + \frac{2}{r}\frac{\partial r}{\partial x}U_1'^2 \tag{7.3.4e}$$

$$+ \frac{*A_{11}}{r^2}\left(\frac{\partial U_2'^1}{\partial x} - \frac{\partial U_1'^1}{\partial \theta}\right) + \frac{\partial U_1'^2}{\partial x} \tag{7.3.4f}$$

$$\Gamma_{22}'^{12} = \frac{\partial U_2'^2}{\partial \theta} + \frac{1}{r}\frac{\partial r}{\partial x}(U_2'^1 + *A^{11}r^2 U_1'^2) \tag{7.3.4g}$$

With the above expressions, the nonsymmetric forms of the equilibrium equations replacing Eqs. (7.2.7) to (7.2.9) of the previous section are determined as

$$\left[\left[\frac{\partial^2 V_{(1)}'}{\partial x^2} + \frac{\partial^2 V_{(2)}'}{\partial x\,\partial \theta}\frac{1+\nu}{2}\frac{\sqrt{*A_{11}}}{r} + \frac{\partial^2 V_{(1)}'}{\partial \theta^2}\frac{1-\nu}{2}\frac{*A_{11}}{r^2}\right.\right.$$

$$+ \frac{\partial V_{(1)}'}{\partial x}\left(\frac{\partial h/\partial x}{h} + \frac{\partial r/\partial x}{r} - \frac{\partial *A_{11}/\partial x}{2*A_{11}}\right) + \frac{\partial V_{(2)}'}{\partial \theta}\left(\nu\frac{\partial h/\partial x}{h}\right.$$

$$\left. - \frac{3-\nu}{2}\frac{\partial r/\partial x}{r}\right)\frac{\sqrt{*A_{11}}}{r} - \frac{\partial V_{(3)}'}{\partial x}\sqrt{*A_{11}}\frac{R_2 + \nu R_1}{R_1 R_2}$$

$$+ V_{(1)}'\left[\nu\frac{\partial h/\partial x}{h}\frac{\partial r/\partial x}{r} + \nu\frac{\partial r^2/\partial x^2}{r}\right.$$

$$\left.\left. - \nu\frac{\partial *A_{11}/\partial x}{2*A_{11}}\frac{\partial r/\partial x}{r} - \left(\frac{\partial r/\partial x}{r}\right)^2\right]\right]$$

$$+ V_{(3)}'\sqrt{*A_{11}}\left(\frac{\partial R_1/\partial x}{R_1^2} + \nu\frac{\partial R_2/\partial x}{R_2^2} - \frac{\partial h/\partial x}{h}\frac{R_2 + \nu R_1}{R_1 R_2}\right)$$

$$
+ \; \frac{1-\nu}{r}\frac{\partial r}{\partial x}\frac{R_1 - R_2}{R_1 R_2}\Bigg) + \frac{F'_{(1)}{}^* A_{11}}{f}\Bigg]\Bigg]
$$

$$
+ \; \frac{n_{(11)}}{f}\Bigg[\Bigg[\frac{\partial^2 V'_{(1)}}{\partial x^2} - \frac{\partial V'_{(1)}}{\partial x}\frac{\partial^* A_{11}/\partial x}{2^* A_{11}} - \frac{\partial V'_{(3)}}{\partial x}\frac{\sqrt{^* A_{11}}}{R_1}
$$

$$
+ \; V'_{(3)}\sqrt{^* A_{11}}\,\frac{\partial R_1/\partial x}{R_1^2}\Bigg]\Bigg] + \frac{n_{(22)}}{f}\Bigg[\Bigg[\frac{\partial^2 V'_{(1)}}{\partial \theta^2}\frac{^* A_{11}}{r^2}
$$

$$
+ \; \frac{\partial V'_{(1)}}{\partial x}\frac{\partial r/\partial x}{r} - \frac{\partial V'_{(2)}}{\partial \theta}\frac{2\sqrt{^* A_{11}}\,\partial r/\partial x}{r} + V'_{(3)}\frac{\sqrt{^* A_{11}}}{R_2}
$$

$$
+ \; V'_{(1)}\Bigg[\frac{\partial r/\partial x}{r}\frac{\partial^* A_{11}/\partial x}{2^* A_{11}} - \frac{\partial^2 r/\partial x^2}{r} - \left(\frac{\partial r/\partial x}{r}\right)^2\Bigg]
$$

$$
+ \; V'_{(3)}\sqrt{^* A_{11}}\left(\frac{2\partial r/\partial x}{r}\frac{R_1 - R_2}{R_1 R_1} - \frac{\partial R_2/\partial x}{R_2^2}\right)\Bigg]\Bigg]
$$

$$
+ \; \frac{F_{(1)}\sqrt{^* A_{11}}}{f}\Bigg[\Bigg[\frac{\partial V'_{(2)}}{\partial \theta}\frac{\sqrt{^* A_{11}}}{r} + V'_{(1)}\frac{\partial r/\partial x}{r}
$$

$$
- \; V'_{(3)}\frac{\sqrt{^* A_{11}}}{R_2}\Bigg]\Bigg] = 0 \qquad\qquad (7.3.5a)
$$

$$
\Bigg[\Bigg[\frac{\partial^2 V'_{(1)}}{\partial x\,\partial \theta}\frac{1+\nu}{2} + \frac{\partial^2 V'_{(2)}}{\partial x^2}\frac{1-\nu}{2}\frac{r}{\sqrt{(A_{11})}} + \frac{\partial^2 V'_{(2)}}{\partial \theta^2}\frac{\sqrt{^* A_{11}}}{r}
$$

$$
+ \; \frac{\partial V'_{(1)}}{\partial \theta}\left(\frac{\partial h/\partial x}{h}\frac{1-\nu}{2} + \frac{3-\nu}{2}\frac{\partial r/\partial x}{r}\right)
$$

$$
+ \; \frac{\partial V'_{(2)}}{\partial x}\frac{1-\nu}{2}\frac{r}{\sqrt{^* A_{11}}}\left(\frac{\partial h/\partial x}{h} - \frac{\partial^* A_{11}/\partial x}{2^* A_{11}} + \frac{\partial r/\partial x}{r}\right)
$$

$$
- \; V'_{(2)}\frac{1-\nu}{2}\frac{r}{\sqrt{^* 2A_{11}}}\left(\frac{\partial^2 r}{\partial x^2} + \frac{\partial r^2/\partial x}{r^2} + \frac{\partial r/\partial x}{r}\frac{\partial h/\partial x}{h}\right.
$$

$$
- \; \frac{\partial r/\partial x}{r}\frac{\partial^* A_{11}/\partial x}{2^* A_{11}}\Bigg) - \frac{\partial V'_{(3)}}{\partial \theta}\sqrt{^* A_{11}}\,\frac{R_1 + \nu R_2}{R_1 R_2}
$$

$$
+ \; \frac{F'_{(2)}r\sqrt{^* A_{11}}}{f}\Bigg]\Bigg] + \frac{n_{(11)}}{f}\Bigg[\frac{\partial^2 V'_{(2)}}{\partial x^2}\frac{r}{\sqrt{^* A_{11}}}
$$

$$- \frac{\partial V'_{(2)}}{\partial x} \frac{r}{\sqrt{^*A_{11}}} \frac{\partial^* A_{11}/\partial x}{2^* A_{11}} - V'_{(2)} \frac{r}{\sqrt{^*A_{11}}} \left(\frac{\partial^2 r/\partial x^2}{r} \right.$$

$$\left. - \frac{\partial^* A_{11}/\partial x}{2^* A_{11}} \frac{\partial r/\partial x}{r} \right) + \frac{\partial V'_{(3)}}{\partial} \frac{\sqrt{^*A_{11}}}{R_1} \Bigg] \Bigg]$$

$$+ \frac{n_{(22)}}{f} \Bigg[\Bigg[\frac{\partial^2 V'_{(2)}}{\partial \theta^2} \frac{\sqrt{^*A_{11}}}{r} + \frac{\partial V'_{(1)}}{\partial \theta} \frac{2 \partial r/\partial x}{r} $$

$$+ \frac{\partial V'_{(2)}}{\partial x} \frac{r}{\sqrt{^*A_{11}}} - V'_{(2)} \frac{(\partial r/\partial x)^2}{r\sqrt{^*A_{11}}} - \frac{\partial V'_{(3)}}{\partial \theta} \frac{\sqrt{^*A_{11}}}{R_2} \Bigg] \Bigg] = 0$$

$$(7.3.5b)$$

$$\Bigg[\Bigg[\frac{\partial V'_{(1)}}{\partial x} \frac{R_2 + \nu R_1}{R_1 R_2} + \frac{\partial V'_{(2)}}{\partial \theta} \frac{\sqrt{^*A_{11}}}{r} \frac{R_1 + \nu R_2}{R_1 R_2} $$

$$+ V'_{(1)} \frac{\partial r/\partial x}{r} \frac{R_1 + \nu R_2}{R_1 R_2}$$

$$- V'_{(3)} \frac{\sqrt{^*A_{11}}}{R_1 R_2} \left(\frac{R_2 + \nu R_1}{R_1} + \frac{R_1 + \nu R_2}{R_2} \right)$$

$$+ \frac{F'_{(3)} \sqrt{^*A_{11}}}{f} \Bigg] \Bigg] + \frac{n_{(11)}}{f} \Bigg[\Bigg[\frac{\partial^2 V'_{(3)}/\partial x^2}{\sqrt{^*A_{11}}} $$

$$+ \frac{\partial V'_{(1)}/\partial x}{R_1} - \frac{\partial V'_{(2)}}{\partial \theta} \frac{\sqrt{^*A_{11}}}{r R_1} - \frac{\partial V'_{(3)}/\partial x}{\sqrt{^*A_{11}}} \frac{\partial^* A_{11}/\partial x}{2^* A_{11}} $$

$$- \frac{V'_{(1)}}{R_1} \left(\frac{\partial R_1/\partial x}{R_1} + \frac{\partial r/\partial x}{r} \right) + V'_{(3)} \frac{\sqrt{^*A_{11}}}{R_1 R_2} \Bigg] \Bigg]$$

$$+ \frac{n_{(22)}}{f} \Bigg[\Bigg[\frac{\partial^2 V'_{(3)}}{\partial \theta^2} \frac{\sqrt{^*A_{11}}}{r R_2} - \frac{\partial V'_{(1)}/\partial x}{R_2} $$

$$- \frac{\partial V'_{(2)}}{\partial \theta} \frac{\sqrt{^*A_{11}}}{r R_2} + \frac{\partial V'_{(3)}/\partial x}{\sqrt{^*A_{11}}} \frac{\partial r/\partial x}{r}$$

$$+ \frac{V'_{(1)}}{R_1} \frac{1}{r} \frac{\partial r}{\partial x} + V'_{(3)} \frac{\sqrt{^*A_{11}}}{R_1 R_2} \Bigg] \Bigg] = 0$$

$$(7.3.5c)$$

in which

$$f = \frac{Eh}{1 - \nu^2} \qquad (7.3.5d)$$

In general Eqs. (7.3.5) have the form

$$[[\ldots \text{classic membrane eqs.} \ldots]] + \frac{n_{(11)}}{f} [[\ldots \text{displ.}$$

$$\text{quantities} \ldots]] + \frac{n_{(22)}}{f} [[\ldots \text{displ. quantities} \ldots]] = 0 \quad (7.3.6)$$

in which the terms within the first pair of double brackets correspond to the classic linear membrane equilibrium equations expressed in terms of displacements, and the terms within the other pairs of double brackets are the effects of the initial stress state obtained via the prestressing phase. Note that Eq. (7.3.5a) contains terms multiplied by $F_{(1)}$, the tangential force during the prestressing phase. This came about because the equilibrium equations of the prestressing phase were used to eliminate and condense terms in Eq. (7.3.5a).

If a special example of the previous equilibrium equations is considered, it is possible to obtain a measure of the relative contribution of each group of bracketed terms. Consider a constant thickness sphere of radius R with a constant initial internal pressure, i.e., $n_{(11)} = n_{(22)}$. Let x be the angle θ the normal to the membrane makes with the axis of revolution. After elimination and specialization of terms, Eqs. (7.3.5) become

$$\left[\left[\frac{\partial^2 V'_{(1)}}{\partial x^2} + \frac{\partial^2 V'_{(2)}}{\partial x \, \partial \theta} \frac{1 + \nu}{2 \sin \phi} + \frac{\partial^2 V'_{(1)}}{\partial \theta^2} \frac{1 - \nu}{2 \sin^2 \phi} + \frac{\partial V'_{(1)}}{\partial x} \cot \phi \right. \right.$$

$$- \frac{\partial V'_{(2)}}{\partial \theta} \frac{3 - \nu}{2} \frac{\cot \phi}{\sin \phi} - \frac{\partial V'_{(3)}}{\partial x} (1 + \nu) - V'_{(1)}(\nu + \cot^2 \phi)$$

$$\left. + \frac{F'_{(1)} R^2}{f} \right] \right] + \frac{n_{(11)}}{f} \left[\left[\frac{\partial^2 V'_{(1)}}{\partial x^2} + \frac{\partial^2 V'_{(1)}/\partial \theta^2}{\sin^2 \phi} \right. \right.$$

$$\left. + \frac{\partial V'_{(1)}}{\partial x} \cot \phi - V'_{(2)} \frac{2 \cot \phi}{\sin \phi} + V'_{(1)}(1 - \cot^2 \phi) \right] \right]$$

$$+ \frac{F_{(1)} R}{f} \left[\left[\frac{\partial V'_{(2)}/\partial \theta}{\sin \phi} + V'_{(1)} \cot \phi - V'_{(3)} \right] \right] = 0 \quad (7.3.7a)$$

$$\left[\left[\frac{\partial^2 V'_{(1)}}{\partial x \, \partial \theta} \frac{1 + \nu}{2} + \frac{\partial^2 V'_{(2)}}{\partial x^2} \frac{1 - \nu}{2} \sin \phi + \frac{\partial^2 V'_{(2)}/\partial \theta^2}{\sin \phi} \right. \right.$$

$$+ \frac{\partial V'_{(1)}}{\partial \theta} \frac{3 - \nu}{2} \cot \phi + \frac{\partial V'_{(2)}}{\partial \theta} \frac{1 - \nu}{2} \cos \phi$$

$$- V'_{(2)} \frac{1 - \nu}{2} (\cot^2 \phi - 1) \sin \phi - \frac{\partial V'_{(3)}}{\partial \theta} (1 + \nu)$$

$$+ \left. \frac{F'_{(2)} R^2 \sin \phi}{f} \right] \right] + \frac{n_{(11)}}{f} \left[\left[\frac{\partial^2 V'_{(2)}}{\partial x^2} \sin \phi \right. \right.$$

$$+ \frac{\partial^2 V'_{(2)} \partial \theta^2}{\sin \phi} + \frac{\partial V'_{(2)}}{\partial x} \cos \phi$$

$$+ \frac{2 \partial V'_{(1)}}{\partial \theta} \cot \phi - V'_{(2)} (\cot^2 \phi - 1) \sin \phi \left. \left. \right] \right] = 0$$

$$(7.3.7b)$$

$$\left[\left[\frac{\partial V'_{(1)}}{\partial x} (1 - \nu) + \frac{\partial V'_{(2)}}{\partial \theta} \frac{1}{\sin \phi} + V'_{(1)} (1 - \nu) \cot \phi - 2 V'_{(3)} (1 - \nu) \right. \right.$$

$$+ \left. \frac{F'_{(3)} R^2}{f} \right] \right] + \frac{n_{(11)}}{f} \left[\left[\frac{\partial^2 V'_{(3)}}{\partial x^2} + \frac{\partial^2 V'_{(3)} / \partial \theta^2}{\sin^2 \phi} \right. \right.$$

$$+ \frac{\partial V'_{(3)}}{\partial x} \cot \phi + 2 V'_{(3)} \left. \left. \right] \right] = 0 \qquad (7.3.7c)$$

It can be seen for each of Eqs. (7.3.7a and b) that the terms within each of the double brackets are of the same order of magnitude as the terms of the other bracketed expressions within the same equation. Therefore, the relative contributions of the bracketed expressions in Eqs. (7.3.7a and b) are dependent on the ratio

$$\frac{n_{(11)}}{f} = \frac{\sigma_\phi (1 - \nu^2)}{E}$$

in which σ_ϕ = initial stress due to prestressing = $n_{(11)}/(2h)$. For example, if representative values for commercial plastics are used, this ratio has the approximate limits

$$0.03 \leqslant \frac{n_{(11)}}{f} \leq 0.05$$

If only Eqs. (7.3.7a and b) are considered, the initial stress contributes from 3 to 5 percent of the behavior of the inflated sphere when additional in-service loads are applied.

From examination of Eq. (7.3.7c), it can be seen that the two double-bracketed expressions are not necessarily of the same order of magnitude. The terms due to changes in the normal curvature, including $\partial^2 V'_{(3)}/\partial x^2$, occur only in the expression multiplied by $n_{(11)}/f$. It can be shown that this is in agreement with the equations for classic membrane behavior without initial stress. That is, if $n_{(11)}$ were zero, the sixth-order system posed by Eq. (7.3.5) would reduce to the fifth-order system of classic linear membrane theory. In the neighborhood of load and stiffness discontinuities and of nonmembrane-type boundaries, the changes in the normal curvature can be quite large compared to the other displacement quantities. Therefore, the contribution of the initial stress state to Eq. (7.3.7c) is important in those regions and cannot be neglected.

The preceding analysis suggests a simplifying assumption. It seems reasonable that close approximations to the solution for the true in-service behavior of initially stressed membranes can be obtained using a more compact set of equations in which the effects of $n_{(11)}$ and $n_{(22)}$ are neglected in Eqs. (7.3.5a and b) but still included in Eq. (7.3.5c). This approximation was investigated for various sample problems and will be described subsequently.

7.3.1   Solution procedure

For a general membrane of revolution, all the coefficients of the displacement variables and their derivatives in Eqs. (7.3.5) are independent of θ. Therefore, if it is assumed that the force resultants and displacements are expressible as Fourier series in θ, then Eqs. (7.3.5) are reducible to a set of six ordinary differential equations with x as the independent variable. Equations (7.3.5) admit solutions of the form

$$V'_{(1)}(x, \theta) = \Sigma V''^{n}_{(1)}(x) \cos n\theta \qquad (7.3.8a)$$

$$V'_{(2)}(x, \theta) = \Sigma V''^{n}_{(2)}(x) \sin n\theta \qquad (7.3.8b)$$

$$V'_{(3)}(x, \theta) = \Sigma V''^{n}_{(3)}(x) \cos n\theta \qquad (7.3.8c)$$

In addition, let

$$F'_{(1)}(x, \theta) = \Sigma F''^{n}_{(1)}(x) \cos n\theta \qquad (7.3.9a)$$

$$F'_{(2)}(x, \theta) = \Sigma F''^{n}_{(2)}(x) \sin n\theta \qquad (7.3.9b)$$

$$F'_{(3)}(x, \theta) = \Sigma F''^{n}_{(3)}(x) \cos n\theta \qquad (7.3.9c)$$

If Eqs. (7.3.8) and (7.3.9) are substituted into Eqs. (7.3.5), each equation has the form

$$\Sigma[A_n(x) \sin n\theta + B_n(x) \cos n\theta] = 0$$

Therefore each $A_n(x)$ and $B_n(x)$ must vanish independently of the others.

The expanded ordinary differential equations of equilibrium are

$$\frac{\partial^2 V''^n_{(1)}(f + n_{(11)})}{\sqrt{*A_{11}}}$$

$$+ \frac{\partial V''^n_{(1)}}{\partial x} \frac{\left[f\left(\dfrac{\partial h/\partial x}{h} + \dfrac{\partial r/\partial x}{r} - \dfrac{\partial *A_{11}/\partial x}{2*A_{11}} \right) - n_{(11)} \dfrac{\partial *A_{11}/\partial x}{2*A_{11}} + n_{(22)} \dfrac{\partial r/\partial x}{r} \right]}{\sqrt{*A_{11}}}$$

$$- \frac{\partial V''^n_{(3)}}{\partial x}\left(f\frac{R_2 + \nu R_1}{R_1 R_2} + \frac{n_{(11)}}{R_1} - \frac{n_{(22)}}{R_2} \right) + n\frac{\partial V''^n_{(2)}}{\partial x}\frac{\left(\dfrac{1 + \nu}{2}f \right)}{r}$$

$$+ \frac{V''^n_{(1)}}{\sqrt{*A_{11}}}\left\{ f\left[\nu \frac{\partial h/\partial x}{h}\frac{\partial r/\partial x}{r} + \nu\frac{\partial^2 r/\partial x^2}{r} - \nu\frac{\partial *A_{11}/\partial x}{2*A_{11}} - \left(\frac{\partial r/\partial x}{r} \right)^2 \right.\right.$$

$$\left. - n^2 \frac{1 - \nu}{2}\frac{A_{11}}{r^2} \right] + n_{(22)}\left[\frac{\partial *A_{11}/\partial x}{2*A_{11}}\frac{\partial r/\partial x}{r} - \frac{\partial^2 r/\partial x^2}{r} + \left(\frac{\partial r/\partial x}{r} \right)^2 \right.$$

$$\left.\left. - n^2 \frac{*A_{11}}{r^2} \right] + F_{(1)}\sqrt{*A_{11}}\frac{\partial r/\partial x}{r} \right\} + V''^n_{(3)}\left[f\left(\frac{\partial R_1/\partial x}{R_1^2} + \nu\frac{\partial R_2/\partial x}{R_2^2} \right.\right.$$

$$- \frac{\partial h/\partial x}{h}\frac{R_2 + \nu R_1}{R_1 R_2} + \frac{1 - \nu}{r}\frac{\partial r}{\partial x}\frac{R_1 - R_2}{R_1 R_2} \Bigg) + n_{(11)}\frac{\partial R_1/\partial x}{R_1^2}$$

$$+ n_{(22)}\left(\frac{2\partial r/\partial x}{r}\frac{R_1 - R_2}{R_1 R_2} - \frac{\partial R_2/\partial x}{R_2^2} \right) - \frac{F_{(1)}\sqrt{*A_{11}}}{R_2} \Bigg]$$

$$+ nV''^n_{(2)}\frac{\left[f\left(\nu\dfrac{\partial h/\partial x}{h} - \dfrac{3 - \nu}{2}\dfrac{\partial r/\partial x}{r} \right) - n_{(22)}\dfrac{2\partial r/\partial x}{r} + F_{(1)}\sqrt{*A_{11}} \right]}{r}$$

$$+ \sqrt{*A_{11}}\,F''^n_{(1)} = 0 \tag{7.3.10a}$$

$$\frac{\partial^2 V''^n_{(2)}}{\partial x^2}\left(f\frac{1 - \nu}{2} + n_{(11)} \right)\frac{r}{*A_{11}} + \frac{\partial V''^n_{(2)}}{\partial x}\left[f\frac{1 - \nu}{2}\left(\frac{\partial h/\partial x}{h} \right.\right.$$

$$- \frac{\partial *A_{11}/\partial x}{2*A_{11}} + \frac{\partial r/\partial x}{r} \bigg) - n_{(11)}\frac{\partial *A_{11}/\partial x}{2*A_{11}}$$

$$\left. + n_{(22)}\frac{\partial r/\partial x}{r} \right]\frac{r}{*A_{11}} - \frac{n\dfrac{\partial V''^n_{(1)}}{\partial x}\left(f\dfrac{1 + \nu}{2} \right)}{\sqrt{*A_{11}}}$$

$$- nV_{(1)}'' \frac{\left[f\left(\dfrac{1 - \nu}{2} \dfrac{\partial h/\partial x}{h} + \dfrac{3 - \nu}{2} \dfrac{\partial r/\partial x}{r} \right) + n_{(22)} \dfrac{2\partial r/\partial x}{r} \right]}{\sqrt{*A_{11}}}$$

$$- V_{(2)}'' \left\{ f \frac{1 - \nu}{2} \left[\frac{\partial^2 r/\partial x^2}{r} + \left(\frac{\partial r/\partial x}{r} \right)^2 + \frac{\partial h/\partial x}{h} \frac{\partial r/\partial x}{r} \right. \right.$$

$$- \frac{\partial r/\partial x}{r} \frac{\partial^* A_{11}/\partial x}{2^* A_{11}} \left. \right] + n^2 \frac{{}^* A_{11}}{r^2} f + n_{(11)} \left(\frac{\partial^2 r/\partial x^2}{r} \right.$$

$$- \frac{\partial^* A_{11}/\partial x}{2^* A_{11}} \frac{\partial r/\partial x}{r} \right) + n_{(22)} \left[n^2 \frac{{}^* A_{11}}{r^2} + \left(\frac{\partial r/\partial x}{r} \right)^2 \right] \left. \right\} \frac{r}{{}^* A_{11}}$$

$$+ nV_{(3)}'' \left(f \frac{R_1 + \nu R_2}{R_1 R_2} - \frac{n_{(11)}}{R_1} + \frac{n_{(22)}}{R_2} \right)$$

$$+ F_{(2)}'' r = 0 \tag{7.3.10b}$$

$$\frac{\partial^2 V_{(3)}''}{\partial x^2} \left(\frac{n_{(11)}}{{}^* A_{11}} \right) + \frac{\partial V_{(1)}''}{\partial x} \frac{\left(f \dfrac{R_2 + \nu R_1}{R_1 R_2} + \dfrac{n_{(11)}}{R_1} - \dfrac{n_{(22)}}{R_2} \right)}{\sqrt{*A_{11}}}$$

$$+ \frac{\partial V_{(3)}''}{\partial x} \frac{\left(n_{(22)} \dfrac{\partial r/\partial x}{r} - n_{(11)} \dfrac{\partial^* A_{11}/\partial x}{2^* A_{11}} \right)}{{}^* A_{11}}$$

$$+ V_{(1)}'' \frac{\left[f \dfrac{\partial r/\partial x}{r} \dfrac{R_1 + \nu R_2}{R_1 R_2} - \dfrac{n_{(11)}}{R_1} \left(\dfrac{\partial R_1/\partial x}{R_1} + \dfrac{\partial r/\partial x}{r} \right) + \dfrac{n_{(22)}}{R_1} \dfrac{\partial r/\partial x}{r} \right]}{\sqrt{*A_{11}}}$$

$$- V_{(3)}'' \left[f \left(\frac{R_2 + \nu R_1}{R_1} + \frac{R_1' + \nu R_2}{R_2} \right) - n_{(11)} - n_{(22)} \right.$$

$$+ n^2 \frac{n_{(22)}}{r^2} R_1 R_2 \left. \right] R_1 R_2 + nV_{(2)}'' \frac{\left(f \dfrac{R_1 + \nu R_2}{R_1 R_2} - \dfrac{n_{(11)}}{R_1} + \dfrac{n_{(22)}}{R_2} \right)}{r}$$

$$+ F_{(3)}'' = 0 \tag{7.3.10c}$$

The solutions to the boundary-value problem posed by Eqs. (7.3.10) can be obtained for any particular membrane of revolution if six membrane-type boundary conditions for each value of n are correctly prescribed. For a membrane of revolution continuous at the apex there should be three conditions imposed at the apex for each value of n. In most cases, the edge conditions can be readily obtained using Fourier expansions.

The descriptions of the boundary conditions at the apex are of some interest since the apex is a regular singular point of Eqs. (7.3.10). The adopted coordinate system is not regular at the apex, i.e., there is no one-to-one correspondence between real points on C and coordinate points (x, θ) at $x = 0$. Because the solutions $V''_{(1)}$, $V''_{(2)}$, $V''_{(3)}$ are analytic at the apex, it is possible to obtain solutions to Eqs. (7.3.10) using series expansions in the neighborhood of $x = 0$. When these series expansions are substituted into the equilibrium equations, which are then evaluated at $x = 0$ with the singularity removed, certain relations between the constants in the series expansions will be obtained. Because the constants of the expansions correspond to the derivatives of the displacements evaluated at the apex, the relations between the constants may be considered as boundary conditions.

Let x be the angle ϕ the normal to the shell makes with the axis of revolution. In this case, $A_{11} = R_1^2$, $\partial r / \partial x = R_1 \cos \phi$, and $r = R_2 \sin \phi$. The series expansions in terms of ϕ are

$$V''_{(1)} = \Sigma a_m^n \phi^m \tag{7.3.11a}$$

$$V''_{(2)} = \Sigma b_m^n \phi^m \tag{7.3.11b}$$

$$V''_{(3)} = \Sigma c_m^n \phi^m \tag{7.3.11c}$$

If the membrane of revolution is continuous at the apex, the apex is an umbilical (spherical) point, and in the region $\phi = 0$ the equilibrium equations Eqs. (7.3.10) have the following form ($R_1 = R_2 = R =$ constant, $n_{(11)} = n_{(22)}$, and h is assumed constant in the region of $\phi = 0$):

$$\frac{\partial^2 V''_{(1)}}{\partial x^2} (f + n_{(11)}) + \frac{\partial V''_{(1)}}{\partial x} (f + n_{(11)}) \cot \phi - \frac{\partial V''_{(3)}}{\partial x} [f(1 + \nu)]$$

$$+ \frac{n V''_{(2)} \left(f \dfrac{1 + \nu}{2} \right)}{\sin \phi} - V''_{(1)} \left[f \left(\nu + \cot^2 \phi + n^2 \dfrac{1 - \nu}{2} \dfrac{1}{\sin^2 \phi} \right) \right.$$

$$\left. + n_{(11)} \left(\cot^2 \phi - 1 + \dfrac{n^2}{\sin^2 \phi} \right) - F_{(1)} R \cot \phi \right] - V''_{(3)} (F_{(1)} R)$$

$$- n V''_{(2)} \frac{\left(f \dfrac{3 - \nu}{2} \cot \phi + 2 n_{(11)} \cot \phi - F_{(1)} R \right)}{\sin \phi} + F''_{(1)} R^2 = 0$$

$$\tag{7.3.12a}$$

$$\frac{\partial^2 V''_{(2)}}{\partial x^2} \left(f \frac{1 - \nu}{2} + n_{(11)} \right) \sin \phi + \frac{\partial V''_{(2)}}{\partial x} \left(f \frac{1 - \nu}{2} + n_{(11)} \right) \cos \phi$$

$$- \frac{n \partial V''^n_{(1)}}{\partial x} \left(f \frac{1 + \nu}{2} \right) - V''^n_{(2)} \left[\left(f \frac{1 - \nu}{2} + n_{(11)} \right) (\cot^2 \phi - 1) \sin \phi \right.$$

$$+ (f + n_{(11)}) \frac{n^2}{\sin \phi} \right] + nV''^n_{(3)} [(1 + \nu)f] - nV''^n_{(1)} \left(\frac{3 - \nu}{2} \right) f$$

$$+ 2n_{11} \Bigg) \cot \phi + F''^n_{(2)} R^2 \sin \phi = 0 \qquad\qquad (7.3.12b)$$

$$\frac{\partial^2 V''^n_{(3)}}{\partial x^2} (n_{(11)}) + \frac{\partial V''^n_{(1)}}{\partial x} [f(1 + \nu)] + \frac{\partial V''^n_{(3)}}{\partial x} [n_{(11)} \cot \phi]$$

$$+ V''^n_{(1)} \left[f(1 + \nu) \cot \phi + nV''^n_{(2)} \left[\frac{(1 + \nu)}{\sin \phi} \right] - V''^n_{(3)} \left[2f(1 + \nu) - 2n_{(11)} \right.$$

$$+ \frac{n^2 n_{(11)}}{\sin^2 \phi} \right] + F''^n_{(3)} R^2 = 0 \qquad\qquad (7.3.12c)$$

Equations (7.3.12) become analytic if Eq. (7.3.12a) is multiplied by $\sin^2 \phi$ and if Eqs. (7.3.12b and c) are multiplied by $\sin \phi$. If Eqs. (7.3.11) and the series expansion of $\sin \phi$ and $\cos \phi$ are substituted into the analytic forms of Eqs. (7.3.12) and if terms are collected in powers of ϕ, the vanishing of all terms independently of ϕ results in a set of relations between the constants of Eqs. (7.3.11), which must then be satisfied for each value of n. The solutions to these relationships for the values $n = 0$, 1, 2 are

$$n = 0 \text{ (axisymmetric) } (a_1^0, b_1^0, c_0^0 \text{ arbitrary)}$$

$$a_0^0 = b_0^0 = c_1^0 = b_2^0 = 0 \qquad\qquad (7.3.13a)$$

$$a_2^0 = -F'^0_{(1)} \frac{R^2}{3} (f + n_{(11)}) \qquad\qquad (7.3.13b)$$

$$c_2^0 = \frac{2f(1 + \nu)(c_0^0 - a_1^0) - 2n_{(11)} c_0^0 - F'^0_{(3)} R^2}{4n_{(11)}} \qquad\qquad (7.3.13c)$$

$$n = 1 \text{ (antisymmetric) } (a_2^1, b_0^1, c_1^1 \text{ arbitrary)}$$

$$c_0^1 = a_1^1 = b_1^1 = 0 \qquad\qquad (7.3.14a)$$

$$a_0^1 = -b_0^1 \qquad\qquad (7.3.14b)$$

$$b_2^1 = \frac{\left(\frac{5 + \nu f}{2} + 2n_{(11)} \right) a_2^1 - (1 + \nu)fc_1^1 + F'^1_{(1)} R^2}{\frac{1 - 3\nu f}{2} + 2n_{(11)}} - \frac{b_0^1}{2} \qquad (7.3.14c)$$

$$c_2^1 = -\frac{F'^2_{(3)} R^2}{3n_{(11)}} \qquad\qquad (7.3.14d)$$

$$n = 2 \ (a_3^2, \ b_1^2, \ c_2^2 \ \text{arbitrary})$$

$$a_0^2 = b_0^2 = c_0^2 = c_1^2 = 0 \qquad\qquad (7.3.15a)$$

$$a_1^2 = -b_1^2 \qquad\qquad (7.3.15b)$$

$$a_2^2 = \frac{2F_{(1)}'^2 R^2[(1 - 3\nu)f + 4n_{(11)}] + F_{(2)}'^2 R^2[(5 + 3\nu)f + 2n_{(11)}]}{15(f + n_{(11)})[f(1 - \nu) + 2n_{(11)}]}$$

$$\qquad\qquad (7.3.15c)$$

$$b_2^2 = -\frac{3}{2} a_2^2 + \frac{2F_{(2)}'^2 R^2 + F_{(1)}'^2}{6(f + n_{(11)})} \qquad\qquad (7.3.15d)$$

$$b_3^2 = \frac{[(3 + \nu)f + 2n_{(11)}]}{n_{(11)} - \nu f}\left(\frac{a_3^2}{2} + \frac{b_1^2}{12}\right) - \frac{(1 + \nu)fc_2^2}{2(n_{(11)} - \nu f)} \qquad (7.3.15e)$$

$$c_3^2 = -\frac{f(1 + \nu)}{5n_{(11)}}(3a_2^2 + 2b_2^2) \qquad\qquad (7.3.15f)$$

For each value of n, the displacements in the neighborhood of $\phi = 0$ could be determined if in each case values are assigned to three, and only three, of the constants in Eqs. (7.3.13). This agrees with the fact that there should be three membrane-type conditions imposed at the edge of the shell. Note that Eqs. (7.3.13) to (7.3.15) satisfy the requirements that the strains be finite at the apex, i.e., $V_{(1)} + \partial V_{(2)}/\partial\theta = 0$ and $\partial V_{(1)}/\partial\theta - V_{(2)} = 0$, and that the slope be continuous at the apex.

7.3.2 Numerical integration scheme

The solution to the boundary-value problem posed by Eqs. (7.3.10) may be obtained from the solution to a set of initial-value problems formed by assuming independent values of the unknown derivatives of the displacements at one boundary, preferably the apex. The unknown displacement derivatives at the apex are equal to the arbitrary coefficients [given in Eqs. (7.3.13) to (7.3.15) for $n \leq 2$] multiplied by known constants, e.g.,

$$\left.\frac{d^m V_{(1)}''^n m}{d\phi^m}\right|_{\phi=0} = m!a_m^n$$

Because for each value of n, there are three arbitrary constants, and four independent solutions to the initial-value problem can be obtained. The boundary-value problem solution is a linear combination of the four initial-value problem solutions, the constants of the combination being determined from the boundary conditions at the edge of the membrane.

The numerical integration scheme chosen for the initial-value problems has been described previously for axisymmetric loading and remains essentially the same for the asymmetric equations; more variables and more equations are considered.

In the neighborhood of $\phi = 0$, special care must be taken in the numerical integration of the initial-value problems. Because the apex is a regular singular point of Eqs. (7.3.15), the method is not strictly applicable, and consistent Taylor series approximations are required. For each value of n sufficient terms in the series, expansions of the derivatives must be taken such that none of the arbitrary constants are omitted in the approximations.

For values of $n \leqslant 2$ the integration scheme for the increments in the neighborhood of $\phi = 0$ should be

1. For each displacement function assume that the nonzero derivative immediately after the highest-order arbitrary derivative is constant throughout an increment and assign a value to that constant.

2. The lower-order derivatives of the displacements at the end of an increment are approximated by Taylor series expansions about the beginning of the increment.

3. Better approximations of the second derivatives at the end of the increment can be obtained by substituting the Taylor series approximations into Eqs. (7.3.12).

4. If the new values of the second derivatives when compared to the previous values do not satisfy some convergence criteria, new values for the derivatives immediately after the highest-order arbitrary derivative are calculated on the basis of the new second derivatives.

5. The process is repeated by returning to step 2.

7.3.3 Sample problems

The numerical solution method described previously has been tested on various problems. The in-service asymmetric behavior of the same ring-supported sphere (see Fig. 7.2.1) used to demonstrate the symmetrically loaded in-service phase solution is now considered.

The boundary conditions at the sphere apex are given by Eqs. (7.3.13) to (7.3.15). For the ring support at $\phi = \phi_0$, it is assumed that once the desired shape has been attained via the prestressing phase, the membrane is completely clamped at the base. The boundary conditions are

$$V_{(1)}^{\prime n} = V_{(2)}^{\prime n} = V_{(3)}^{\prime n} = 0 \qquad \text{at } \phi = \phi_0$$

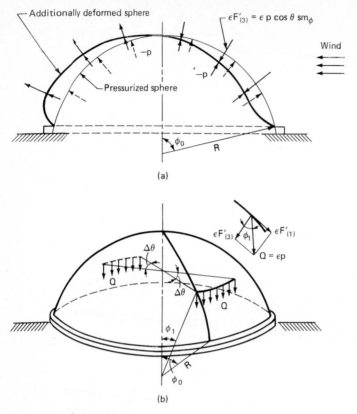

Figure 7.3.1   Sphere with (a) wind load; (b) partial ring load.

The first sphere problem chosen to demonstrate the asymmetric solution method is that of a wind loading (see Fig. 7.3.1a), a crude approximation to which can be expressed as [79]

$$F'_{(3)} = \varepsilon p \sin \phi \cos \theta$$

in which p = internal prestressing, and ε = small parameter. The entries in Table 7.3.1, when multiplied by either $\cos \theta$ or $\sin \theta$, are the added displacements and stress resultants due to a wind load with a maximum intensity of 40 psf (1.9 kPa) superposed on a prestressed sphere (Fig. 7.3.1a) in which p = 400 psf (19 kPa) and $\phi_0 = \pi/4$. For example,

$$\varepsilon V'_{(3)}(\phi, \theta) = (\text{entries in column 4, Table 7.3.1}) \cos \theta$$

$$\varepsilon n'_{(22)}(\phi, \theta) = (\text{entries in column 6, Table 7.3.1}) \cos \theta$$

$$\varepsilon n'_{(12)}(\phi, \theta) = (\text{entries in column 7, Table 7.3.1}) \sin \theta$$

As shown previously, an approximate set of equations is derivable in which the contributions of the initial stress are neglected in the first two

TABLE 7.3.1   Sphere with Wind Loads (sin θ and cos θ Terms)

ϕ, rad (1)	$\varepsilon V'_{(1)} \times 10^3$ (2)	$\varepsilon V'_{(2)} \times 10^3$ (3)	$\varepsilon V'_{(3)} \times 10^3$ (4)	$\varepsilon n'_{(11)},$ lb/ft (5)	$\varepsilon n'_{(22)},$ lb/ft (6)	$\varepsilon n'_{(12)},$ lb/ft (7)
0.000	−6.791	6.971	0.000	0.00	0.00	0.00
0.079	−6.673	6.718	2.894	−7.13	−21.31	−6.68
0.157	−6.324	6.499	5.699	−14.11	−42.17	−13.35
0.236	−5.755	6.133	8.318	−20.82	−62.08	−19.95
0.314	−4.985	5.624	10.631	−27.03	−80.26	−26.43
0.393	−4.047	4.974	12.477	−32.56	−95.65	−32.66
0.471	−2.991	4.188	13.620	−37.18	−106.51	−38.43
0.550	−1.895	3.278	13.704	−40.63	−110.14	−43.39
0.628	−0.881	2.257	12.160	−42.53	−101.91	−46.96
0.707	−0.145	1.150	8.082	−42.48	−74.34	−48.14
0.785	0.000	0.000	0.000	−39.97	−14.92	−45.27

NOTE: 1 lb/ft = 14.6 N/m.

of the three in-service phase equilibrium equations. The sphere example just given was used to test the accuracy of this approximation. Portions of the results of the test are shown in Fig. 7.3.2 where the displacements are compared for the exact and approximate formulations. Also shown in Fig. 7.3.2 are the solutions for a wind-loaded sphere using the classic membrane theory as described in Sec. 6.4.3, wherein effects of initial stress are completely neglected. It can be seen that the approximate solution is in close agreement with the exact solution. Also, both of these solutions are significantly different from the solution obtained from the classic membrane theory.

The second sphere problem considered is that of two ring segments supporting line loads oriented in the axial direction (see Fig. 7.3.1b). This load system has been chosen to demonstrate the solution procedure for problems in which values of $n = 0, 1, 2$, etc., must be considered. A practical illustration of this type of load system is that of a platform suspended from two ring segments attached to the sphere.

The axially oriented ring loads can be expanded in Fourier series with respect to θ, and expressions for the load components to be used in Eqs. (7.3.12) are found to be (see Fig. 7.3.1b)

$$\varepsilon F'^{n}_{(i)} = 0 \qquad n \geqslant 0, \phi \neq \phi_1, i = 1, 2, 3$$

$$\varepsilon F'^{0}_{(1)}(\phi_1) = \frac{Q\,\Delta\theta}{\pi} \sin \phi_1$$

$$\varepsilon F'^{n}_{(1)}(\phi_1) = \frac{2Q}{n\pi} \sin \frac{n\,\Delta\theta}{2} [(-1)^n + 1] \sin \phi_1 \qquad n \geqslant 1$$

$$\varepsilon F'^{n}_{(2)}(\phi_1) = 0 \qquad n \geqslant 0$$

$$\varepsilon F'^{0}_{(3)}(\phi_1) = \frac{Q\,\Delta\theta}{\pi}\cos\phi_1$$

$$\varepsilon F'^{n}_{(3)}(\phi_1) = \frac{2Q}{n\pi}\sin\frac{n\,\Delta\theta}{2}\,[(-1)^n + 1]\cos\phi_1 \qquad n \geqslant 1$$

where ϕ_1 = angular location of a ring segment
  $\Delta\theta$ = central azimuth angle of ring segment
  Q = ring load, lb/ft (N/m)

Table 7.3.2 gives the $n = 0,\ 1,\ 2$ results for a partial ring loading in which $Q = 40$ lb/ft (584 N/m), $\Delta\theta = \pi/4$, $\phi_1 = \pi/8$, and $\phi_0 = \pi/4$.

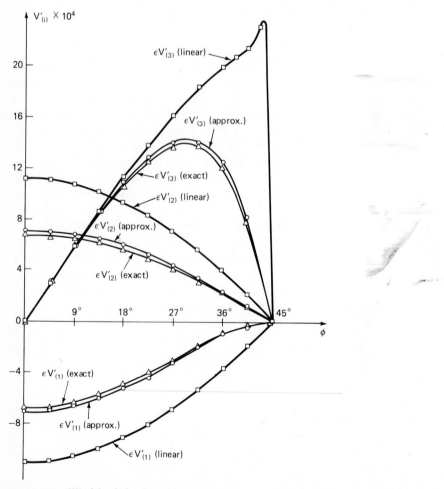

Figure 7.3.2 Wind-loaded sphere comparisons.

TABLE 7.3.2 Sphere with Ring Load (sin $n\theta$ and $n\theta$ Terms)

Row (1)	ϕ, rad (2)		0.000 (3)	0.079 (4)	0.157 (5)1	0.236 (6)	0.314 (7)	0.393 (8)	0.471 (9)	0.550 (10)	0.628 (11)	0.707 (12)	0.785 (13)
2	$\varepsilon V'^{n}_{(1)} \times 10^3$	$n = 0$	0.000	0.002	0.012	0.041	0.104	0.199	0.250	0.219	0.146	0.065	0.000
3		$n = 1$	0.000	0.000	0.000	0.000	0.000	0.000	0.000	0.000	0.000	0.000	0.000
4		$n = 2$	0.000	−0.101	−0.182	−0.217	−0.169	−0.025	0.092	0.113	0.083	0.039	0.000
5	$\varepsilon V'^{n}_{(2)} \times 10^3$	$n = 0$	0.000	0.000	0.000	0.000	0.000	0.000	0.000	0.000	0.000	0.000	0.000
6		$n = 1$	0.000	0.000	0.000	0.000	0.000	0.000	0.000	0.000	0.000	0.000	0.000
7		$n = 2$	0.000	0.103	0.201	0.284	0.343	0.358	0.304	0.226	0.150	0.076	0.000
8	$\varepsilon V'^{n}_{(3)} \times 10^3$	$n = 0$	0.222	0.277	0.436	0.739	1.247	1.650	0.834	0.144	−0.158	−0.195	0.000
9		$n = 1$	0.000	0.000	0.000	0.000	0.000	0.000	0.000	0.000	0.000	0.000	0.000
10		$n = 2$	0.000	0.096	0.410	1.018	2.062	3.015	1.837	0.797	0.281	0.045	−0.000
11	$\varepsilon n^{n}_{(11)}$, lb/ft	$n = 0$	−3.2	−3.3	−3.6	−4.2	−5.2	−10.4	−9.0	−8.6	−7.9	−7.1	−6.6
12		$n = 1$	0.0	0.0	0.0	0.0	0.0	0.0	0.0	0.0	0.0	0.0	0.0
13		$n = 2$	−13.9	−8.8	−8.8	−8.8	−8.8	−16.0	−11.3	−9.2	−7.1	−5.4	−4.2
14	$\varepsilon n^{n}_{(22)}$, lb/ft	$n = 0$	−3.2	−3.5	−4.6	−6.7	−10.3	−14.4	−5.4	−1.3	0.3	−0.2	−2.5
15		$n = 1$	0.0	0.0	0.0	0.0	0.0	0.0	0.0	0.0	0.0	0.0	0.0
16		$n = 2$	−5.2	8.0	5.8	1.3	−6.6	−16.9	−7.1	−1.2	0.5	0.1	−1.6
17	$\varepsilon n^{n}_{(12)}$, lb/ft	$n = 0$	0.0	0.0	0.0	0.0	0.0	0.0	0.0	0.0	0.0	0.0	0.0
18		$n = 1$	−0.0	0.0	0.0	0.0	0.0	0.0	0.0	0.0	0.0	0.0	0.0
19		$n = 2$	4.1	7.9	6.9	5.0	1.8	−3.1	−6.0	−5.5	−4.4	−3.5	−3.1

NOTE: 1 lb/ft = 14.6 N/m.

7.4 Dynamic Response of Initially Stressed Membranes

The dynamic behavior of extremely thin membranes which have attained a given stress state due to previous loads is presented in this section [42]. The response of these structural components to the action of further in-service dynamic loads will be studied and the effects of the initial stress state determined. The equations of motion for a thin membrane with a general configuration and a prescribed initial stress state are derived. A solution procedure for the asymmetric modes of free vibration of a membrane of revolution is derived and illustrated by means of sample problems. The forced response of these same problems is also studied.

7.4.1 Equations of motion

The assumptions made in the derivation of the equations of motion for the superposition of small oscillations about the initially stressed configuration of a general membrane are outlined in Sec. 7.3. If we add an inertial term to the static equations given by Eqs. (7.1.14) and (7.1.15), the equations of motion are found to be

$$\left(n'^{\alpha\gamma}\big|_\alpha + F'^\gamma - h\rho\,\frac{\partial^2 V'^\gamma}{\partial t^2}\right) + \left(*n^{\rho\gamma}\Gamma'^\alpha_{\rho\alpha} + *n^{\alpha\rho}\Gamma'^\gamma_{\rho\alpha}\right) = 0 \quad (7.4.1a)$$

$$\left(n'^{\alpha\gamma}*B_{\alpha\gamma} + F'^3 - h\rho\,\frac{\partial^2 V'^3}{\partial t^2}\right) + \left(n^{\alpha\gamma}B'_{\alpha\gamma}\right) = 0 \quad (7.4.1b)$$

in which covariant differentiation is denoted by a vertical line used as a subscript and ρ = mass density.

Equations (7.4.1) can be expressed entirely in terms of displacement functions by substituting the relations between the force resultants and the displacements. Thus

$$n'^{\alpha\gamma} = \frac{Eh}{1+\nu}\left[\,*A^{\alpha\rho}U'^\gamma_\rho + *A^{\gamma\rho}U'^\alpha_\rho + \frac{2\nu}{(1-\nu)}\,*A^{\alpha\gamma}U'^\rho_\rho\right] - *n^{\alpha\gamma}U'^\rho_\rho$$

The general equations of motion can then be specialized for a membrane of revolution with an arbitrary meridional contour. Asymmetric oscillations about an axially symmetric initial stress state have been considered. Therefore, derivatives with respect to θ of all quantities associated with the initially stressed membrane of revolution are identically zero. The physical forms of Eqs. (7.4.1) specialized for a general membrane of revolution are

$$\frac{\partial^2 V'_{(1)}}{\partial x^2}(1+K_1) + \frac{\partial^2 V'_{(2)}}{\partial x\,\partial\theta}\left(\frac{1+\nu}{2}\right)\frac{\sqrt{*A_{11}}}{r} + \frac{\partial^2 V'_{(1)}}{\partial\theta^2}\left(\frac{1-\nu}{2}+K_2\right)\frac{*A_{11}}{r^2}$$

$$+\frac{\partial V'_{(1)}}{\partial x}\left(\frac{\partial h/\partial x}{h} - \frac{\partial *A_{11}/\partial x}{2*A_{11}} + \frac{\partial r/\partial x}{r} - K_1\frac{\partial *A_{11}/\partial x}{2*A_{11}} + K_2\frac{\partial r/\partial x}{r}\right)$$

$$+\frac{\partial V'_{(2)}}{\partial\theta}\left(\frac{\partial h/\partial x}{h}\nu - \frac{3-\nu}{2}\frac{\partial r/\partial x}{r} - K_2\frac{2\partial r/\partial x}{r} + \frac{F_{(1)}}{K}\sqrt{*A_{11}}\right)\frac{\sqrt{*A_{11}}}{r}$$

$$-\frac{\partial V'_{(3)}}{\partial x}\left(\frac{R_2+\nu R_1}{R_1 R_2} + \frac{K_1}{R_1} - \frac{K_2}{R_2}\right)\sqrt{*A_{11}} + V'_{(1)}\left\{\frac{\partial h/\partial x}{h}\frac{\partial r/\partial x}{r}\nu\right.$$

$$+\nu\frac{\partial r/\partial x}{r} - \left(\frac{\partial r/\partial x}{r}\right)^2 - \nu\frac{\partial *A_{11}/\partial x}{2*A_{11}}\frac{\partial r/\partial x}{r} - K_2\left[\frac{\partial^2 r/\partial x^2}{r} + \left(\frac{\partial r/\partial x}{r}\right)^2\right.$$

$$\left.-\frac{\partial *A_{11}/\partial x}{2*A_{11}}\frac{\partial r/\partial x}{r}\right] + \frac{F_{(1)}}{K}\sqrt{*A_{11}}\frac{\partial r/\partial x}{r}\right\} + V'_{(3)}\left[\frac{\partial R_1/\partial x}{R_1^2}\right.$$

$$+\nu\frac{\partial R_2/\partial x}{R_2^2} - \frac{\partial h/\partial x}{h}\frac{R_2+\nu R_1}{R_1 R_2} + \frac{1-\nu}{r}\frac{\partial r}{\partial x}\frac{R_1-R_2}{R_1 R_2} + K_1\frac{\partial R_1/\partial x}{R_1^2}$$

$$+K_2\left(\frac{2\partial r/\partial x}{r}\frac{R_1-R_2}{R_1 R_2} - \frac{\partial R_2/\partial x}{R_2^2}\right) - \frac{F_{(1)}\sqrt{*A_{11}}}{KR_2}\right]\sqrt{*A_{11}} + \frac{F'_{(1)}}{K}*A_{11}$$

$$-\frac{h\rho}{K}*A_{11}\frac{\partial^2 V'_{(1)}}{\partial t^2} = 0 \tag{7.4.2a}$$

$$\frac{\partial^2 V'_{(1)}}{\partial x\,\partial\theta}\left(\frac{1+\nu}{2}\right) + \frac{\partial^2 V'_{(2)}}{\partial x^2}\left(\frac{1-\nu}{2}+K_1\right)\frac{r}{\sqrt{A*_{11}}} + \frac{\partial^2 V'_{(2)}}{\partial\theta^2}(1+K_2)\frac{\sqrt{*A_{11}}}{r}$$

$$+\frac{\partial V'_{(1)}}{\partial\theta}\left(\frac{1-\nu}{2}\frac{\partial h/\partial x}{h} + \frac{\partial r/\partial x}{r} + K_2 2\frac{\partial r/\partial x}{r}\right) + \frac{\partial V'_{(2)}}{\partial x}\left[\frac{1-\nu}{2}\left(\frac{\partial h/\partial x}{h}\right.\right.$$

$$\left.\left.-\frac{\partial *A_{11}/\partial x}{2*A_{11}} + \frac{\partial r/\partial x}{r}\right) - K_1\frac{\partial *A_{11}/\partial x}{2*A_{11}} + K_2\frac{\partial r/\partial x}{r}\right]\frac{r}{\sqrt{*A_{11}}}$$

$$-V'_{(2)}\left\{\frac{1-\nu}{2}\left[\frac{\partial^2 r/\partial x^2}{r} + \left(\frac{\partial r/\partial x}{r}\right)^2 + \frac{\partial r/\partial x}{r}\frac{\partial h/\partial x}{h} - \frac{\partial r/\partial x}{r}\frac{\partial *A_{11}/\partial x}{2*A_{11}}\right]\right.$$

$$\left.+K_1\left(\frac{\partial^2 r/\partial x^2}{r} - \frac{\partial *A_{11}/\partial x}{2*A_{11}}\frac{\partial r/\partial x}{r}\right) + K_2\left(\frac{\partial r/\partial x}{r}\right)^2\right\}\frac{r}{\sqrt{*A_{11}}}$$

$$- \frac{\partial V'_{(3)}}{\partial \theta} \left(\frac{R_1 + \nu R_2}{R_1 R_2} - \frac{K_1}{R_1} + \frac{K_2}{R_2} \right) \sqrt{*A_{11}} + \frac{F'_{(2)}}{K} r \sqrt{*A_{11}}$$

$$- \frac{h\rho}{K} r \sqrt{*A_{11}} \frac{\partial^2 V'_{(2)}}{\partial t^2} = 0 \qquad\qquad (7.4.2b)$$

$$\frac{\partial^2 V'_{(3)}}{\partial x^2} \left(\frac{K_1}{\sqrt{*A_{11}}} \right) + \frac{\partial^2 V'_{(3)}}{\partial \theta^2} \left(\frac{K_2}{r} \right) \frac{\sqrt{*A_{11}}}{r} + \frac{\partial V'_{(1)}}{\partial x} \left(\frac{R_2 + \nu R_1}{R_1 R_2} + \frac{K_1}{R_1} - \frac{K_2}{R_2} \right)$$

$$+ \frac{\partial V'_{(2)}}{\partial \theta} \left(\frac{R_1 + \nu R_2}{R_1 R_2} - \frac{K_1}{R_1} + \frac{K_2}{R_2} \right) \frac{\sqrt{*A_{11}}}{r} - \frac{\partial V'_{(3)}}{\partial x} \left(K_1 \frac{\partial * A_{11}/\partial x}{2 * A_{11}} \right.$$

$$\left. - K_2 \frac{\partial r/\partial x}{r} \right) \frac{1}{\sqrt{*A_{11}}} + V'_{(1)} \left[\frac{\partial r/\partial x}{r} \frac{R_1 + \nu R_2}{R_1 R_2} - \frac{K_1}{R_1} \left(\frac{\partial R_1/\partial x}{R_1} + \frac{\partial r/\partial x}{r} \right) \right.$$

$$\left. + \frac{K_2}{R_2} \frac{\partial r/\partial x}{r} \right] - V'_{(3)} \left(\frac{R_2 + \nu R_1}{R_1} + \frac{R_1 + \nu R_2}{R_2} - K_1 - K_2 \right) \frac{\sqrt{*A_{11}}}{R_1 R_2}$$

$$+ \frac{F'_{(3)}}{K} \sqrt{*A_{11}} - \frac{h\rho}{K} \sqrt{*A_{11}} \frac{\partial^2 V'_{(3)}}{\partial t^2} = 0 \qquad\qquad (7.4.2c)$$

in which

$$K = \frac{Eh}{1 - \nu^2} \qquad\qquad (7.4.3a)$$

$$K_1 = \frac{n_{(11)}}{K} \qquad\qquad (7.4.3b)$$

$$K_2 = \frac{n_{(22)}}{K} \qquad\qquad (7.4.3c)$$

An approximate formulation of Eqs. (7.4.2) is presented in the previous section. In that approximate theory for initially stressed membranes, it was shown that the effects of the initial stress can be neglected in the first two of the equations of motion. However, in the third equation of motion, Eq. (7.4.2c), the initial stress components should be retained in order to adequately describe the membrane behavior in the neighborhood of discontinuities and non-membrane-type boundaries. The approximate formulation of the equations of motion for a general membrane of revolution is

$$\frac{\partial^2 V'_{(1)}}{\partial x^2} + \frac{\partial^2 V'_{(2)}}{\partial x\,\partial\theta}\frac{1+\nu}{2}\frac{\sqrt{^*A_{11}}}{r} + \frac{\partial^2 V'_{(1)}}{\partial\theta^2}\frac{1-\nu}{2}\frac{^*A_{11}}{r^2} + \frac{\partial V'_{(1)}}{\partial x}\left(\frac{\partial h/\partial x}{h}\right.$$

$$\left. - \frac{\partial^*A_{11}/\partial x}{2^*A_{11}} + \frac{\partial r/\partial x}{r}\right) + \frac{\partial V'_{(2)}}{\partial\theta}\left[\nu\left(\frac{\partial h/\partial x}{h} - \frac{3-\nu}{2}\frac{\partial r/\partial x}{r}\right)\right]\frac{\sqrt{^*A_{11}}}{r}$$

$$- \frac{\partial V'_{(3)}}{\partial x}\frac{R_2 + \nu R_1}{R_1 R_2}\sqrt{^*A_{11}} + V'_{(1)}\left[\nu\left(\frac{\partial h/\partial x}{h}\frac{\partial r/\partial x}{r} + \frac{\partial^2 r/\partial x^2}{r}\right.\right.$$

$$\left. - \frac{\partial^*A_{11}/\partial x}{2^*A_{11}}\frac{\partial r/\partial x}{r}\right) - \left(\frac{\partial r/\partial x}{r}\right)^2\right] + V'_{(3)}\left(\frac{\partial R_1/\partial x}{R_1^2} + \nu\frac{\partial R_2/\partial x}{R_2^2}\right.$$

$$\left. - \frac{\partial h/\partial x}{h}\frac{R_1 + \nu R_1}{R_1 R_2} + \frac{1-\nu}{r}\frac{\partial r}{\partial x}\frac{R_1 - R_2}{R_1 R_2}\right)\sqrt{^*A_{11}} + \frac{F'_{(1)}}{K}{^*A_{11}}$$

$$- \frac{h\rho}{K}{^*A_{11}}\frac{\partial^2 V'_{(1)}}{\partial t^2} = 0 \tag{7.4.4a}$$

$$\frac{\partial^2 V'_{(1)}}{\partial x\,\partial\theta}\frac{1+\nu}{2} + \frac{\partial^2 V'_{(2)}}{\partial x^2}\frac{1-\nu}{2}\frac{r}{\sqrt{^*A_{11}}} + \frac{\partial^2 V'_{(2)}}{\partial\theta^2}\frac{\sqrt{^*A_{11}}}{r}$$

$$+ \frac{\partial V'_{(1)}}{\partial\theta}\left(\frac{1-\nu}{2}\frac{\partial h/\partial x}{h} + \frac{\partial r/\partial x}{r}\right) + \frac{\partial V'_{(2)}}{\partial x}\frac{1-\nu}{2}\left(\frac{\partial h/\partial x}{h} + \frac{\partial r/\partial x}{r}\right.$$

$$\left. - \frac{\partial^*A_{11}/\partial x}{2^*A_{11}}\right)\frac{r}{\sqrt{^*A_{11}}} - V'_{(2)}\frac{1-\nu}{2}\left[\frac{\partial^2 r/\partial x^2}{r} + \left(\frac{\partial r/\partial x}{r}\right)^2 + \frac{\partial r/\partial x}{r}\frac{\partial h/\partial x}{h}\right.$$

$$\left. - \frac{\partial r/\partial x}{r}\frac{\partial^*A_{11}/\partial x}{2^*A_{11}}\right]\frac{r}{\sqrt{^*A_{11}}} - \frac{\partial V'_{(3)}}{\partial\theta}\frac{R_1 + \nu R_2}{R_1 R_2}\sqrt{^*A_{11}} + \frac{F'_{(2)}}{K}r\sqrt{^*A_{11}}$$

$$- \frac{h\rho}{K}r\sqrt{^*A_{11}}\frac{\partial^2 V'_{(2)}}{\partial t^2} = 0 \tag{7.4.4b}$$

$$\frac{\partial^2 V'_{(3)}}{\partial x^2}\frac{K_1}{\sqrt{^*A_{11}}} + \frac{\partial^2 V'_{(3)}}{\partial\theta^2}\frac{K_2}{r} + \frac{\partial V'_{(1)}}{\partial x}\left(\frac{R_2 + \nu R_1}{R_1 R_2} + \frac{K_1}{R_1} - \frac{K_2}{R_2}\right)$$

$$+ \frac{\partial V'_{(2)}}{\partial\theta}\left(\frac{R_1 + \nu R_2}{R_1 R_2} - \frac{K_1}{R_1} + \frac{K_2}{R_2}\right)\frac{\sqrt{^*A_{11}}}{r} - \frac{\partial V'_{(3)}/\partial x}{\sqrt{^*A_{11}}}\left(K_1\frac{\partial^*A_{11}/\partial x}{2^*A_{11}}\right.$$

$$\left. - K_2\frac{\partial r/\partial x}{r}\right) + V'_{(1)}\left[\frac{\partial r/\partial x}{r}\frac{R_1 + \nu R_2}{R_1 R_2} - \frac{K_1}{R_1}\left(\frac{\partial R_1/\partial x}{R_1} + \frac{\partial r/\partial x}{r}\right)\right.$$

$$+ \left. \frac{K_2}{R_2} \frac{\partial r / \partial x}{r} \right] - V'_{(3)} \left(\frac{R_2 + \nu R_1}{R_1} + \frac{R_1 + \nu R_2}{R_2} - K_1 - K_2 \right) \frac{\sqrt{*A_{11}}}{R_1 R_2}$$

$$+ \frac{F'_{(3)}}{K} \sqrt{*A_{11}} - \frac{h\rho}{K} \sqrt{*A_{11}} \frac{\partial^2 V'_{(3)}}{\partial t^2} = 0 \qquad (7.4.4c)$$

It was shown previously that the approximate static equations equivalent to Eqs. (7.4.4) gave solutions in close agreement with the solutions to the more exact formulation. Also, both of those solutions were significantly different from the solutions to the equivalent classic membrane equations where the effects of initial stress are ignored. In a succeeding section, the solutions to both the approximate and the exact equations, Eqs. (7.4.4) and (7.4.2), respectively, will be compared for sample dynamic problems.

7.4.2   Solution technique for free vibration

In the consideration of the natural frequencies and associated mode shapes for the asymmetric superposed equations of motion, all the forcing functions are assumed to be zero and the displacements to be harmonic functions:

$$V'_{(1)}(x, \theta, t) = V'^*_{(1)}(x, \theta) \exp (i\omega t) \qquad (7.4.5a)$$

$$V'_{(2)}(x, \theta, t) = V'^*_{(2)}(x, \theta) \exp (i\omega t) \qquad (7.4.5b)$$

$$V'_{(3)}(x, \theta, t) = V'^*_{(3)}(x, \theta) \exp (i\omega t) \qquad (7.5.5c)$$

If in addition, the functions are expanded in Fourier series, it can be seen that Eqs. (7.4.2) and (7.4.4) admit solutions of the form (torsional vibrations are ignored)

$$V'^*_{(1)}(x, \theta) = \sum_{n=0}^{\infty} V'^n_{(1)}(x) \cos n\theta \qquad (7.4.6a)$$

$$V'^*_{(2)}(x, \theta) = \sum_{n=0}^{\infty} V'^n_{(2)}(x) \sin n\theta \qquad (7.4.6b)$$

$$V'^*_{(3)}(x, \theta) = \sum_{n=0}^{\infty} V'^n_{(3)}(x) \cos n\theta \qquad (7.4.6c)$$

When trial ω is substituted into the equations of motion, the combined initial-value and boundary-value problem is converted into an equivalent boundary-value problem for which one method of solution has been presented in the previous section. The trial frequencies are tested by calculating a determinant associated with the equivalent static solution. Because the elements of this determinant (described in the following paragraphs) are continuous functions of ω, a change in sign of the determinant for two different trial values of ω implies the existence of a natural frequency between those two trial values. Then, using binary search techniques, one can obtain better approximations to the natural frequency.

For an assumed value of ω, the sixth-order boundary-value problem is transformed into three initial-value problems by assuming three independent sets of homogeneous boundary conditions at one edge of the membrane, and the solution to each homogeneous initial-value problem is propagated along a typical meridian. In order that the homogeneous boundary conditions at the edge be satisfied, the linear homogeneous equations formed from the corresponding portions of the partial solutions must have a nontrivial solution. In other words, the determinant of the coefficient matrix for the partial solutions corresponding to the boundary conditions must be zero.

$$
\left\{ \begin{array}{c} \text{True} \\ \text{boundary} \\ \text{conditions} \end{array} \right\} = \left[\begin{array}{c} \text{boundary values} \\ \text{for} \\ \text{partial solutions} \end{array} \right] \left\{ \begin{array}{c} \alpha_1 \\ \alpha_2 \\ \alpha_3 \end{array} \right\} = \left\{ \begin{array}{c} 0 \\ 0 \\ 0 \end{array} \right\}
$$

Therefore

$$
\det \left[\begin{array}{c} \text{boundary values} \\ \text{for} \\ \text{partial solutions} \end{array} \right] = 0
$$

in which α_i = the constants of the combination. If the determinant is not zero, the trial value of ω is not a natural frequency. After two values of ω are found for which the signs of the corresponding determinants are opposite, a half-interval search is instituted to find the bracketed value of ω corresponding to a natural frequency.

If the starting point for the numerical integration of the initial-value problems is the apex, special methods must be used in the neighborhood of the apex. Because it is not possible to consider all terms in the Taylor series expansions, it is necessary to use a successive approximation scheme for propagating solutions in the neighborhood of the apex. It was found that in cases where the geometry of the membrane changes rapidly near the apex, e.g., a paraboloid, more rapid convergence of the successive approximations was possible if, instead of using Taylor expansions of the displacement functions ($V'_{(1)}$, $V'_{(2)}$, $V'_{(3)}$), Taylor expansions of $V'_{(1)}/R_1$, $V'_{(2)}/r$, and $V'_{(3)}$ were used.

7.4.3   Sample solutions: free vibrations

In order to test the solution procedures described in the previous section for both the exact and the approximate equations of motion, several illustrative examples were solved for their natural frequencies and mode shapes, only the $n = 0, 1, 2$ harmonics being considered. A computer program was developed which was capable of handling a membrane of revolution with a general meridional configuration. The program was tested on several sample configurations including (1) a sphere continuous at the apex and fixed or simply supported at the base; and (2) a paraboloid of revolution continuous at the apex and fixed or simply supported at the base. The results for these problems are presented in this section.

The first example considered was an initially stressed spherical segment fixed at its base as shown in Fig. 7.4.1. The initial stress state was attained via internal pressure $\bar{p}$. Several numerical problems were considered. Figures 7.4.2 and 7.4.3 depict the first three natural displacement and stress mode shapes for each of the first two harmonics of a sphere with an opening angle of $\phi_0 = 45°$. The properties of the initially stressed membrane considered are $E = 300,000$ psi (20.7 GPa), $\nu = 0.4$, $h = 0.02$ ft (6.1 mm), $R = 100$ ft (30.5 m), and $n_{(11)} = n_{(22)} = 20,000$ lb/ft (2.92×10^5 N/m). When the approximate formulation was used, results were obtained which were in close agreement to the more exact solution. See Table 7.4.1 for a comparison of solutions for a sphere with an opening angle of 45°.

It is of interest to note that for a steep membrane vibrating in its lowest axisymmetric mode there exists a node, $V'_{(3)} = 0$, at an interior point. The location of the node varies with the opening angle considered. Therefore,

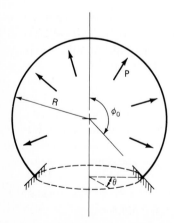

Figure 7.4.1 Sphere continuous at apex and fixed at base.

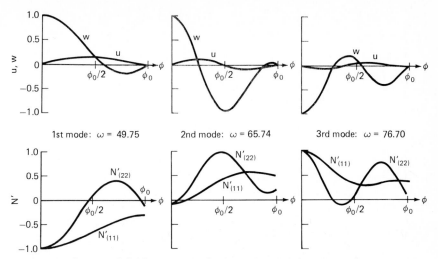

Figure 7.4.2 Stress and displacement mode shapes ($n = 0$) for sphere.

a uniform, dynamic prestress loading cannot be adequately represented by only one mode when the modal analysis method for forced vibrations (described in a following section) is used. In Fig. 7.4.4 the normal displacements for the first axisymmetric mode shapes of spheres with various opening angles are plotted. For extremely steep membranes, note that the node occurs near $\phi_0/2$.

The effects of the thickness and initial stress parameters were also studied. For a membrane of constant thickness, both the thickness h and the

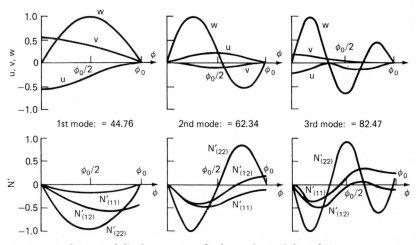

Figure 7.4.3 Stress and displacement mode shapes ($n = 1$) for sphere.

TABLE 7.4.1 Comparison of Exact and Approximate Formulations for Sphere Fixed at $\phi_0 = 45°$

	$n = 0$				$n = 1$			
ϕ^0	$V_{(3)}$ (exact)	$V_{(3)}$ (approx.)	$n'_{(11)}$ (exact)	$n'_{(11)}$ (approx.)	$V_{(3)}$ (exact)	$V_{(3)}$ (approx.)	$n'_{(11)}$ (exact)	$n'_{(11)}$ (approx.)
	ω_1 (exact) = 49.75	ω_1 (approx.) = 49.75	ω_1 (approx.) = 49.13	ω_1 (approx.) = 49.13	ω_1 (exact) = 44.76	ω_1 (approx.) = 44.76	ω_1 (exact) = 43.93	ω_1 (approx.) = 43.93
0	1.000	1.000	−100.0	−100.0	0.000	0.000	−0.0	−0.0
9	0.815	0.815	−93.1	−93.1	0.611	0.609	−56.8	−55.7
18	0.383	0.382	−75.7	−77.7	0.967	0.965	−93.5	−92.5
27	−0.028	−0.030	−54.6	−57.6	0.930	0.930	−97.5	−97.6
36	−0.175	−0.176	−37.4	−40.3	0.541	0.544	−68.5	−69.3
45	0.000	0.000	−29.4	−30.7	0.000	0.000	−16.5	−14.8
	ω_2 (exact) = 65.74		ω_2 (approx.) = 65.43		ω_2 (exact) = 62.34		ω_2 (approx.) = 61.76	
0	1.000	1.000	−14.3	−12.3	0.000	0.000	−0.0	−0.0
9	0.337	0.338	8.8	8.6	0.922	0.920	−75.0	−71.2
18	−0.730	−0.722	57.6	54.2	0.745	0.747	−100.0	−100.0
27	−0.837	−0.818	93.1	90.5	−0.166	−0.160	−73.0	−80.7
36	−0.185	−0.168	99.0	99.5	−0.546	−0.542	−35.8	−45.2
45	−0.000	−0.000	90.0	91.0	0.000	0.000	−26.0	−29.3
	ω_3 (exact) = 76.69		ω_3 (approx.) = 76.43		ω_3 (exact) = 82.47		ω_3 (approx.) = 82.02	
0	−1.000	−1.000	100.0	100.0	0.000	0.000	0.0	0.0
9	−0.415	−0.414	79.5	81.5	0.970	0.990	−82.0	−75.2
18	0.211	0.205	44.0	47.5	−0.155	−0.155	−21.5	−21.5
27	−0.127	−0.135	30.8	32.4	−0.528	−0.532	85.0	78.0
36	−0.382	−0.383	36.7	37.4	0.376	0.373	92.2	94.8
45	0.000	0.000	37.9	40.3	0.000	0.000	73.5	73.6

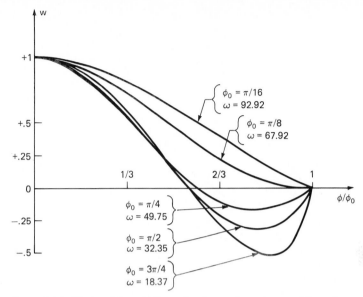

Figure 7.4.4 First axisymmetric mode shapes for spheres with various opening angles.

initial stress parameter $*n_{(11)}$ are contained within the single parametric expression:

$$K_1 = \frac{*n_{(11)}(1 - \nu^2)}{Eh}$$

Therefore, for a given initial stress, the effect of a change in thickness can be obtained by maintaining a constant thickness and changing the initial stress. Figure 7.4.5 shows the effect of initial stress on the values of the natural frequencies of a sphere with an opening angle of 45°. It was also found that the initial stress had no appreciable effect on the shapes of the natural modes for cases in which the ratio of initial stress to Young's modulus is small. In cases with large ratios of initial stress to Young's modulus, that ratio has been found [15, 42] to have a significant effect on the modal shape, e.g., in the development and location of node points.

In order to demonstrate the solution procedure on a sample problem with a more complex geometry than that of a sphere, the paraboloid of revolution shown in Fig. 7.4.6 was considered. Plotted in Fig. 7.4.7 are the first three displacement mode shapes for the first two harmonics of a paraboloid continuous at the apex and fixed at $\phi_0 = 45°$. The data for the particular paraboloid considered in Fig. 7.4.6 are $E = 300,000$ psi (20.7 GPa), $\nu = 0.4$, $h = 0.011$ ft, $a = 50$ ft (15.25 m), and $\bar{p} = 500$ psf (23.95

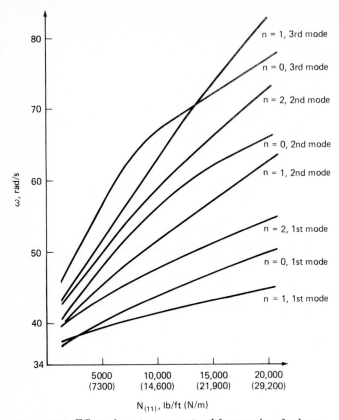

Figure 7.4.5 Effect of prestress on natural frequencies of spheres.

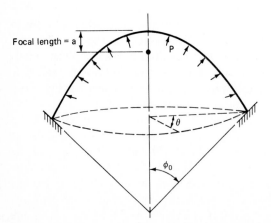

Figure 7.4.6 Paraboloid of revolution.

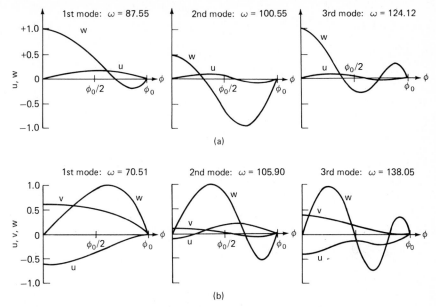

Figure 7.4.7 Mode shapes for paraboloid of revolution. (a) $n = 0$; (b) $n = 1$.

kPa), where $\bar{p}$ = the internal pressure used to calculate the initial stress resultants.

7.4.4 Forced vibrations

Once the natural frequencies and associated mode shapes of free vibration for an initially stressed membrane of revolution are calculated, it is possible to determine the response of the membrane to time-dependent loading conditions by means of the modal analysis method. In this method it is assumed the transient displacements can be represented as an infinite, convergent sum of the mode shapes of free vibration, each mode shape being multiplied by a time-dependent participation factor. The problem is then reduced to the determination of a sufficient number of mode participation factors to adequately represent the forced response of the initially stressed membrane.

The equations of motion can be written as

$$L_1(V'_{(1)}, V'_{(2)}, V'_{(3)}) + F'_{(1)}\sqrt{*A_{11}} - h\rho\sqrt{*A_{11}}\,\frac{\partial^2 V'_{(1)}}{\partial t^2}$$

$$- hC\sqrt{*A_{11}}\,\frac{\partial V'_{(1)}}{\partial t} = 0 \quad (7.4.8a)$$

$$L_2(V'_{(1)}, V'_{(2)}, V'_{(3)}) + F'_{(2)}r - h\rho r \frac{\partial^2 V'_{(2)}}{\partial t^2} - hCr \frac{\partial V'_{(2)}}{\partial t} = 0 \quad (7.4.8b)$$

$$L_3(V'_{(1)}, V'_{(2)}, V'_{(3)}) + F'_{(3)} - h\rho \frac{\partial^2 V'_{(3)}}{\partial t^2} - hC \frac{\partial V'_{(3)}}{\partial t} = 0 \quad (7.4.8c)$$

in which $L_i(V'_{(1)}, V'_{(2)}, V'_{(3)})$ = the differential operator related to the equivalent static problem; and C = the coefficient of viscous damping. Note that Eqs. (7.4.8) are a slight generalization of Eqs. (7.4.2) or (7.4.4) in that a first approximation to the effect of viscous damping has been included. It is assumed that the membrane is underdamped, i.e., $C < C_{cr}$, in which C_{cr} = the coefficient of critical damping.

Let the displacements be expanded in terms of the natural modes as follows:

$$V'_{(1)}(x, \theta, t) = \sum_{n=0}^{\infty} \sum_{i=1}^{\infty} V''^n_{(1)i}(x^1) \cos n\theta f_{ni}(t) \quad (7.4.9a)$$

$$V'_{(2)}(x, \theta, t) = \sum_{n=1}^{\infty} \sum_{i=1}^{\infty} V''^n_{(2)i}(x^1) \sin n\theta f_{ni}(t) \quad (7.4.9b)$$

$$V'_{(3)}(x, \theta, t) = \sum_{n=0}^{\infty} \sum_{i=1}^{\infty} V''^n_{(3)i}(x^1) \cos n\theta f_{ni}(t) \quad (7.4.9c)$$

in which $V''^n_{(1)i}$, $V''^n_{(2)i}$, $V''^n_{(3)i}$ = the modal displacements for the ith ordered mode of the nth harmonic; and $f_{ni}(t)$ = the corresponding time-dependent mode participation factor.

Equations (7.4.8) are combined as follows [42]: (1) Eqs. (7.4.9) are substituted into Eqs. (7.4.8); (2) the equations of motion for free vibration are used to eliminate each of the L_i values; (3) the equations are multiplied by $V''^n_{(1)i} \cos n\theta$, $V''^n_{(2)i} \sin n\theta$, $V''^n_{(3)i} \cos n\theta$, respectively, and then added together; (4) the result is then integrated over the surface area of the initially stressed middle surface; and (5) orthogonality properties of the natural modes are used to eliminate terms. When the subject steps are completed, the ordinary differential equation

$$\ddot{f}_{ni} + 2\omega_{ni}\beta\dot{f}_{ni} + f_{ni} = Q_{ni}(t) \quad (7.4.10a)$$

is obtained, in which

$$Q_{ni}(t) = \frac{1}{\rho K_{ni}} \int_a^b (V''^n_{(1)i}F''^n_{(1)} + V''^n_{(2)i}F''^n_{(2)}$$

$$+ V''^n_{(3)i}F''^n_{(3)}) \frac{r\sqrt{*A_{11}}}{h} dx \quad (7.4.10b)$$

$$K_{ni} = \int_a^b [(V''^n_{(1)i})^2 + (V''^n_{(2)i})^2 + (V''^n_{(3)i})^2] r \sqrt{*A_{11}} \, dx \qquad (7.4.10c)$$

In the preceding equations, $f_{ni} = df_{ni}/dt$; $\omega_{ni} =$ the ith ordered natural frequency for the nth harmonic; $\beta =$ the ratio of the coefficient of damping to the coefficient of critical damping; $K_{ni} =$ the modal normalization constant; and $F'_{(i)}(x, \theta, t)$ have been expanded as Fourier series in θ.

The problem of determining the transient response of initially stressed membranes for which natural frequencies and mode shapes are available has been reduced to that of solving the classic initial-value problem posed by Eqs. (7.4.10). The initial-value problem can be integrated by means of Duhamel integral techniques if the transient load vector is expressible in terms of a single forcing function, or by means of Newmark's method. These techniques have been described for multiple-degree-of-freedom systems in Secs. 4.3 and 5.1.

To treat a particular forced vibration problem, using as a basis for solution the superposition theory presented herein, it is necessary to select a value for ε, the small nondimensional parameter defined by the superposed load vector. If the dynamic load vector superposed is denoted by $\Delta \overline{F}(t)$, a reasonable measure for ε is

$$\varepsilon = \frac{\displaystyle\int_0^{2\pi} \int_a^b \Delta \overline{F}_{max}(t) r \sqrt{*A_{11}} \, dx \, d\theta}{\displaystyle\int_0^{2\pi} \int_a^b \overline{F} r \sqrt{*A_{11}} \, dx}$$

in which $\overline{F} =$ the static load vector on the initially stressed membrane.

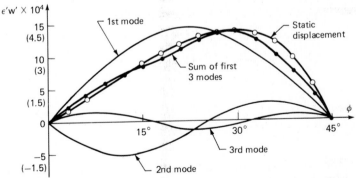

Figure 7.4.8 Comparison of static and dynamic results of wind-loaded sphere.

To illustrate the method, consider the same wind-loaded sphere as considered in Sec. 7.3.3. For a dynamic wind [$\varepsilon\bar{p}$ = 40 psf (1.9 kPa)] load on a sphere continuous at the apex and fixed at ϕ_0 = 45°, the first three partial sums (at $t = \infty$) of the normal modes multiplied by the participation factors are compared in Fig. 7.4.8 to the equivalent static solution. It can be seen that only three modes need be considered in order to give answers in close agreement with the static solution.

Nonlinear Behavior of Membranes

There are few nonlinear solution techniques for membrane structures that do not require numerical solutions on digital computers. The perturbation method described in Sec. 8.2 could be used for hand calculations of membranes of revolution during the prestressing phase. The method consists of superposing an infinite set of asymptotically convergent linear solutions to the nonlinear equations of a membrane with the deformed surface used as reference. Also, in Sec. 7.5 of Ref. 5, an approximate nonlinear solution is developed which is useful in the region adjoining the end closures of inflated circular tubes. That solution method can be used to account for the significant deviation of linear results for momentless shells from those predicted by nonlinear theory at edges of membranes. Such deviations were noted previously in Chap. 7.

A bibliography of nonlinear methods relying on computer simulations is given in Ref. 14 for works reported prior to 1973. More recent efforts directed specifically at membrane structures have relied on the use of finite element methodology and will be the primary focus of this chapter.

Based on the nonlinear field equations for membranes developed in Sec. 8.1, a superparametric finite element model for strongly curved membranes is presented in Sec. 8.3 which can be used for nonlinear static analysis during both the prestressing and in-service phases. It is often difficult to predict an initial reference shape which is both in equilibrium and has sufficient stiffness to enable further incremental analysis. The viscous relaxation technique [16, 89, 90] is an effective scheme for establishing an equilibrium configuration in such cases.

Dynamic response predictions are computationally intensive for membrane structures if nonlinear effects are included. Linear analyses (as described in Sec. 8.4.3) are often sufficient for purposes of initial analysis, but a nonlinear analysis is required for final validation of the structural design and for cases of extreme loading. An isoparametric finite element for the nonlinear dynamic analysis of membranes is presented in Sec. 8.4.

8.1 Nonlinear Field Equations for Prestressing of Membranes

This section is concerned with thin elastic membranes whose final desired shapes are obtained by prestressing and which undergo considerable displacement because of the prestressing. The behavior of membranes during the prestressing phase is such that classic linear theory [79, 119, 122] is inadequate for proper analysis. We will develop here an adequate nonlinear theory following that of Refs. 26 and 172. In a subsequent section (Sec. 8.3) we will develop equations for time-varying problems using a general lagrangian description of the motion with an arbitrarily located reference surface. In this section it will be tacitly assumed that the prestressed configuration $*C$ is the reference configuration for all calculations of both stresses and displacements. That is, the displacements are those from the initial state to the deformed state, but they are expressed in a coordinate system embedded in the deformed configuration.

The same notations and conventions as used in Chap. 7 are adopted here. All quantities on the deformed membrane in static equilibrium are denoted by using a leading superscript "$*$." Parameters of the undeformed membrane are denoted by using a leading superscript "^{0}." Greek subscripts and superscripts take on the values 1, 2; lowercase Latin subscripts and superscripts, the values 1, 2, 3. A repeated index indicates summation over the range of the index. Whenever a summation sign is used, the summation convention does not apply with respect to the script under the summation sign. Physical components of tensors are denoted by subscripts enclosed in parentheses. See Appendix B for a summary of tensor calculus.

8.1.1 Differential geometry

It is possible to form the field equations of a membrane in terms of the middle surface of the membrane and the unit normal to this surface. Because the desired membrane shape corresponds to the loaded membrane, the deformed middle surface $*C$ has been chosen as the reference surface. The thickness of the deformed membrane is denoted by $*h$. Let x^i be curvilinear coordinates embedded in $*C$, in which x^α = nondimensional middle surface coordinates, and x^3 = the distance along the unit normal $*\hat{N}$ from the deformed middle surface $*C$.

The position vector of a point on $*C$ is denoted by $*\overline{R}$. The covariant base vectors of $*C$ are (see Appendix B)

$$*\overline{A}_\alpha = \frac{\partial *\overline{R}}{\partial x^\alpha} \tag{8.1.1}$$

and the element of length on $*C$ is

$$(d*s)^2 = *A_{\alpha\gamma}\, dx^\alpha\, dx^\gamma \tag{8.1.2}$$

in which $*A_{\alpha\gamma}$ = the nondimensional metric tensor of C_R. The nondimensional curvature tensors of C_R are

$$*B_{\alpha\gamma} = \frac{1}{2}\left(*\overline{A}_\alpha \cdot \frac{\partial *\hat{N}}{\partial x^\gamma} + *\overline{A}_\gamma^- \cdot \frac{\partial *\hat{N}}{\partial x^\alpha}\right) \tag{8.1.3a}$$

$$*B_\alpha^\gamma = *A^{\gamma\rho}*B_{\alpha\rho} \tag{8.1.3b}$$

Covariant differentiation on $*C$ is denoted by a vertical line and expressed as

$$T^\alpha|_\gamma = \frac{\partial T^\alpha}{\partial x^\gamma} + *\Gamma^\alpha_{\rho\gamma} T^\rho \tag{8.1.4}$$

in which the Christoffel symbols of the second kind are

$$*\Gamma^\gamma_{\alpha\rho} = \frac{1}{2}\left(\frac{\partial *A_{\sigma\alpha}}{\partial x^\rho} + \frac{\partial *A_{\sigma\rho}}{\partial x^\alpha} - \frac{\partial *A_{\alpha\rho}}{\partial x^\sigma}\right)*A^{\gamma\sigma} \tag{8.1.5}$$

The gaussian curvature of C_R is defined as

$$*K = \frac{1}{*A}\,(*B_{11}*B_{22} - *B_{12}*B_{21}) \tag{8.1.6}$$

in which $*A$ = the determinant of $*A_{\alpha\gamma}$.

The position vector of a general point is

$$*\overline{R}(x^i) = [*\overline{R}(x^\alpha) + x^3*\hat{N}] \tag{8.1.7}$$

The covariant base vectors of a general point are

$$*\overline{G}_3 = *\hat{N} \qquad *\overline{G}_\alpha = (\delta_\alpha^\gamma - x^3*B_\alpha^\gamma)\,*\overline{A}_r \tag{8.1.8}$$

in which $\delta_\alpha^\gamma =$ the Kronecker delta. The metric tensor components at a general point are

$$*G_{\alpha\gamma} = *\overline{G}_\alpha \cdot *\overline{G}_\gamma = [*A_{\alpha\gamma} - 2x^{3}*B_{\alpha\gamma} + (x^3)^{2}*B_\alpha^\rho*B_{\gamma\rho}] \quad (8.1.9)$$

8.1.2 Equilibrium

The equilibrium of a shell element can now be considered. First the components of the three-dimensional stress tensor are replaced by an equivalent system of stress resultants acting on $*C$. The force resultant vectors for a thin membrane with negligible moments and transverse shears are

$$*\overline{n}^\alpha = *n^{\alpha\gamma} \frac{*\overline{A}_\gamma}{\sqrt{*A^{\alpha\alpha}}} \quad (8.1.10)$$

(no sum on α) in which $*n^{\alpha\gamma} =$ the force resultant tensors in the $*\overline{A}_\alpha$ direction which are given by

$$*n^{\alpha\gamma} = \int_{-*n/2}^{+*n/2} \sigma^{\alpha\rho}(\delta_\rho^\gamma - x_3*B_\rho^\gamma) \sqrt{\frac{G}{*A}}\, dx^3 \quad (8.1.11)$$

in which $G =$ the determinant of $*G_{\alpha\gamma}$, and $\sigma^{ij} =$ the components of the symmetric Cauchy stress tensor.

The Cauchy stress tensor is the natural measure of stress; it is referred to base vectors in $*C$ and measured per unit area of the deformed state. An alternative definition [18] which will be convenient to use in a succeeding section is that of the Kirchhoff stress tensor, τ^{ij}, which is referred to base vectors in ^0C and is measured per unit area of the undeformed state. The two terms are related by

$$\tau^{ij} = \sqrt{I_3}\, \sigma^{ij}$$

where $I_3 =$ third strain invariant $= \sqrt{*G/^0G}$, ^0G being the determinant of the metric tensor $^0G_{\alpha\gamma}$ of ^0C. Even if the material were incompressible, i.e., $*G = {}^0G$, the physical components of τ^{ij} differ from those of σ^{ij} in that base vectors in different coordinate systems are used to define them.

The load vector $*\overline{F}$ is the force on $*C$ per unit area of $*C$ and is given by

$$*\overline{F} = (*F^{\gamma}*\overline{A}_\gamma + *F^3*\hat{N}) \quad (8.1.12)$$

in which $*F^\gamma$ and $*F^3 =$ the components in the $*\overline{A}_\gamma$ and $*\hat{N}$ directions, respectively. The equations of force equilibrium are therefore given by

$$*n^\alpha|_\alpha^\gamma + *F^\gamma = 0 \quad (8.1.13a)$$

$$*n^{\rho\gamma}*B_{\gamma\rho} + *F^3 = 0 \quad (8.1.13b)$$

in which a vertical line denotes covariant differentiation (see Appendix B) with respect to coordinates on $*C$.

8.1.3 Strain equations

The strain-displacement relations are derivable by considering the displacement vector $\overline{\mathbf{V}}$ of a general point in the membrane

$$*\overline{\mathbf{V}} = [*\overline{V} + x^3 (*\hat{N} - {}^0_*\hat{N})] \tag{8.1.14}$$

in which $*\overline{V}$ = the displacement vector of the corresponding point on the middle surface of $*C$, and ${}^0_*\hat{N}$ = the unit normal to the undeformed surface ${}^0C$ but referred to coordinates and base vectors in $*C$. We hereafter adopt the notations that a leading subscript on a variable denotes the reference surface if it is different from the surface on which the variable is defined (denoted by the leading superscript).

From the general theory of strain [172],

$$2*\gamma_{ij} = *\overline{G}_i \cdot \frac{\partial *\overline{V}}{\partial x^j} + *\overline{G}_j \cdot \frac{\partial *\overline{V}}{\partial x^i} - \frac{\partial *\overline{V}}{\partial x^i} \cdot \frac{\partial *\overline{V}}{\partial x^j} \tag{8.1.15a}$$

Note that $*\overline{\mathbf{V}}$ is referred to a coordinate system embedded in $*C$. If we were to use instead (as in a succeeding section) a displacement vector ${}^*_0\overline{\mathbf{V}}$ referred to a coordinate system embedded in ${}^0C$, then in Eq. (8.1.15) we would change the sign preceding the last (nonlinear) term. Three strain invariants can be formed by $*\gamma_{ij}$:

$$I_1 = {}^0G^{rs}*G_{rs} \tag{8.1.15b}$$

$$I_2 = *G^{rs} \, {}^0G_{rs} \, I_3 \tag{8.1.15c}$$

$$I_3 = \frac{*G}{{}^0G} \tag{8.1.15d}$$

where ${}^0G_{rs} = *G_{rs} - 2*\gamma_{rs}$ = metric tensor in undeformed state ${}^0C$.

The strain components $*\gamma_{\alpha\gamma}$ become

$$2*\gamma_{\alpha\gamma} = \left[\left(*\overline{A}_\alpha \cdot \frac{\partial *\overline{V}}{\partial x^\gamma} + *\overline{A}_\gamma \cdot \frac{\partial *\overline{V}}{\partial x^\alpha} - \frac{\partial *\overline{V}}{\partial x^\alpha} \cdot \frac{\partial *\overline{V}}{\partial x^\gamma} \right) - 2x^3 \, (*B_{\alpha\gamma} - {}^0_*B_{\alpha\gamma}) \right.$$
$$\left. + (x^3)^2 (*B^\sigma_\gamma *B_{\alpha\sigma} - {}^0_*B^\rho_\gamma \, {}^0_*B_{\alpha\sigma}) + \cdots \right] \tag{8.1.16}$$

in which ${}^0_*B_{\alpha\gamma}$ = the curvature tensor of ${}^0C$ referred to $*C$.

Let

$$*\overline{V} = *V^{\gamma*}\overline{A}_\gamma + *V^{3*}\hat{N} \tag{8.1.17}$$

in which $*V^i$ = the tensor components of $*\overline{V}$ in the directions of the surface base vectors $*\overline{A}_\gamma$, $*\hat{N}$. Then

$$\frac{\partial *\overline{V}}{\partial x^\alpha} = \left[\frac{\partial *\overline{V}}{\partial x^\alpha} - \left(*B_\alpha^{\gamma*}\overline{A}_\gamma + \frac{\partial^0_* \hat{N}}{\partial x^\alpha} \right) \right] \tag{8.1.18a}$$

$$\frac{\partial *\overline{V}}{\partial x^3} = (*\hat{N} - {}^0_*\hat{N}) \tag{8.1.18b}$$

$$\frac{\partial^0_* \hat{N}}{\partial x^\alpha} = - {}^0_* B_\alpha^{\gamma 0}_* \overline{A}_\gamma \tag{8.1.18c}$$

$$\frac{\partial *\overline{V}}{\partial x^\alpha} = U_\alpha^{\gamma*}\overline{A}_\gamma + U_\alpha^{3*}\hat{N} \tag{8.1.18d}$$

in which

$$U_\alpha^\gamma = *V^\gamma|_\alpha - *V^{3*}B_\alpha^\gamma \tag{8.1.19a}$$

$$U_\alpha^3 = *V^3|_\alpha + *V^{\gamma*}B_{\gamma\alpha} \tag{8.1.19b}$$

Because membranes are extremely thin, it is assumed that the ratio of thickness to the smallest radius of curvature is small compared to unity. Therefore, if Eqs. (8.1.18) and (8.1.19) are substituted into Eq. (8.1.16),

$$*\gamma_{\alpha\gamma} = *\gamma_{\alpha\gamma} - x^{3*}\kappa_{\alpha\gamma} \tag{8.1.20a}$$

in which

$$*\gamma_{\alpha\gamma} = \tfrac{1}{2}(U_\alpha^{\rho*}A_{\rho\gamma} + U_\gamma^{\rho*}A_{\rho\alpha} - U_\alpha^\sigma U_\gamma^{\rho*}A_{\sigma\rho} - U_\alpha^3 U_\gamma^3) \tag{8.1.20b}$$

$$*\kappa_{\alpha\gamma} = (*B_{\alpha\gamma} - {}^0_* B_{\alpha\gamma}) \tag{8.1.20c}$$

8.1.4   Perfectly elastic behavior

When the strains to which a shell is subjected do not exceed certain physical constants, e.g., the yield limit, the behavior can be considered perfectly elastic. From the principle of virtual work,

$$\sigma_{ij} = \frac{1}{2}\left(\frac{\partial *U}{\partial *\gamma_{ij}} + \frac{\partial *U}{\partial *\gamma_{ji}} \right) \tag{8.1.21}$$

where $*U$ = strain energy density per unit volume of $*C$.

Various forms for the strain energy density $*U$ can be postulated [18, 50, 172, 187]. It can be expressed as a function of the strain invariants I_1, I_2, I_3 in Eqs. (8.1.15b through d). Thus, taking the indicated derivatives in Eq. (8.1.21a), we obtain

$$\sigma^{ij} = 2\left[{}^0G^{ij}\frac{\partial *U}{\partial I_1} + ({}^0G^{ij}I_1 - {}^0G^{ir0}G^{js}*G_{rs})\frac{\partial *U}{\partial I_2} + I_3*G^{ij}\frac{\partial *U}{\partial I_3} \right]$$

(8.1.22)

For an incompressible ($I_3 = 1$) rubber like material, the following isotropic linearized form of $*U$ is commonly adopted [50, 187]:

$$*U = C_1(I_1 - 3) + C_2(I_2 - 3)$$

with C_1 and C_2 as material constants. Then

$$\sigma^{ij} = 2C_1{}^0G^{ij} + 2C_2D^{ij} + p*G^{ij}$$

(8.1.23a)

with

$$D^{ij} = {}^0G^{ij}I_1 - {}^0G^{ir0}G^{js}*G_{rs}$$

(8.1.23b)

and $p = 2\dfrac{\partial *U}{\partial I_3}$ = hydrostatic pressure determinable from plane stress condition for the membrane:

$$p = -2({}^0_*\lambda)^2\left(C_1 + C_2\frac{*A}{{}^0A} \right)$$

(8.1.23c)

where ${}^0_*\lambda = \dfrac{{}^0h}{*h}$ = ratio of undeformed to deformed thickness and ${}^0_*A$ = determinant of ${}^0_*A_{\alpha\beta}$.

If it is assumed that the material is homogeneous and initially unstrained, $*U$ can be expressed as a power series in the strains. If the material is assumed linear (hookean) and isotropic in the final state, the stress-strain relations become

$$\sigma^{\alpha\gamma} = \left[\frac{E\nu}{(1 + \nu)(1 - 2\nu)} *A^{\alpha\gamma}*A^{\tau\rho} \right.$$

$$\left. + \frac{E}{2(1 + \nu)}(*A^{\alpha\tau}*A^{\gamma\rho} + *A^{\alpha\tau}*A^{\gamma\tau}) \right]*\gamma_{\tau\rho}$$

(8.1.24)

in which E = Young's modulus of elasticity, and ν = Poisson's ratio.

The relations between the force resultants and the displacements can be obtained by substituting Eqs. (8.1.20) and (8.1.24) into Eq. (8.1.11).

Then the thinness assumption can be used to eliminate all terms involving $*h^n$, for $n \geqslant 3$, compared to terms of order $*h$. This implies that σ^{33} is of the same order of magnitude or smaller than $n^{\alpha\gamma}$. Therefore, for a hookean isotropic material

$$
*n^{\alpha\gamma} = \frac{E*h}{2(1+\nu)} \left\{ \left[*A^{\alpha\rho} U_\rho^\gamma + *A^{\gamma\rho} U_\rho^\alpha + \frac{2\nu}{(1-\nu)} *A^{\alpha\gamma} U_\rho^\rho \right] \right.
$$
$$
- \frac{1}{2} \left[*A^{\alpha\rho}* A^{\gamma\tau} + *A^{\gamma\rho}* A^{\alpha\tau} + \frac{2\nu}{(1-\nu)} *A^{\alpha\gamma}* A^{\tau\rho} \right] \left(*A_{\beta\sigma} U_\tau^\beta U_\rho^\sigma \right.
$$
$$
\left. \left. - U_\tau^3 U_\rho^3 \right) \right\}
\tag{8.1.25}
$$

in which U_γ^α and U_α^3 are given by Eqs. (8.1.17).

8.1.5 Physical components, energy, and boundary conditions

The physical components of the force resultant tensors are the magnitudes of the force resultant vectors in the directions of the unit vectors $*\overline{A}_\gamma / \sqrt{*A_{\gamma\gamma}}$ (no sum):

$$
\overline{n}^\alpha = \sum_\gamma *n_{(\alpha\gamma)} \frac{*\overline{A}_\gamma}{\sqrt{*A_{\gamma\gamma}}}
\tag{8.1.26a}
$$

in which the physical components are

$$
*n_{(\alpha\gamma)} = *n^{\alpha\gamma} \sqrt{\frac{*A_{\gamma\gamma}}{*A^{\alpha\alpha}}} \qquad \text{(no sum)} \tag{8.1.26b}
$$

The physical components of the force and displacement vectors of $*C$ are the magnitudes in the direction of the unit vectors $*\overline{A}_\gamma / \sqrt{*A_{\gamma\gamma}}$ (no sum) and $*\hat{N}$, i.e.,

$$
\overline{V} = \sum_\gamma *V_{(\gamma)} \frac{*\overline{A}_\gamma}{\sqrt{*A_{\gamma\gamma}}} + *V_{(3)} *\hat{N}
\tag{8.1.27a}
$$

$$
\overline{F} = \sum_\gamma *F_{(\gamma)} \frac{*\overline{A}_\gamma}{\sqrt{*A_{\gamma\gamma}}} + *F_{(3)} *\hat{N}
\tag{8.1.27b}
$$

in which the physical components of displacement and force in the directions of the natural base vectors of the deformed surface $*C$ are

$$
*V_{(\gamma)} = *V^\gamma \sqrt{*A_{\gamma\gamma}} \qquad \text{(no sum)} \tag{8.1.27c}
$$

$$*V_{(3)} = *V_3 \qquad\qquad (8.1.27d)$$

$$*F_{(\gamma)} = *F^\gamma \sqrt{*A_{\gamma\gamma}} \qquad \text{(no sum)} \qquad (8.1.27e)$$

$$*F_{(3)} = *F^3 \qquad\qquad (8.1.27f)$$

To this point, only the field equations have been considered. In order to solve completely any particular membrane problem, it is necessary to specify the conditions prevailing at the boundaries of the membrane. Only membrane boundary conditions are considered in this study.

The principle of virtual work during a small virtual motion $\delta\overline{V}$ from the equilibrium state on $*C$ of a thin membrane is stated as

$$0 = \int_{*C}\int \overline{F} \cdot \delta\overline{V} \sqrt{*A}\, dx_1\, dx_2 - \int_{*C}\int *n^{\alpha\beta}\, \delta\gamma_{\alpha\beta} \sqrt{*A}\, dx_1\, dx_2 \tag{8.1.28}$$

where $\delta\overline{V}$ = virtual displacement vector with corresponding virtual strain $\delta\gamma_{\alpha\beta}$. Equation (8.1.28), or its equivalent expressed in terms of quantities referred to ^0C, will serve as the basis for the finite element models considered in subsequent sections. If $\delta\gamma_{\alpha\beta}$ in Eq. (8.1.28) were replaced by Eqs. (8.1.16) and then integrated by parts, the equations of equilibrium given by Eqs. (8.1.13) would be recovered plus additional terms involving contour integrals around $*C$. To satisfy the principle of virtual work these additional terms must be zero. Thus,

$$\sum_\alpha \sum_\rho *A_{\rho 1}\eta_{(\alpha)} \frac{*\eta_{(\alpha\rho)}}{\sqrt{*A_{\rho\rho}*A_{11}}}\, \delta V_{(1)}$$

$$+ \sum_\alpha \sum_\rho *A_{\rho 2}\eta_{(\alpha)} \frac{*\eta_{(\alpha\rho)}}{\sqrt{*A_{\rho\rho}*A_{22}}}\, \delta V_{(2)} = 0 \qquad (8.1.29a)$$

on the bounding curve $*S$ of the membrane, where $\eta_{(\alpha)}$ are the physical components of the unit outward normal to $*S$ on the surface $*C$,

$$\hat{\eta} = \eta_\alpha *\overline{A}^\alpha \tag{8.1.29b}$$

and $\delta V_{(\alpha)}$ are the physical components in the directions of the base vectors of $*C$ of the virtual displacement vector

$$\delta\overline{V} = \delta V^\gamma\, \overline{A}_\gamma + \delta V^3\, \hat{N} \tag{8.1.29c}$$

In this section the field equations and boundary conditions have been presented for a nonlinear theory of thin membrance shells. Only five assumptions have been made: (1) membrane behavior; (2) extreme thinness; (3) elastic behavior (hyperelastic or hookean); (4) homogeneous, initially unstrained material; and (5) isotropic material.

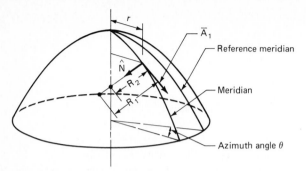

Figure 8.2.1   Geometry of membranes of revolution.

8.2   Prestressed Membranes of Revolution

In this section the field equations are specialized for shells of revolution loaded axisymmetrically, and a method of solution is presented for these equations. A surface of revolution is shown in Fig. 8.2.1. Let x^1 be some regular coordinate along the meridian of the shell. Let x^2 be the azimuth angle θ which a meridian plane makes with a reference meridional plane. The radius of curvature of a meridional line is R_1; the radius of curvature of a cross-sectional circle is R_2. In addition, r is the perpendicular distance from the axis of revolution to a point of $^*C$. Because the shell is axially symmetric, all quantities are independent of θ, and derivatives with respect to θ are zero. The important geometric quantities are

$$^*A_{11} = {}^*A_{11}(x^1) \qquad ^*A_{22} = (r)^2 \qquad ^*A_{12} = {}^*A_{21} = 0 \quad (8.2.1a)$$

$$^*B_{22} = \frac{(r)^2}{R_2} \qquad ^*B_{11} = \frac{A_{11}}{R_1} \qquad ^*B_{12} = {}^*B_{21} = 0 \qquad (8.2.1b)$$

Note that the unit normal is directed toward the axis of revolution.

For a shell of revolution, the physical components of the force resultant tensor, of the load vector, and of the displacement vector are

$$n_{(11)} = {}^*n^{11*}A_{11} \qquad n_{(22)} = {}^*n^{22}(r)^2 \qquad ^*n_{(12)} = {}^*n_{(21)} = 0 \quad (8.2.2a)$$

$$^*F_{(1)} = {}^*F^1\sqrt{^*A_{11}} \qquad ^*F_{(2)} = 0 \qquad ^*F_{(3)} = {}^*F^3 \qquad (8.2.2b)$$

and

$$^*V_{(1)} = {}^*V^1\sqrt{^*A_{11}} \qquad ^*V_{(2)} = 0 \qquad ^*V_{(3)} = {}^*V^3 \qquad (8.2.2c)$$

$$*n_{(22)} = \frac{R_2}{r*IR_1} \left[-R_1 r* I* F_{(3)} + \int \left(*F_{(3)} R_2 \frac{dr}{dx^1} \right. \right.$$

$$\left. \left. + *F_{(1)} r \sqrt{*A_{11}} \right) *I \, dx^1 + C_2 \right] \tag{8.2.6b}$$

in which C_2 is to be evaluated from the boundary conditions.

8.2.1 Perturbation method

A closed-form solution to Eqs. (8.2.4) for the displacements in terms of familiar functions does not seem likely in general. However, Eqs. (8.2.4) do lend themselves to an infinite series solution which is convergent if certain parametric inequalities are satisfied. By means of the implicit function theorems [190], Eqs. (8.2.4) can be reduced to a normal form:

$$\frac{d}{dx^1} *V_{(1)} = \epsilon f_1(*V_{(1)}, *V_{(3)}; x^1)$$

$$\frac{d}{dx^1} *V_{(3)} = \epsilon f_2(*V_{(1)}, *V_{(3)}; x^1)$$

in which $\epsilon = $ the small nondimensional parameter defined by

$$\epsilon = \frac{(1 - \nu^2)*P}{E} \tag{8.2.7}$$

with $*P = $ characteristic value of surface tractions.

It has been shown [172, 188, 191] that it is possible to apply a perturbation method to these equations and obtain solutions in the form

$$*V_{(1)} = \sum_{i=1}^{\infty} \epsilon^i * V_{(1)}^{(i)} \tag{8.2.8a}$$

$$*V_{(3)} = \sum_{i=1}^{\infty} \epsilon^i * V_{(3)}^{(i)} \tag{8.2.8b}$$

in which the $i = 0$ terms have been eliminated because, as $\epsilon \rightarrow 0$, $E \rightarrow \infty$. This formulation leads to sets of equations in which the left members of each set have the same coefficients as every other set. Only the right-hand side will vary with the script i on $*V_{(1)}^{(i)}$ and $*V_{(3)}^{(i)}$.

The displacement vector is assumed to be expanded in an absolutely convergent series in ϵ with a nonzero radius of convergence. It is further assumed that the term-by-term derivatives of the series expansions are also absolutely convergent. It has been shown for problems in nonlinear

Therefore, the physical forms of the membrane equilibrium equation

$$\frac{d}{dx^1}\left(\frac{*n_{(11)}}{*A_{11}}\right) + \left(\frac{1}{*A_{11}}\frac{d*A_{11}}{dx^1} + \frac{1}{r}\frac{dr}{dx^1}\right)\left(\frac{*n_{(11)}}{*A_{11}}\right)$$

$$-\frac{1}{r*A_{11}}*n_{(22)}\frac{dr}{dx^1} + \frac{*F_{(1)}}{\sqrt{*A_{11}}} = 0 \quad (8.2.2)$$

$$\left(\frac{*n_{(11)}}{R_1}\right) + \left(\frac{*n_{(22)}}{R_2}\right) + *F_{(3)} = 0 \quad (8.2.3)$$

It is possible to cast Eq. (8.1.25) completely into physical components

$$\left(\frac{1}{\sqrt{*A_{11}}}\frac{d}{dx^1}*V_{(1)} - \frac{1}{R_1}*V_{(3)}\right) + \nu\left(\frac{1}{\sqrt{*A_{11}}}\frac{1}{r}\frac{dr}{dx^1}*V_{(1)} - \frac{1}{R_2}*V_{(3)}\right)$$

$$-\frac{1}{2}\left(\frac{1}{\sqrt{*A_{11}}}\frac{d}{dx^1}*V_{(1)} - \frac{1}{R_1}*V_{(3)}\right)^2$$

$$+\frac{1}{2*A_{11}}\left(\frac{d}{dx^1}*V_{(3)} + \frac{\sqrt{*A_{11}}}{R_1}*V_{(1)}\right)^2 = \frac{*n_{(11)}(1-\nu^2)}{Eh} \quad (8.2.4a)$$

$$\left(\frac{1}{\sqrt{*A_{11}}}\frac{1}{r}\frac{dr}{dx^1}*V_{(1)} - \frac{1}{R_2}*V_{(3)}\right) + \nu\left(\frac{1}{\sqrt{*A_{11}}}\frac{d}{dx^1}*V_{(1)} - \frac{1}{R_1}*V_{(3)}\right)$$

$$-\frac{1}{2}\left(\frac{1}{\sqrt{*A_{11}}}\frac{1}{r}\frac{dr}{dx^1}*V_{(1)} - \frac{1}{R_2}*V_{(3)}\right)^2 = \frac{*n_{(22)}(1-\nu^2)}{Eh}$$

$$(8.2.4b)$$

Because we have assumed the deformed state $*C$ of the membrane is the reference upon which all geometric values are defined, we can develop a closed-form solution for the stress resultants. The equilibrium equations can be solved using an integrating factor,

$$*I = \exp\left(\int \frac{1}{r}\frac{R_2}{R_1}\frac{dr}{dx^1}dx^1\right) \quad (8.2.5)$$

The solutions are

$$*n_{(11)} = \frac{-1}{r*I}\left[\int\left(*F_{(3)}R_2\frac{dr}{dx^1} + *F_{(1)}r\sqrt{*A_{11}}\right)*I\,dx^1 + C_2\right] \quad (8.2.6a)$$

elasticity [192] that each term of the series solution exists and is unique if the corresponding linear solution exists and is unique.

The loads are known functions of x^1, and closed-form solutions for the force resultants have been determined. However, the loads and force resultants have been expanded in powers of ϵ in order to handle problems in which the two necessary boundary conditions are both of the displacement type. Let

$$*n_{(11)} = \sum_{i=1}^{\infty} \epsilon^{i*} n_{(11)}^{(i)} \qquad *n_{(22)} = \sum_{i=1}^{\infty} \epsilon^{i*} n_{(22)}^{(i)} \qquad (8.2.9a)$$

$$*F_{(1)} = \sum_{i=1}^{\infty} \epsilon^{i*} F_{(1)}^{(i)} \qquad *F_{(3)} = \sum_{i=1}^{\infty} \epsilon^{i*} F_{(3)}^{(i)} \qquad (8.2.9b)$$

in which

$$*n_{(11)}^{(i)} = \frac{-1}{r^*I} \left[\int \left(*F_{(3)}^{(i)} R_2 \frac{dr}{dx^1} + *F_{(1)}^{(i)} r \sqrt{*A_{11}} \right) *I \, dx^1 + C_2^{(i)} \right]$$

$$(8.2.10a)$$

$$*n_{(22)}^{(i)} = \frac{R_2}{r^*IR_1} \left[-r^*IR_1 *F_{(3)}^{(i)} + \int \left(*F_{(3)}^{(i)} R_2 \frac{dr}{dx^1} \right. \right.$$

$$\left. \left. + *F_{(1)}^{(i)} r \sqrt{*A_{11}} \right) *I \, dx^1 + C_2^{(i)} \right] \qquad (8.2.10b)$$

and $*F_{(1)}^{(i)} = *F_{(3)}^{(i)} = 0$ for $i > 2$. It is now possible to expand Eqs. (8.2.4) in powers of ϵ. If terms are then collected in powers of ϵ, both equations have the form

$$\epsilon\phi_1(x^1) + \epsilon^2\phi_2(x^1) + \cdots + \epsilon^n\phi_n(x^1) = 0$$

in which $\phi_1(x^1), \phi_2(x^1), \ldots, \phi_n(x^1)$ are independent of ϵ. Because the displacements are functions of ϵ, and because Eqs. (8.2.4) must be satisfied for any value of ϵ in the range $0 \leq \epsilon < \epsilon_0$, the coefficients of $\epsilon, \epsilon^2, \ldots, \epsilon^n$ must vanish independently, where ϵ_0 is some bounding value of the parameter ϵ given by Eq. (8.2.7).

When this expansion and collection of terms is performed, the following infinite set of linear differential equations results:

$$\left(\frac{1}{\sqrt{*A_{11}}} \frac{d}{dx^1} *V_{(1)}^{(n)} - \frac{1}{R_1} *V_{(3)}^{(n)} \right)$$

$$+ \nu \left(\frac{1}{\sqrt{*A_{11}}} \frac{1}{r} \frac{dr}{dx^1} *V_{(1)}^{(n)} - \frac{1}{R_2} *V_{(3)}^{(n)} \right) = \Psi^{(n)} \qquad (8.2.11a)$$

$$\left(\frac{1}{\sqrt{*A_{11}}} \frac{1}{r} \frac{dr}{dx^1} * V_{(1)}^{(n)} - \frac{1}{R_2} * V_{(3)}^{(n)} \right)$$

$$+ \nu \left(\frac{1}{\sqrt{*A_{11}}} \frac{d}{dx^1} * V_{(1)}^{(n)} - \frac{1}{R_1} * V_{(3)}^{(n)} \right) = \Lambda^{(n)} \quad (8.2.11b)$$

in which $\Psi^{(n)}$ and $\Lambda^{(n)}$ are the right sides of the nth set of linear equations:

$$\Psi^{(n)} = \Psi^{(n)} \left(* V_{(1)}^{(m)}, * V_{(3)}^{m}; x^1 \right) \qquad (8.2.12a)$$

$$\Lambda^{(n)} = \Lambda^{(n)} \left(* V_{(1)}^{(m)}, * V_{(3)}^{(m)}; x^1 \right) \qquad (8.2.12b)$$

in which $m = 1, 2, \ldots, (n-1)$. The first few values of $\Psi^{(n)}$ and $\Lambda^{(n)}$ are

$$\Psi^{(1)} = -\frac{(1-\nu^2)}{*hE*Ir} \left[\int \left(*F_{(3)}^{(1)} R_2 \frac{dr}{dx^1} + *F_{(1)}^{(1)} r \sqrt{*A_{11}} \right) *I \, dx^1 + C_2^{(1)} \right]$$

$$\Lambda^{(1)} = -\frac{(1-\nu^2)}{*hE*Ir} \left[R_1 r* I* F_{(3)}^{(1)} - \int \left(*F_{(3)}^{(1)} R_2 \frac{dr}{dx^1} \right. \right. \qquad (8.2.13a)$$

$$\left. \left. + *F_{(1)}^{(1)} r \sqrt{*A_{11}} \right) *I \, dx^1 + C_2^{(1)} \right] \quad (8.2.13b)$$

$$\Psi^{(2)} = \frac{1}{2} \left[\left(\frac{1}{\sqrt{*A_{11}}} \frac{d}{dx^1} * V_{(1)}^{(1)} - \frac{1}{R_1} * V_{(3)}^{(1)} \right)^2 \right.$$

$$\left. + \frac{1}{*A_{11}} \left(\frac{d}{dx^1} * V_{(3)}^{(1)} + \frac{\sqrt{*A_{11}}}{R_1} * V_{(1)}^{(1)} \right)^2 \right] - \frac{(1-\nu^2)}{*hEr*I} C_2^{(2)} \quad (8.2.14a)$$

$$\Lambda^{(2)} = \frac{1}{2} \left(\frac{1}{\sqrt{*A_{11}}} \frac{1}{r} \frac{dr}{dx^1} * V_{(1)}^{(1)} - \frac{1}{R_2} * V_{(3)}^{(1)} \right)^2 + \frac{(1-\nu^2)}{*hEr*I} \frac{R_2}{R_1} C_2^{(2)} \quad (8.2.14b)$$

$$\Psi^{(3)} = \left(\frac{1}{\sqrt{*A_{11}}} \frac{d}{dx^1} * V_{(1)}^{(1)} - \frac{1}{R_1} * V_{(3)}^{(1)} \right) \left(\frac{1}{\sqrt{*A_{11}}} \frac{d}{dx^1} * V_{(1)}^{(2)} \right.$$

$$\left. - \frac{1}{R_2} * V_{(3)}^{(2)} \right) + \frac{1}{*A_{11}} \left(\frac{d}{dx^1} * V_{(3)}^{(1)} + \frac{\sqrt{*A_{11}}}{R_1} V_{(1)}^{(1)} \right) \left(\frac{d}{dx^1} * V_{(3)}^{(2)} \right.$$

$$\left. + \frac{\sqrt{*A_{11}}}{R_1} * V_{(1)}^{(2)} \right) - \frac{(1-\nu^2)}{*hEr*I} C_2^{(3)} \quad (8.2.15a)$$

$$\Lambda^{(3)} = \left(\frac{1}{\sqrt{*A_{11}}} \frac{1}{r} \frac{dr}{dx^1} * V_{(1)}^{(1)} - \frac{1}{R_2} * V_{(3)}^{(1)} \right) \left(\frac{1}{\sqrt{*A_{11}}} \frac{1}{r} \frac{dr}{dx^1} * V_{(1)}^{(2)} \right.$$

$$\left. - \frac{1}{R_2} * V_{(3)}^{(2)} \right) + \frac{(1 - \nu^2)}{*hEr*I} \frac{R_2}{R_1} C_2^{(3)} \tag{8.2.15b}$$

The rest of the $\Psi^{(n)}$ and $\Lambda^{(n)}$, $n > 3$, can be obtained by completing the expansion of Eqs. (8.2.4).

An important conclusion can be drawn from the above expansion technique: nonlinear material properties would have no effect on the determination of the displacements until the third terms in the expansions are considered. Therefore, because ϵ is small compared to unity for the usual choice of materials for membranes, physically reasonable material nonlinearities have little effect on the displacements in comparison to the influence of the geometric nonlinearities.

In the perturbation method, each approximation is not independent of the other terms in the expansions. However, the solution for each term depends only on the preceding terms, which will have already been determined. That is, the solutions for $* V_{(1)}^{(n)}$ and $* V_{(3)}^{(n)}$ depend only on $* V_{(1)}^{(i)}$ and $* V_{(3)}^{(i)}$, where $i = 1, 2, \ldots , (n - 1)$.

Equations (8.2.11) can be solved using an integrating factor. The nth terms in the expansions are

$$* V_{(1)}^{(n)} = \frac{*I}{(1 - \nu^2)} \left\{ \int \frac{\sqrt{*A_{11}}}{R_1} \left[\Psi^{(n)} (R_1 + \nu R_2) \right. \right.$$

$$\left. \left. - \Lambda^{(n)}(\nu R_1 + R_2) \right] \frac{dx^1}{*I} + C_3^{(n)} \right\} \tag{8.2.16a}$$

$$* V_{(3)}^{(n)} = - \frac{R_2}{(1 - \nu^2)} \left(\Lambda^{(n)} - \nu \Psi^{(n)} \right) + \frac{R_2}{\sqrt{*A_{11}}} \frac{1}{r} \frac{dr}{dx^1} * V_{(1)}^{(n)} \tag{8.2.16b}$$

in which the first few values of $\Psi^{(n)}$ and $\Lambda^{(n)}$ are given by Eqs. (8.2.13) to (8.2.15). It is possible to solve Eqs. (8.2.16) repeatedly for $n = 1, 2, 3, \ldots$ until the desired accuracy is obtained. The complete solutions for the displacements are then

$$* V_{(1)} = \left(\epsilon * V_{(1)}^{(1)} + \epsilon^2 * V_{(1)}^{(2)} + \epsilon^3 * V_{(1)}^{(3)} + \ldots \right) \tag{8.2.17a}$$

$$* V_{(3)} = \left(\epsilon * V_{(3)}^{(1)} + \epsilon^2 * V_{(3)}^{(2)} + \epsilon^3 * V_{(3)}^{(3)} + \ldots \right) \tag{8.2.17b}$$

in which ϵ is given by Eq. (8.2.7). Note that the first terms in the expansions correspond to the classic linear solutions for the displacements.

The solutions developed in this section are valid in regions where there are no discontinuities in load, thickness, stiffness, or curvature. In the neighborhood of discontinuities, special methods must be used. The solutions, Eqs. (8.2.16), exist if $\sqrt{*A_{11}}$, $1/r$, dr/dx^1, R_1, and R_2 are bounded continuous functions of x^1.

8.2.2 Sample problem

In order to demonstrate the method of solution derived for the prestressing phase, a sphere problem is considered in this section. The method of solution has been tested on more complex shell problems [15, 171, 188], but this simple problem has been chosen because a compact solution exists and it is possible to prove convergence for this example.

For the purpose of this study the thickness is assumed constant, and a constant internal pressure, with no tangential loads, is used as the load system. A fictitious, but fairly realistic, material is assumed, with the properties $E = 300,000$ psi (2.07 GPa) and $\nu = 0.4$. These properties represent those of various commercial plastics.

The example considered is the spherical segment shown in Fig. 8.2.2. The x^1 coordinate is chosen as ϕ, the angle the normal to the meridian forms with the axis of revolution. Therefore,

$$R_1 = R_2 = R \qquad *A_{11} = R^2$$

$$r = R \sin \phi \qquad \frac{dr}{dx^1} = R \cos \phi \qquad *I = \sin \phi$$

in which $R =$ the radius of the deformed shell. The segment is restrained at the edge $\phi = \phi_0$ such that no vertical displacement Δ_v is allowed. Therefore, $*n_{(11)}$ and $*n_{(22)}$ are finite at $\phi = 0$, and

$$\Delta_{v|\phi = \phi_0} = (*V_{(1)} \sin \phi_0 + *V_{(3)} \cos \phi_0) = 0 \qquad (8.2.18)$$

From Eqs. (8.2.9),

$$p = \sum_{n=1}^{\infty} \epsilon^{n*} F_{(3)}^{(n)} \qquad 0 = \sum_{m=1}^{\infty} \epsilon^{n*} F_{(1)}^{(n)} \qquad (8.2.19)$$

in which $p =$ the applied normal pressure, and in this case the magnitude of $*P$ is equal to that of p. Thus,

$$*F_{(3)}^{(1)} = \frac{E}{(1 - \nu^2)} \qquad (8.2.20a)$$

$$*F_{(1)}^{(n)} = 0 \text{ for } n \geqslant 1 \qquad *F_{(3)}^{(n)} = 0 \text{ for } n \geqslant 2 \qquad (8.2.20b)$$

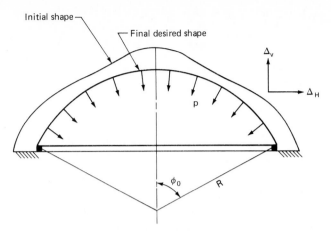

Figure 8.2.2   Prestressing of a spherical segment.

The above are substituted into Eqs. (8.2.10), (8.2.13), and (8.2.16), the integration is performed, and Eq. (8.2.18) is satisfied. This yields

$$*n^{(1)}_{(11)} = *n^{(1)}_{(22)} = \frac{R}{2} \frac{E}{(1 - \nu^2)} \tag{8.2.21a}$$

$$*V^{(1)}_{(1)} = R^2 \frac{\sin \phi \cos \phi_0}{2*h(1 - \nu^2)} \tag{8.2.21b}$$

$$*V^{(1)}_{(3)} = - \frac{R^2}{2*h(1 + \nu)} (1 - \cos \phi \cos \phi_0) \tag{8.2.21c}$$

If a similar procedure is followed for the second approximation, it can be shown that

$$*n_{(11)} = *n_{(22)} = - \frac{pR}{2} \tag{8.2.22a}$$

$$*V_{(1)} = - \frac{R^2 p \sin \phi \cos \phi_0}{2*hE} (1 - \nu) \left\{ 1 - \left[\frac{Rp}{4*hE} \left(\frac{1 - \nu}{1 + \nu} \right) \right] \right.$$

$$\left. + 2 \left[\frac{Rp}{4*hE} \left(\frac{1 - \nu}{1 + \nu} \right) \right]^2 + \cdots \right\} \tag{8.2.22b}$$

$$*V_{(3)} = + \frac{R^2 p [1 - \cos \phi \cos \phi_0]}{2*hE} (1 - \nu) \left\{ 1 \right.$$

$$\left. - \left[\frac{Rp}{4*hE} \left(\frac{1 - \nu}{1 + \nu} \right) \right] + 2 \left[\frac{Rp}{4*hE} \left(\frac{1 - \nu}{1 + \nu} \right) \right]^2 + \cdots \right\} \tag{8.2.22c}$$

TABLE 8.2.1 Pressurization of a Sphere

ϕ, rad (1)	$*n_{(11)}$, lb/ft (2)	$*n_{(22)}$, lb/ft (3)	$(*V_{(1)}/R) \times 10^4$				$(*V_{(3)}/R) \times 10^4$			
			ϵ term (4)	ϵ^2 term (5)	Columns (4) + (5) (6)	Exact (7)	ϵ term (8)	ϵ^2 term (9)	Columns (8) + (9) (10)	Exact (11)
0.000	2000.0	2000.0	0.000	0.000	0.000	0.000	−40.680	−0.202	−40.881	−40.883
0.039	2000.0	2000.0	3.856	0.019	3.875	3.875	−40.755	−0.202	−40.957	−40.959
0.079	2000.0	2000.0	7.705	0.038	7.744	7.744	−40.982	−0.203	−41.186	−41.187
0.118	2000.0	2000.0	11.543	0.057	11.601	11.601	−41.360	−0.205	−41.565	−41.567
0.157	2000.0	2000.0	15.363	0.076	15.440	15.440	−41.889	−0.208	−42.097	−42.100
0.196	2000.0	2000.0	19.160	0.095	19.255	19.255	−42.567	−0.211	−42.778	−42.779
0.236	2000.0	2000.0	22.926	0.114	23.040	23.041	−43.393	−0.215	−43.608	−43.610
0.275	2000.0	2000.0	26.658	0.132	26.490	26.791	−44.367	−0.220	−44.587	−44.589
0.314	2000.0	2000.0	30.348	0.151	30.499	30.500	−45.486	−0.226	−45.712	−45.714
0.353	2000.0	2000.0	33.992	0.169	34.161	34.162	−46.750	−0.232	−46.982	−46.983
0.393	2000.0	2000.0	37.583	0.186	37.769	37.771	−48.155	−0.239	−48.394	−48.396
0.432	2000.0	2000.0	41.116	0.204	41.650	41.322	−49.701	−0.247	−49.947	−49.949
0.471	2000.0	2000.0	44.586	0.221	44.807	44.809	−51.384	−0.255	−51.639	−51.640
0.511	2000.0	2000.0	47.987	0.238	48.225	48.227	−53.202	−0.264	−53.466	−53.467
0.550	2000.0	2000.0	51.314	0.255	51.569	51.571	−55.152	−0.274	−55.425	−55.427
0.589	2000.0	2000.0	54.562	0.271	54.833	54.835	−57.231	−0.284	−57.515	−54.517
0.628	2000.0	2000.0	57.726	0.286	58.012	58.014	−59.436	−0.295	−59.731	−59.733
0.668	2000.0	2000.0	60.801	0.302	61.102	61.105	−61.763	−0.606	−62.070	−62.072
0.707	2000.0	2000.0	63.782	0.316	64.098	64.100	−64.210	−0.319	−64.528	−64.531
0.746	2000.0	2000.0	66.665	0.331	66.995	66.998	−66.772	−0.331	−76.103	−67.105
0.785	2000.0	2000.0	69.444	0.344	69.789	69.789	−69.444	−0.344	−69.789	−69.791

NOTE: 1 lb/ft = 47.9 Pa.

It is possible to continue to solve for the successive approximations. However, for this particular example, an expression for the sum of the approximation can be derived if it is assumed that the material is linearly elastic. By mathematical induction it can be shown that Eqs. (8.2.22) sum to

$$^*V_{(1)} = R(1 + \nu) \sin \phi \cos \phi_0 \left[1 - \sqrt{\frac{Rp}{4^*hE} \left(\frac{1 - \nu}{1 + \nu} \right)} \right] \qquad (8.2.23a)$$

$$^*V_{(3)} = - R(1 + \nu)(1 - \cos \phi \cos \phi_0) \left[1 - \sqrt{1 + \frac{Rp}{4^*hE} \left(\frac{1 - \nu}{1 + \nu} \right)} \right]$$

$$(8.2.23b)$$

when

$$\left| \frac{Rp}{4^*hE} \left(\frac{1 - \nu}{1 + \nu} \right) \right| < \frac{1}{4} \qquad (8.2.23c)$$

Table 8.2.1 shows results for the pressurization of a sphere in which E = 300,000 psi (2.07 GPa), ν = 0.4, h = 0.002 ft (0.61 mm), p = -400 psf (-19.2 kPa), the smallest characteristic length is the sphere radius, R = 10 ft (3.05 m), and ϕ_0 = $\pi/4$. It can be seen from Table 8.2.1 that the sum of the first two approximations gives results in close agreement with the closed-form solutions.

8.3 Finite Element Analysis of Nonlinear Static Behavior

When membranes with more general geometries than axisymmetric membranes are considered, closed-form solutions are impractical and a numerical model is needed to discretize the differential equations into algebraic forms. The finite element method can be used to develop such a model. The fundamentals of the finite element method were described in Sec. 3.1. Here we will use that methodology to discretize the governing field equations given in Sec. 8.1 for general curved membranes.

Early investigations [143, 193, 194] of nonlinear membranes adopted flat triangular elements: numerous elements were required to represent strongly curved membranes. In order to avoid spurious discontinuities in slope across element boundaries, a curved element should be used. In this section a superparametric element is adopted to improve slope continuity. A superparametric element is one in which the geometry is approximated by polynomials of higher order than those used for the displacement variables. We will see that such elements have some inherent drawbacks but do provide accurate answers in certain classes of problems. In a subsequent section where time-varying problems are considered, we will develop an isoparametric element for membranes which avoids those drawbacks.

Because in this and the next section we will be using various configurations of the membrane as the reference surface under varying conditions, let us adopt a special notation to aid in identifying those conditions. Recall that a leading *super*script on a variable denotes the surface on which the variable is defined, but a leading *sub*script denotes the reference surface for measurement of magnitudes and orientation of the variable. If there is only a single leading script, the reference surface coincides with the definition surface. The static equilibrium surface is denoted by $*C$, the unstrained surface by ^0C. We now adopt the notation $^@C$ to denote a general reference surface. In some instances $^@C$ will coincide with $*C$ (updated lagrangian coordinates), or with ^0C (total lagrangian coordinates).

The following derivation is a synthesis of those reported in Refs. 41, 134–136, and 195, and the examples used are drawn from those works. We here adopt the notation of using capital Latin letters as scripts to denote nodal numbers on an element. Commas used as subscripts denote partial differentiations with respect to the coordinate component with the succeeding subscript, for example, $\partial f_i/\partial x_\alpha = f_{i,\alpha}$. The convention of using repeated indices to indicate summations is also used for nodal numbers.

In a fashion similar to that used for finite element models of cables, we now assume a discrete representation of the continuously distributed components of displacement in a typical membrane element as

$$\underset{@}{*}V^i = {}^@\Phi_N(x_1, x_2)\underset{@}{*}d_N^i \tag{8.3.1}$$

in which $\underset{@}{*}d_N^i$ are nodal values of the components $\underset{@}{*}V^i$ of displacement to $*C$ from the selected reference surface $^@C$ and $^@\Phi_N(x_1, x_2)$ represents a row matrix of polynomial interpolation functions of x_1, x_2 defined on $^@C$. Note that $\underset{@}{*}d_N^i$ are nodal displacement components in the directions of the base vectors $^@\overline{A}_\alpha$, $^@\hat{N}$ of the surface $^@C$. They are not cartesian components. The interpolation functions must satisfy the criteria the $\underset{@}{*}V^{\alpha i}(x_{1M}, x_{2M}) = \underset{@}{*}d_M^i$, and so forth (see Sec. 5.3).

Upon substitution of Eqs. (8.3.1) into Eqs. (8.1.20), we obtain the strain tensor in the form

$$\underset{@}{*}\gamma_{\alpha\beta} = {}^@e_{N\alpha\beta i}\underset{@}{*}d_N^i \pm {}^@\eta_{NM\alpha\beta ij}\underset{@}{*}d_N^i\underset{@}{*}d_M^j \tag{8.3.2a}$$

in which

$$^@e_{N\alpha\beta\lambda} = \tfrac{1}{2}({}^@A_{\lambda\beta}{}^@\Phi_{N,\alpha} + {}^@A_{\rho\beta}{}^@\Gamma^\rho_{\alpha\lambda}{}^@\Phi_N + {}^@A_{\lambda\alpha}{}^@\phi_{N,\beta}$$

$$+ {}^@A_{\rho\alpha}{}^@\Gamma^\rho_{\lambda\beta}{}^@\Phi_N) \tag{8.3.2b}$$

$$^{@}e_{N\alpha\beta3} = -^{@}\Phi_N\,^{@}B_{\alpha\beta} \tag{8.3.2c}$$

$$^{@}\eta_{MN\alpha\beta\gamma\rho} = \tfrac{1}{2}(^{@}\phi_{M,\alpha}\,^{@}A_{\lambda\rho}\,^{@}\phi_{N,\beta} + ^{@}\phi_M\,^{@}\Gamma^{\mu}_{\alpha\lambda}\,^{@}A_{\mu\rho}\,^{@}\phi_{N,\beta}$$

$$+ ^{@}\phi_{M,\alpha}\,^{@}A_{\mu\rho}\,^{@}\Gamma^{\mu}_{\beta\lambda}\,^{@}\phi_N + ^{@}\phi_M\,^{@}\Gamma^{\mu}_{\lambda\alpha}\,^{@}\Gamma^{\gamma}_{\rho\beta}\,^{@}A_{\mu\gamma}\,^{@}\phi_N$$

$$+ ^{@}\phi_M\,^{@}B_{\lambda\alpha}\,^{@}B_{\beta\rho}\,^{@}\phi_N) \tag{8.3.2d}$$

$$^{@}\eta_{MN\alpha\beta\lambda3} = \tfrac{1}{4}(-^{@}\phi_M\,^{@}B_{\alpha\rho}\,^{@}\Gamma^{\rho}_{\beta\lambda}\,^{@}\phi_N - ^{@}\phi_M\,^{@}B_{\alpha\lambda}\,^{@}\phi_{N,\beta} - ^{@}\phi_M\,^{@}B_{\beta\rho}\,^{@}\Gamma^{\rho}_{\alpha\lambda}\,^{@}\phi_N$$

$$- ^{@}\Lambda_M\,^{@}B_{\beta\lambda}\,^{@}\phi_{N,\alpha} + ^{@}\phi_{M,\beta}\,^{@}B_{\alpha\lambda}\,^{@}\phi_N + ^{@}\phi_{M,\alpha}\,^{@}B_{\beta\lambda}\,^{@}\phi_N)$$

$$\tag{8.3.2e}$$

$$^{@}\eta_{MN\alpha\beta33} = \tfrac{1}{2}(^{@}\phi_M\,^{@}B^{\lambda}_{\alpha}\,^{@}B^{\rho}_{\beta}\,^{@}A_{\lambda\rho}\,^{@}\phi_N + ^{@}\phi_{M,\alpha}\,^{@}\phi_{N,\beta}) \tag{8.3.2f}$$

The negative sign before the nonlinear term in Eq. (8.3.2a) is adopted if we use the deformed surface $^*C$ as the origin and reference $^{@}C$ for displacements. The positive sign is adopted if we use the undeformed surface ^0C as the origin and reference. The terms $^{@}A_{\alpha\beta}$, $^{@}B_{\alpha\beta}$, $^{@}\Gamma^{\gamma}_{\alpha\beta}$ in Eqs. (8.3.2b to f) are geometric properties of the selected reference surface (^0C or $^*C$) and can vary continuously over the element as x_1, x_2 vary. Thus, we must specify, exactly or approximately, how those quantities vary in order to carry out any required integrations. The stresses are obtained by substitution of Eqs. (8.3.2) into Eqs. (8.1.23) for a hyperelastic material or into Eq. (8.1.24) for a hookean material.

8.3.1 Superparametric element

We noted above that variations over the element of geometric properties, $^{@}A_{\alpha\beta}$ and so forth, of the reference surface need to be specified. Toward that end we can adopt an approximate technique [196] of surface patching so as to provide a curved element with continuous slopes across element interfaces.

A curved quadrilateral membrane element can be defined by four space curves and four nodal points on the middle surface of the actual membrane. Each element is assigned its own local curvilinear coordinates x_1, x_2 which vary continuously from zero to one along the boundaries. These coordinate lines need not be orthogonal and need not lie along the lines of principal curvature of the true membrane.

In terms of these coordinates an interpolation surface for the element can be defined by

$$y^i(x_1, x_2) = \{^{@}f(x_1)\}[^{@}B^i]\{^{@}f(x_2)\} \tag{8.3.3}$$

where y^i are components of the elemental position vector $^{@}\overline{R}$ referred to a global cartesian coordinate system; $\{^{@}f(x_{\alpha})\}$ is a vector of interpolating functions—called "blending functions" by Coons [196]—which form a basis for the element geometry; $[^{@}B]$ is the element boundary geometry matrix containing information with respect to the global cartesian coordinate system about the location, slopes, twists, and higher derivatives of the corner nodal points; the sizes of $\{^{@}f\}$ and $[^{@}B]$ are dependent on the order or interpolation chosen for the element.

If the order of interpolation $= r = 2$, the interpolated element has continuity only of location across the element boundaries. If $r = 4$, slope continuity can also be enforced across element boundaries. If $r = 6$, continuity of twist between adjacent elements can be enforced. Different orders or interpolation could be used in the two different boundary directions. Such a choice is feasible only if the membrane has a clearly defined geometry in one global direction. It was decided to use $r = 4$ in the current work and to assume bicubic interpolating functions.

The element boundary geometry matrix is given by

$$[^{@}B^i] = \begin{bmatrix} ^{@}y^i(1) & ^{@}y^i(4) & ^{@}y^i_{,2}(1) & ^{@}y^i_{,2}(4) \\ ^{@}y^i(2) & ^{@}y^i(3) & ^{@}y^i_{,2}(2) & ^{@}y^i_{,2}(3) \\ ^{@}y^i_{,1}(1) & ^{@}y^i_{,1}(4) & ^{@}y^i_{,12}(1) & ^{@}y^i_{,12}(4) \\ ^{@}y^i_{,1}(2) & ^{@}y^i_{,1}(3) & ^{@}y^i_{,12}(2) & ^{@}y^i_{,12}(3) \end{bmatrix} \qquad (8.3.4)$$

where $^{@}y^i(M) = i$th component of the element position vector $^{@}\overline{R}$ at corner node M

$$^{@}y^i_{,\alpha}(M) = \frac{\partial ^{@}y^i(M)}{\partial x_\alpha} = i\text{th component of } ^{@}\overline{R}_{,\alpha} \text{ at corner node } M$$

$$^{@}y^i_{,\alpha\beta}(M) = \frac{\partial^2 ^{@}y^i(M)}{\partial x_\alpha \, \partial x_\beta} = i\text{th component of } ^{@}\overline{R}_{,\alpha\beta} \text{ at corner node } M.$$

If a cubic basis is chosen for the blending function $\{^{@}f(x_\alpha\mu)\}$, if it is assumed that

$$[^{@}f_1(x_\alpha)\,^{@}f_2(x_\alpha)\,^{@}f_3(x_\alpha)\,^{@}f_4(x_\alpha)] = [x_\alpha^3 x_\alpha^2 x_\alpha \ 1] \begin{bmatrix} 2 & -2 & 1 & 1 \\ -3 & 3 & -2 & -1 \\ 0 & 0 & 1 & 0 \\ 1 & 0 & 0 & 0 \end{bmatrix}$$

and if Eqs. (8.3.4) and (8.3.5) are substituted into Eq. (8.3.3), it can be shown that (1) for this choice of $\{^{@}f(x_\alpha)\}$ the element is continuous with all neighboring elements; and (2) slope continuity is enforced across all boundaries if $^{@}y^i(M)$, $^{@}y^i_{,\alpha}(M)$, and $^{@}y^i_{,\alpha\beta}(M)$ for each element coincide for common nodes in adjacent elements.

The coefficients of $[^{@}B^i_N]$ remain to be specified. There are several alternative methods for approximating these coefficients:

1. *Exact specification.* If the mathematical descriptions of the membrane geometry are sufficiently well-defined such that the first and sec-

ond derivatives of the position vector can be determined explicity, e.g., a cylinder, one can simply input the slopes and twists of the surface at each node into the appropriate coefficients of $[^@B^i]$.

2. *Interpolation of surrounding elements.* The cartesian coordinates of the surface nodes can be input to calculate the upper left submatrix of $[^@B^i]$, and the slopes and twists of the surface at the corner nodes of one element can be obtained from interpolation of the cartesian coordinates of nodes in surrounding elements. The coefficients of $[^@B^i]$ corresponding to each of the four nodes on an element can then be calculated from weighted averages of the slopes and twists of the same node on incident elements.

3. *Interpolation within elements.* For a more accurate estimate of the $[^@B^i]$ matrices, 12 additional nodes, 8 on the boundary and 4 interior to the element, can be defined at the intersections of the curvilinear coordinate lines $x_\alpha = 0, \frac{1}{3}, \frac{2}{3}, 1$. The cartesian coordinates of these auxiliary nodes can then be used to interpolate for the slopes and twists at the corner nodes.

In this way the coefficients of $[^@B^i]$ are calculated directly for each element. One can then either average the coefficients of each element $[^@B^i]$ for incident elements at the node considered, or accept the slight discrepancies between coefficients in $[^@B^i]$ for common nodes in adjacent elements.

It was decided to adopt the last technique for specifying the coefficients of $[^@B^i]$ and to accept the small discrepancies in coefficients between $[^@B^i]$ matrices for adjacent elements. These errors imply slight discontinuities in slope across element boundaries, but this disadvantage was offset by considerable savings in computer storage and computational time. The auxiliary nodal points are selected by inspection (see Fig. 8.3.1 for a description of the element geometry). It is emphasized that these additional nodes are introduced solely for the purposes of interpolating the element geometry and need not be used to approximate the deformation pattern of the shell. Evaluating Eqs. (8.3.3) to (8.3.5) at all 16 geometry nodes allows one to calculate the entries for $[^@B^i]$ as shown in Table 8.3.1.

Once the actual shell geometry has been approximated by Eq. (8.3.3), the first and second fundamental forms $^@A_{\alpha\beta}$, $^@B_{\alpha\beta}$ and the Christoffel symbols $^@\Gamma^\lambda_{\alpha\beta}$ for the typical element can be calculated (see Appendix B).

First fundamental form (metric tensor):

$$^@A_{\alpha\beta} = \frac{^@\partial y^i}{\partial x_\alpha} \cdot \frac{^@\partial y^i}{\partial x_\beta} \tag{8.3.6a}$$

Second fundamental form (curvature tensor):

$$^@B_{\alpha\beta} = {^@\overline{A}}_{\alpha,\beta} \cdot {^@\hat{N}} \tag{8.3.6b}$$

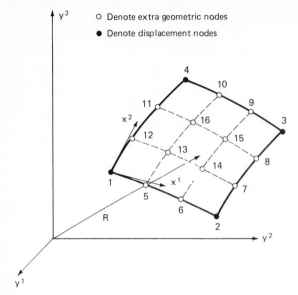

Figure 8.3.1 Extra nodes for superparametric element.

Christoffel symbols:

$$^{@}\Gamma^{\gamma}_{\alpha\beta} = \frac{^{@}A_{\gamma\lambda}}{2} \left({}^{@}A_{\lambda\alpha,\beta} + {}^{@}A_{\lambda\beta,\alpha} - {}^{@}A_{\alpha\beta,\lambda} \right) \qquad (8.3.6c)$$

where $^{@}\overline{A}_i = \dfrac{^{@}\partial\overline{R}}{\partial x_i}$ = covariant base vectors of the element. Thus we have

established the relation between the local curvilinear coordinates used as the basis of the typical element and the global cartesian coordinates of its nodal points.

8.3.2 Semi-inverse solution to prestressing phase [134]

One class of problems for tension structures is that in which we assume that the final prestressed shape has a known geometry. In that instance we can use the final shape $*C$ as the reference $^{@}C$ and calculate the initial shape required to obtain the desired final shape after loading. Here the principle of virtual work is used to derive the stiffness equations from which these displacements can be determined.

Because the final equilibrium state of the membrane is the reference, we write the virtual work [Eq. (8.1.28)] due to small virtual displacements from that state as

TABLE 8.3.1 Coefficients of $[^{\oplus}B^i]$ Using Auxiliary Interior Element Nodes

$^{\oplus}y^{,i}(N)$	1	2	3	4	5	6	7	8	9	10	11	12	13	14	15	16
$^{\oplus}y^i_{,1}(1) =$	-5.50	1.00			9.00	-4.50										
$^{\oplus}y^i_{,1}(2) =$	-1.00	5.50			4.50	-9.00										
$^{\oplus}y^i_{,1}(3) =$			5.50	-1.00					-9.00	4.50						
$^{\oplus}y^i_{,1}(4) =$			1.00	-5.50					-4.50	9.00						
$^{\oplus}y^i_{,2}(1) =$	-5.50			1.00							-4.50	9.00				
$^{\oplus}y^i_{,2}(2) =$		-5.50	1.00				9.00	-4.50								
$^{\oplus}y^i_{,2}(3) =$		-1.00	5.50				4.50	-9.00								
$^{\oplus}y^i_{,2}(4) =$	-1.00			5.50							-9.00	4.50				
$^{\oplus}y^i_{,12}(1) =$	30.25	-5.50	1.00	5.50	-49.50	24.75	9.00	-4.50	-4.50	9.00	24.75	-49.50	81.00	-40.50	20.25	-40.50
$^{\oplus}y^i_{,12}(2) =$	5.50	-1.00	5.50	-30.25	-9.00	4.50	4.50	-9.00	-24.75	49.50	49.50	-24.75	40.50	-81.00	40.50	-20.25
$^{\oplus}y^i_{,12}(3) =$	1.00	-5.50	30.25	5.50	-4.50	9.00	24.75	-49.50	-49.50	24.75	9.00	-4.50	20.25	-40.50	81.00	-40.50
$^{\oplus}y^i_{,12}(4) =$	5.50	-1.00	5.50	-30.25	-9.00	4.50	4.50	-9.00	-24.75	49.50	49.50	-24.75	40.50	-20.25	40.50	81.00

NOTE: jth coefficient = Σ (entries in row i) $\cdot\ ^{\oplus}y^{,i(K)}$.

$$0 = \int \int_{C_R} (^*F^i \, \delta^* V_i - ^*n^{\alpha\beta} \, \delta^* \gamma_{\alpha\beta}) \sqrt{^*A} \; dx_1 \, dx_2 \qquad (8.3.7)$$

where $\delta^* \gamma_{\alpha\beta}$ = small virtual strain corresponding to linear part of eulerian strain tensor $^*\gamma_{\alpha\beta}$ in Eq. (8.3.2a) with $^* V_i$ replaced by
$\delta^* V_i$

$^*n^{\alpha\beta}$ = eulerian stress resultant tensor, Eq. (8.1.11)

$^*F^i$ = contravariant components of the surface traction

$$\overline{F} = {}^*F^\alpha \, {}^*\overline{A}_\alpha + {}^*F^3 {}^*\hat{N}$$

Consider now a hookean material with $^*n^{\alpha\beta}$ given by Eq. (8.1.25). By substitution of Eqs. (8.3.1) (8.3.2), and (8.1.25) into Eq. (8.3.7) and after taking indicated variations, we obtain

$$(^*KL_{MNij} + ^*KN_{MNij}) \, ^*d^i_M = ^*F_{iN} \qquad (8.3.8a)$$

in which

$$^*KL_{MNij} = h \int \int_{C_R} ^*e_{M\alpha\beta i} C^{\alpha\beta\lambda\mu} \, ^*e_{N\lambda\mu j} \sqrt{^*A} \; dx^1 \, dx^2 \qquad (8.3.8b)$$

$$^*KN_{MNij} = h \int_{^*C} \int \eta_{MS\alpha\beta ik} \, ^*d^k_S C^{\alpha\beta\lambda\mu} \, ^*e_{N\beta\mu j} \sqrt{^*A} \; dx^1 \, dx^2 \qquad (8.3.8c)$$

$$^*F_{jN} = (^*A_{ji})|_{\text{at } N} \int \int_{^*C} ^*F^{i} {}^*\Phi_N \sqrt{^*A} \; dx_1 \, dx_2 \qquad (8.3.8d)$$

and $C^{\alpha\beta\lambda\mu}$ are the hookean coefficients:

$$C^{\alpha\beta\lambda\mu} = \frac{E}{2(1 + \nu)} \left(^*A^{\alpha\lambda} {}^*A^{\beta\rho} + ^*A^{\alpha\rho} {}^*A^{\beta\lambda} + \frac{2\nu}{1 - \nu} {}^*A^{\alpha\beta} {}^*A^{\lambda\rho} \right) \qquad (8.3.8e)$$

In Eq. (8.3.8a) $^*KL_{MNij}$ = the conventional linear stiffness matrix. The
nonlinear part of the stiffness matrix, $^*KN_{MNij}$, is a function of nodal displacements. Note that in this instance, $^*F_{jN}$ are not functions of nodal
diplacements because the components of load on the deformed surface are
assumed known a priori. The surface integrals for each element are carried out numerically using gaussian quadrature and the superparametric
approximations to $^*A_{\alpha\beta}$, $\sqrt{^*A}$, etc., described previously. Note that the
calculated $^*d^i_N$ are contravariant components of nodal displacement and
Eqs. (8.1.27) must be used to convert them to physical components.

The solution techniques described in Sec. 3.3 for nonlinear algebraic
equations can be used with Eq. (8.3.8). Because Eqs. (8.3.8) were obtained
using local coordinates for each element, we need to first transform all
tensor quantities into physical components in a single global coordinate
system using the tensor transformation rules described in Appendix B

before determining the tangent stiffness equations for use in a Newton-Raphson scheme to iteratively solve for the displacements.

In order to test the potential of the curved element and to determine the convergence properties of the assumed displacement functions, the first numerical example considered was that of determining the initial shape required to obtain a perfect ring-supported hemisphere under the action of 1 psi (6.9 kN/m^2) internal pressure. The thickness has been taken as 0.04 in (1 mm) and the material was assumed to be a vinyl-coated nylon for which representative isotropic constants are Young's modulus = 40,000 psi (176,000 kN/m^2) and Poisson's ratio = 0.45.

Numerical results for hemispheres with various radii are shown in Fig. 8.3.2 where the apex deflection as a function of radius is plotted. Also shown for comparison purposes is the exact solution obtained by the Poincare perturbation technique discussed in Sec. 8.2. The deviation of finite element solutions is negligible when the radius is less than 30 ft (9 m). Therefore, it can be seen that a relatively few number of curved elements provide accurate solutions for nonlinear problems solved by the single-step method.

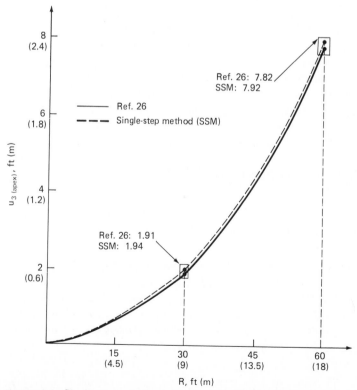

Figure 8.3.2 Single-step solution to hemisphere (Ref. 138).

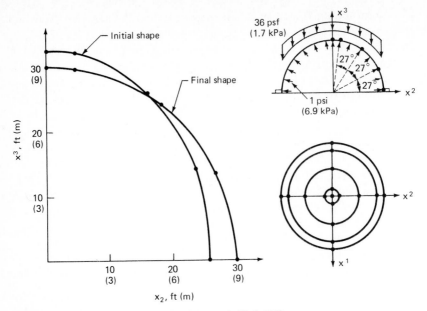

Figure 8.3.3 Single-step solution for formwork (Ref. 138).

One potential method for the erection of concrete domes is to first pressurize a thin membrane and then to spray concrete on its interior [197]. As an example, consider the case in which it is desired that under the combined action of the internal pressure and the dead weight of uncured concrete, the membrane deforms into a perfect hemisphere. The problem considered herein is that of determining the initial shape required to obtain that final hemisphere after pressurization and spraying of concrete.

Figure 8.3.3 shows the initial shape required for a hemispherical membrane (same material as above) of 30-ft (9-m) radius with 1-psi (6.9-kN/m^2) internal pressure and 36-psf (1720 N/m^2) dead load corresponding to 3 in (76 mm) of concrete.

8.3.3   Incremental analysis of hyperelastic membranes [135]

In instances where the final shape is not known a priori, a combined incremental and iterative procedure can be used to determine the displacements from an initially specified surface configuration, ^0C. Toward that end we first must develop the virtual work expression for an incremental displacement process. From the principle of virtual work we can write the virtual work on a deformed configuration as

$$\int_0 \int_C ({}^*_0 F^i \, \delta^*_0 V_i - {}^0 h^*_0 A^{\alpha\beta} \, \delta^*_0 \gamma_{\alpha\beta}) \sqrt{{}^0A} \, dx_1 \, dx_2 = 0 \qquad (8.3.9)$$

where we have adopted the initial unstrained state ^0C as our reference $^@C$. Therefore, $^*_0F^i$ = components of surface traction $^*_0\overline{F} = {}^*_0F^\alpha {}^*_0\overline{A}_\alpha + {}^*_0F^3 {}^*_0\hat{N}$ measured per unit area of undeformed surface ^0C but acting on the deformed surface $^*C$ in the direction of the deformed base vectors $^*_0\overline{A}_\alpha$, $^*_0\hat{N}$. Also, $^*_0n^{\alpha\beta}$ = Kirchhoff stress tensor referred to base vectors $^0\overline{A}_\alpha$ in ^0C and measured per unit area of the undeformed state ^0C (see discussion in Sec. 8.1.2).

We will here assume an isotropic hyperelastic material [50, 187] which is incompressible and isotropic, i.e.,

$$
{}^*_0n^{\alpha\beta} = 2C_1 \left({}^0A^{\alpha\beta} - \frac{{}^*_0\lambda^4}{{}^0A} \, {}^*_0D^{\alpha\beta} \right)
$$

$$
+ \, 2C_2 \left\{ {}^*_0\lambda^{2} \, {}^0A^{\alpha\beta} + \frac{{}^*_0D^{\alpha\beta}}{{}^0A} \, [1 - 2{}^*_0\lambda^4(1 + {}^0A^{\mu\gamma} {}^*_0\gamma_{\mu\gamma})] \right\} \quad (8.3.10a)
$$

$$
{}^*_0D^{\alpha\beta} = {}^0A\,{}^0A^{\alpha\beta} + 2\epsilon^{\alpha\gamma}\epsilon^{\beta\mu} {}^*_0\gamma_{\mu\gamma} \quad\quad\quad (8.3.10b)
$$

in which $^*_0\lambda = {}^*h/{}^0h$ = ratio of deformed to undeformed thickness; and $\epsilon^{\alpha\gamma}$ = two-dimensional perturbation symbol ($\epsilon^{11} = \epsilon^{22} = 0$, $\epsilon^{12} = -\epsilon^{21} = 1$). An alternate approach to the modeling of incompressible materials with strong anisotropy is to assume a bulk modulus much greater than the shear modulus and proceed as for compressible materials. However, results will have a tendency to reproduce pure shear deformations ("locking" or "checkerboarding") if care is not taken in specifying gaussian integration points for the elements. One should select the number of integration points to be less than the number of nodes on the element, e.g., with a four-noded element using one integration gives good results.

The discretized equivalent to Eq. (8.3.9) for a finite element model is obtained by substituting Eqs. (8.3.2) and (8.3.10) into Eq. (8.3.9) and taking indicated variations. Thus, for any arbitrary virtual displacement

$$
{}^*_0P_{Nj} = {}^0h \int_{{}^0C} \int {}^*_0n^{\alpha\beta} \left({}_0e_{N\alpha j} + 2{}_0\eta_{NM\alpha\beta ij} {}^*_0d^i_M\right) \sqrt{{}^0A} \; dx_1 \, dx_2 \quad (8.3.11)
$$

in which $^*_0P_{Nj}$ are the covariant components of the external nodal forces:

$$
{}^*_0P_{Nj} = \int \int_{C_0} {}^*_0F^{i0}\Phi_N {}^0G_{ij}\sqrt{{}^0A} \; dx_1 \, dx_2 \quad\quad (8.3.12)
$$

Equations (8.3.11) and (8.3.12) are the stiffness equations for a hyperelastic membrane element.

Since, as recommended in Sec. 3.3, we are adopting an incremental approach combined with Newton-Raphson iterations to solve this nonlinear algebraic problem, it is first necessary to cast Eqs. (8.3.11) and (8.3.12) into incremental form. Consider the deformations of a typical finite element which carries it first from its initial state ^0C to an equilibrium state $^*C$. Next consider an added increment of displacement from $^*C$ to another equilibrium state $^\#C$. Let $^*_0 d^i_N$, $^*_0 n^{\alpha\beta}$, $^*_0 F^i$, and $^\#_0 d^i_N$, $^\#_0 n^{\alpha\beta}$, $^\#_0 F^i$ denote the nodal displacements, Kirchhoff stresses, and tractions in states $^*C$ and $^\#C$, respectively. Then

$$^\#_0 d^i_N = {}^*_0 d^i_N + w^i_N \tag{8.3.13a}$$

$$^\#_0 n^{\alpha\beta} = {}^*_0 n^{\alpha\beta} + s^{\alpha\beta} \tag{8.3.13b}$$

$$^\#_0 P_{Ni} = {}^*_0 P_{Ni} + r_{Ni} \tag{8.3.13c}$$

in which w^i_N, $s^{\alpha\beta}$, r_{Ni} are incremental changes from $^*_0 d^i_N$, $^*_0 n^{\alpha\beta}$, and $^*_0 P_{Ni}$.

The incremental load r_{Ni} merits special attention. Since $^\#_0 P_{Ni}$ and $^*_0 P_{NI}$ are functions of $^\#_0 d^i_N$ and $^*_0 d^i_N$, respectively, increments in traction $\Delta^\#_0 F^i$ and displacement w^i_N will have two separate effects on r_{Ni}. The derivation of the initial load matrix and incremental load matrix is as follows. From Eq. (8.3.13c)

$$r_{Nj} = {}^\#_0 P_{Nj} - {}^*_0 P_{Nj} = \int_{^0C} \int ({}^\#_0 F^i - {}^*_0 F^i)^0 G_{ij} {}^0 \phi_N \sqrt{{}^0 A}\ dx_1\ dx_2 \tag{8.3.14}$$

For conservative tractions, $({}^\#_0 F^i - {}^*_0 F^i)$ can be evaluated easily since the tractions act in known directions. However, for nonconservative forces, such as pressure which acts in the direction normal to the membrane in both $^*C$ and $^\#C$, further evaluation is needed. If the pressures in $^*C$ and $^\#C$ are $^*_0 P$ and $^\#_0 P = {}^*_0 P + \Delta P$, then

$$^\#F^{i0}\overline{A}_i \sqrt{{}^0 A}\ dx_1\ dx_2 = ({}^*_0 P + \Delta P)\ {}^\#_0 \hat{N} \sqrt{{}^\# A}\ dx_1\ dx_2 \tag{8.3.15a}$$

$$^*F^{i0}\overline{A}_i \sqrt{{}^0 A}\ dx_1\ dx_2 = {}^*_0 P {}^*_0 \hat{N} \sqrt{{}^* A}\ dx_1\ dx_2 \tag{8.3.15b}$$

in which $^\#_0 \hat{N}$ and $^*_0 \hat{N}$ are the unit normals to $^\#C$ and $^*C$, respectively, and are given by

$$^*_0 \hat{N} = \sqrt{\frac{{}^0 A}{{}^* A}}\ (\delta^3_j + {}^0\tilde{E}_{Nkj} {}^*_0 d^K_N + {}^0\tilde{\eta}_{NMikj} {}^*_0 d^i_{N0} {}^*_0 d^K_M)^0 \overline{A}^j \tag{8.3.16a}$$

and

$$^\#_0 \hat{N} = \sqrt{\frac{{}^0 A}{{}^\# A}}\ (\delta^3_j + {}^0\tilde{E}_{Nkj0} d^K_N + {}^0\tilde{\eta}_{NMikj} {}^\#_0 d^i_{N0} {}^\#_0 d^k_M)^0 \overline{A}^j \tag{8.3.16b}$$

in which $^0\tilde{E}_{Nikj}$ and $^0\tilde{\eta}_{NMikj}$ are dependent on the interpolation functions and geometric properties of ^0C and are developed using the procedures of Appendix B and Sec. 8.3.1. (See Ref. 198 for details.)

From Eqs. (8.3.15) and (8.3.16), we obtain

$$({}^{\#}_{0}F^{j} - {}^{*}_{0}F^{j}){}^{0}G_{jr} = P({}^{0}\tilde{E}_{Mkr} + {}^{0}\tilde{\eta}_{NMikr}{}^{*}_{0}d^{i}_{N} + {}^{0}\tilde{\eta}_{MNkir}{}^{*}_{0}d^{i}_{N})w^{k}_{M}$$

$$+ \Delta P(\delta_{3r} + {}^{0}\tilde{E}_{Nkr}{}^{*}_{0}d^{k}_{N} + {}^{0}\tilde{\eta}_{NMikr}{}^{*}_{0}d^{i}_{N}{}^{*}_{0}d^{k}_{M}) \quad (8.3.17)$$

Thus, for pressure loading, Eq. (8.3.14) becomes

$$r_{Nj} = \tilde{R}_{NMij}w^{i}_{M} + q_{Nj} \quad (8.3.18a)$$

in which

$$\tilde{R}_{NMij} = \int_{{}^{0}C} \int p({}^{0}\tilde{E}_{Mij} + {}^{0}\tilde{\eta}_{QMkij}{}^{*}_{0}d^{k}_{Q}$$

$$+ {}^{0}\tilde{\eta}_{MQikj}{}^{*}_{0}d^{k}_{Q}){}^{0}\Phi_{N}\sqrt{{}^{0}A}\ dx_{1}\ dx_{2} \quad (8.3.18b)$$

is the initial load matrix, and

$$q_{Nj} = \int_{{}^{0}C} \int \Delta p(\delta_{3j} + {}^{0}\tilde{E}_{Mkj}{}^{*}_{0}d^{k}_{M}$$

$$+ {}^{0}\tilde{\eta}_{QMikj}{}^{*}_{0}d^{i}_{Q}{}^{*}_{0}d^{k}_{M}){}^{0}\phi_{N}\sqrt{{}^{0}A}\ dx_{1}\ dx_{2} \quad (8.3.18c)$$

is the incremental load matrix.

Now, substituting Eqs. (8.3.13) into Eq. (8.3.11), and then applying the equation of static equilibrium in state $^{*}C$, we obtain (omit higher-order terms)

$$r_{Nj} = \left(\int_{C_{0}} \int 2{}^{0}h{}^{*}_{0}n^{\alpha\beta 0}\eta_{MN\alpha\beta ij}\sqrt{{}^{0}A}\ dx_{1}\ dx_{2} \right) w^{i}_{M} + \int_{C_{0}} \int {}^{0}h\ \Delta n^{\alpha\beta}\ ({}^{0}e_{N\alpha\beta j}$$

$$+ 2{}^{0}\eta_{MN\alpha\beta ij}{}^{*}_{0}d^{i}_{M})\sqrt{{}^{0}A}\ dx_{1}\ dx_{2}$$

in which the incremental stress tensor $\Delta n^{\alpha\beta}$ is

$$\Delta n^{\alpha\beta} = {}^{\#}_{0}n^{\alpha\beta} - {}^{*}_{0}n^{\alpha\beta}$$

$$= \frac{2C_{1}}{{}^{0}A} ({}^{\#}_{0}\lambda^{\#}_{0}D^{\alpha\beta} - {}^{*}_{0}\lambda^{*}_{0}D^{\alpha\beta} + 2C_{2}{}^{0}A^{\alpha\beta}({}^{\#}_{0}\lambda^{2} - {}^{*}_{0}\lambda^{2})$$

$$+ 2\frac{C_{2}}{{}^{0}A} ({}^{\#}_{0}D^{\alpha\beta} - {}^{*}_{0}D^{\alpha\beta}) - \frac{4C_{2}}{{}^{0}A} ({}^{\#}_{0}D^{\alpha\beta}{}^{\#}_{0}\lambda^{4} - {}^{*}_{0}D^{\alpha\beta}{}^{\#}_{0}\lambda^{4})$$

$$- \frac{4C_{2}}{{}^{0}A} ({}^{\#}_{0}D^{\alpha\beta}{}^{\#}_{0}\lambda^{4}\ {}^{0}A^{\mu\gamma}{}^{\#}_{0}\gamma_{\gamma} - {}^{*}_{0}D^{\alpha\beta}{}^{*}_{0}\gamma^{4}\ {}^{0}A^{\mu\gamma}{}^{*}_{0}\gamma_{\mu\gamma}) \quad (8.3.20a)$$

After simplification of Eq. (8.3.20a) using Eqs. (7.1.15), (8.3.2), and (8.3.10), we obtain

$$\Delta n^{\alpha\beta} = {}^{*}_{0}A^{\alpha\beta\xi\eta}({}^{0}e_{M\xi\eta i} + 2^{0}\eta_{MN\xi\eta ij}{}^{*}_{0}d^{j}_{M})w^{i}_{N} \tag{8.3.20b}$$

in which

$$
{}^{*}_{0}A^{\alpha\beta\xi\eta} = \left(2C_1\left(\frac{{}^{*}_{0}T^{\,\xi\eta *}_{0}D^{\alpha\beta}}{{}^{*}A} - \epsilon^{\alpha\xi}\epsilon^{\beta\eta}\right) + C_2\left\{{}^{0}A^{\alpha\beta *}_{0}T^{\xi\eta} + 2\epsilon^{\alpha\xi}\epsilon^{\xi\eta}\left[\left(\frac{{}^{*}A}{{}^{0}A}\right)^2\right.\right.\right.
$$

$$
\left.\left.\left. - 2\right](1 + {}^{0}A^{\mu\gamma *}_{0}\gamma_{\mu\gamma}) + 2^{*}_{0}D^{\alpha\beta}\left(\frac{2^{*}_{0}T^{\xi\eta}}{{}^{*}A} + \frac{2^{*}_{0}T^{\xi\eta}}{{}^{*}A}{}^{0}A^{\mu\gamma *}\gamma_{\mu\gamma} - {}^{0}A^{\xi\eta}\right)\right\}\right)\frac{2^{0}A}{{}^{*}A^2}
\tag{8.3.20c}
$$

and

$$
{}^{*}_{0}T^{\,\alpha\beta} = 2^{0}A^{0}A^{\alpha\beta} + 4\epsilon^{\alpha\mu}\epsilon^{\beta\gamma *}_{0}\gamma_{\mu\gamma} \tag{8.3.20d}
$$

where $\epsilon^{\alpha\mu}$ are the two-dimensional permutation symbols, all quantities are dependent on nodal displacements ${}^{*}_{0}d^{i}_{N}$ in the directions ${}^{*}_{0}\bar{A}_{\alpha}$, ${}^{*}_{0}\hat{N}$, which are assumed to have been previously calculated on $*C$.

Now substitution of Eqs. (8.3.18a) and (8.3.20b) into Eq. (8.3.19) leads to

$$
q_{Nr} = {}^{0}h\int_{{}_{0}C}\int[{}^{*}_{0}n^{\alpha\beta 0}\eta_{NM\alpha\beta kr} + {}^{*}_{0}A^{\alpha\beta\xi\eta}({}^{0}e_{N\alpha\beta r}
$$

$$
+ 2^{0}\eta_{QN\alpha\beta ir}{}^{*}_{0}d^{i}_{R})({}^{0}e_{M\xi\eta k} + 2^{0}\eta_{MQ\xi\eta kj}{}^{*}_{0}d^{j}_{Q})]\sqrt{{}^{0}A}\,dx_1\,dx_2\,w^{k}_{m} \tag{8.3.21a}
$$

which can be rewritten as

$$
[K_{MNkr} + \tilde{K}_{MNkr} - R_{MNkr}]\{w^{k}_{M}\} = \{q_{Nr}\} \tag{8.3.21b}
$$

in which

$$
K_{MNkr} = {}^{0}h\int_{{}_{0}C}\int{}^{*}_{0}n^{\alpha\beta 0}\eta_{MN\alpha\beta ij}\sqrt{{}^{0}A}\,dx_1\,dx_2 \tag{8.3.21c}
$$

$$
\tilde{K}_{MNkr} = {}^{0}h\int_{{}_{0}C}\int{}^{*}_{0}A^{\alpha\beta\xi\eta}({}^{0}e_{N\alpha\beta r} + 2^{0}\eta_{RN\alpha\beta ir}{}^{*}_{0}d^{i}_{R})({}^{0}e_{M\xi\eta k}
$$

$$
+ 2^{0}\eta_{MQ\xi\eta kj}{}^{*}_{0}d^{j}_{Q})\sqrt{{}^{0}A}\,dx_1\,dx_2 \tag{8.3.21d}
$$

Equations (8.3.21) represent the incremental form of the equations for the finite element model of membrane behavior. Matrix $[K]$ is dependent on the stress at $*C$ and is commonly called the "initial stress matrix." $[\tilde{K}]$ is dependent on the displacement to $*C$ and is referred to as the "instantaneous geometric stiffness matrix," and $[R]$ is dependent on the load on $*C$ and is called the "initial load matrix." The matrix $[K + \overset{\vee}{K} - R]$ can also be identified with the instantaneous tangent stiffness matrix.

From Eqs. (8.3.21), which represent the incremental finite element model for membranes, equations for the complete continuum can be obtained by adding the contribution of each element. The equation for the total continuum can be written in incremental form as

$$[K]\{w\} = \{Q\} \tag{8.3.22a}$$

in which $[K]$ = total stiffness matrix, and $\{A\}$ = total load matrix. The solution algorithm after m increments is as follows: (1) evaluate $[K]$ at the mth incremental state (associated with tangent stiffness matrix); (2) solve for $\{w^{(m+1)}\}$ by

$$\{w^{(m+1)}\} = [K]^{-1}\{Q\} \tag{8.3.22b}$$

(3) reevalutate $[K]$ at $(m + 1)$th increment after rth iterate; (4) apply the Newton-Raphson algorithm

$$\{d^{(m+1)}\} = \{d^{(m)}\} + \{w^{(m+1)}\} \tag{8.3.22c}$$

for initial values, or 0th iteration, and

$$\{d^{(m+1,r+1)}\} = \{d^{(m+1,r)}\} - [K]^{-1}\{f(u^{(m+1,r)}, p^{(m+1)}\} \tag{8.3.22d}$$

for the $(r + 1)$th iteration; (5) repeat steps 3 through 4 until convergence to the desired accuracy is obtained; and (6) repeat steps 1 through 5 for next incremental step.

Example membrane problems. The displacement approximation functions in the finite element equations were left in generalized form. For the example problems considered here, only the simplest approximations were taken, i.e., bilinear functions of x_1 and x_2.

To illustrate the finite element model for membrane behavior, two examples are presented here: (1) inflation of an initially flat circular sheet (zero initial curvature); and (2) inflation of a hemispherical segment (double curvature).

In Ref. 199 results were presented for the inflation of flat circular sheets using flat triangular elements. The same example considered in that work was selected for comparison purposes. The membrane sheet is initially 0.05 in (1.3 mm) thick and 8 in (200 mm) in diameter: it has Mooney material constants of $C_1 = 24.0$ psi (170 kN/m^2) and $C_2 = 3.0$ psi (21 kN/m^2). The present finite element model uses 12 elements in a symmetric quarter section with 19 nodes as compared to [199] 38 elements and 27 nodes [134]. Figure 8.3.4 shows the models used and results obtained in

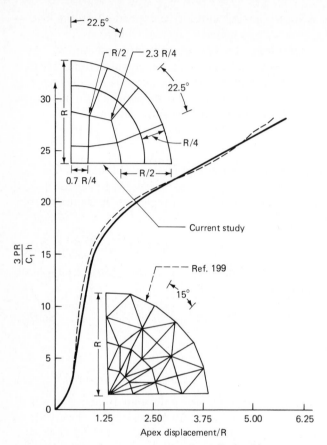

Figure 8.3.4 Inflation of flat circular sheet (Ref. 135).

the two studies. With a lesser number of elements and degrees of freedom, comparable results were obtained in the present study [134]. This is primarily due to the improved capability using curved elements to model the circular boundary. To model that boundary with straight line segments [199] requires many more elements. However, it should be noted that since the initial geometry is flat and only bilinear displacement polynomials were used, the incremented profiles will always remain flat. This is a drawback in the superparametric element: the original geometry approximations is always preserved if lower order displacement approximations are used.

Since the sheet is initially flat, there exists no linear incremental solution at the start of pressurization, no matter how small an increment of pressure is assumed. Therefore, to initiate the solution, an initial displacement was assumed and the Newton-Raphson iteration process was used to correct that trial solution. The incremental process was used with increments of 0.2 psi (1.4 kN/m^2) applied to a maximum of 2.8 psi (19 kN/m^2) with Newton-Raphson iterations performed after every five increments.

The inflation of a hemispherical segment was considered in order to demonstrate the potential of a curved element model for a doubly curved surface. The quadrilateral idealization was found to be inadequate near the apex. The apex was therefore modeled analytically by "cutting" out a 1° sector and replacing the eliminated material with nodal loads equivalent to the umbilical stress resultants.

The finite element model and results are shown in Fig. 8.3.5. The initial thickness of the hemisphere was taken at 0.01 ft (0.003 m), the radius as 50 ft (3 m), and the Mooney material constants as C_1 = 2300 psf (110 kN/m^2) and C_2 = 230 psf (11 kN/m^2). The results obtained using a finite difference modeling of the axisymmetric behavior [200] are also shown in Fig. 8.3.5.

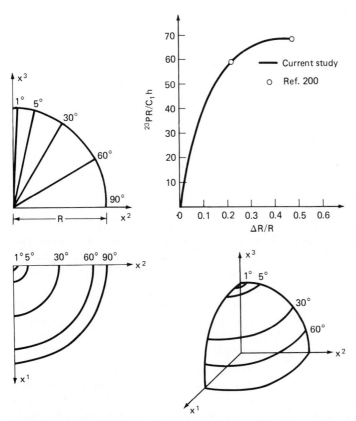

Figure 8.3.5 Inflation of hemisphere (Ref. 135).

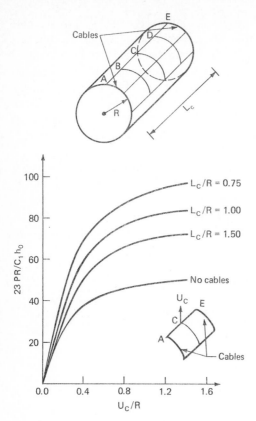

Figure 8.3.6 Inflation of cable-reinforced cylinder (Ref. 136).

Example of cable-reinforced membranes. In Ref. 136, the procedure described in this section is extended to include hyperelastic cables continuously attached to and reinforcing the membrane. A cubic blending function was used to develop a one-dimensional superparametric element for the cable elements which would be compatible with the membrane element. More refined treatments of cable-membrane interaction including slippage effects are treated using finite element techniques in conjunction with interactive computer graphics in the series of Refs. 137 to 141. Straight cable and flat triangular membrane elements are used. The problem of specifying a prestressed configuration is addressed in Refs. 139 and 140. Several techniques are described for solving either for the reference geometry given the prestresses or for the prestresses given the reference geometry, or a combination of stresses and geometry. Least square techniques and iteration of nonlinear algebraic equilibrium equations are two methods employed. In Ref. 141 in-service static problems are treated using nonlinear equations with an interesting inclusion of slip effects between membranes and reinforcing cables. In the above-described works, sophisticated computer hardware is required [137, 138] to implement the interactive computer graphics.

With the model developed here and in Ref. 136, the example of a circumferentially reinforced cylinder was considered. On a right circular cylinder with diameter = 20 ft (6 m), length = 30 ft (9 m), and thickness = 0.01 ft (0.003 m) circumferential cables were placed as shown in Fig. 8.3.6. The cylinder was constrained against longitudinal motion at the terminal circumferential sections. Various diameters and spacing of the cable were considered. The Mooney constants [187] for the cables were taken as C_{1c} = 14,000 psf (670 kN/m^2) and C_{2c} = 4000 psf (190 kN/m^2) and the Mooney constants of the membrane as C_1 = 2300 psf (110 kN/m^2) and C_2 = 280 psf (13 kN/m^2). These are representative of natural rubber and gum rubber, respectively.

In Fig. 8.3.7 is plotted the radial displacement along the longitudinal section of the cylinder at various pressures. The effect of cable spacing on the response of the cylinder can be seen in Fig. 8.3.6, where the radial

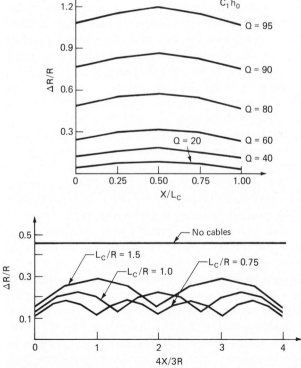

Figure 8.3.7 Radial and longitudinal profile of cable-reinforced cylinders (Ref. 136).

expansions are plotted as a function of internal pressure. For comparison purposes, the radial expansion versus pressure curves for unreinforced membranes are included. The effect of cable spacing is further demonstrated in Fig. 8.3.7 where longitudinal profiles at a pressure of 4 psf (190 N/m²) are plotted for three different spacings.

8.4   Finite Element Analysis of Nonlinear Dynamic Behavior

In this section we consider the nonlinear dynamic behavior of cable-reinforced membranes using finite element analysis techniques. For thin, extremely flexible membranes there are three sources of nonlinearity which are of importance in predicting the dynamic behavior of membrane structures: (1) large dynamic motions of the membranes transverse to their middle surface are possible; (2) fabric and rubberlike materials have nonlinear stress-strain relations; and (3) pressure loadings lead to nonconservative displacment-dependent loads if large rotations of the unit normal to the membrane are possible.

Two families of isoparametric elements for membranes are presented in Sec. 8.4.1. With these elements, discretized equations of motion are developed in Sec. 8.4.2 which are suitable to describe the large dynamic displacements of a curved membrane. We will adopt a general lagrangian coordinate system with curvilinear coordinates embedded in a reference surface $^{@}C$. If we then specify $^{@}C$ to be the static equilibrium state and limit our dynamic motions to be small excursions about $^{*}C$, we will arrive at linear equations of free vibration for a general finite element model of the membrane (see Sec. 8.4.3). Those equations can be solved for modal responses (frequencies and shapes) which can in turn be used for a modal superposition solution for small forced vibrations.

Alternatively, if we specify $^{@}C$ to always be the initial state ^{0}C, we will arrive at a total lagrangian description (see Sec. 8.4.4) of the nonlinear equations of motion at time t on surface $'C$, or at time $t + \Delta t$ on surface $^{#}C$. In the latter case an implicit integration technique, as described in Sec. 5.1 for nonlinear cable systems and in Sec. 8.4.4 for membranes, can be used to develop numerical solutions at discrete times for the nonlinear motions of membranes and cable-reinforced membranes.

A third alternative would be to specify $'C$ to be the surface which is in dynamic equilibrium at time t. This would lead to an updated lagrangian descripton of the nonlinear equations of motion on $^{#}C$ at time $t + \Delta t$. With this alternative, incremental equations describing the nonlinear hydrodynamic interactions with a nonquiescent embedding fluid could be solved iteratively at each discrete time. The dynamic pressures from acoustic or gravity waves in the embedding fluid could be calculated from an updated eulerian model of the fluid coupled to the membrane model.

8.4.1 Isoparametric interpolation for membranes

In Sec. 5.2.1, the procedures and advantages of using isoparametric interpolation for curved cable elements were discussed. The simultaneous interpolation of geometry as well as displacements by the same shape functions can also be used for curved membranes. We will here develop the interpolation equations for location as well as displacements at a general point in a multinoded membrane element in terms of cartesian coordinates and cartesian components of displacement at nodal points on the element. Note that cartesian displacement components are used rather than the components in the directions of the base vectors of the reference surface, as was done in Sec. 8.3 where a superparametric formulation was used. See Refs. 94, 165, 204 for extensive discussions of isoparametric theory.

We specify our curvilinear coordinates x_1, x_2 on the surface of the element to be nondimensional natural coordinates with ranges -1.0 to $+1.0$ and assume the coordinates and displacements at a generic point x_1, x_2 in the element to be given in terms of shape functions $H_N(x_1, x_2)$ by

$$^@y_i(x_1, x_2) = H_N(x_1, x_2)\,^@Y_{iN} \tag{8.4.1a}$$

$$'_@u_i(x_1, x_2) = H_N(x_1, x'_2)'_@D_{iN} \tag{8.4.1b}$$

where $^@Y_{Ni} = i$th cartesian coordinate of node N, and $'_@D_{iN} = i$th cartesian component of displacement of node N from $^@C$ to $'C$. The position and displacement vectors at point x_1, x_2 in terms of $^@y_i$ and $'_@u_i$ are

$$^@\overline{R}(x_1, x_2) = {}^@y_i\hat{e}_i \tag{8.4.2a}$$

$$'_@\overline{V}(x_1, x_2) = '_@u_i\hat{e}_i \tag{8.4.2b}$$

where $\hat{e}_i$ are the unit cartesian base vectors. The shape function $H_N(x_1, x_2)$ has the property that its numerical value is unity at node N and is zero at other nodes. We will adopt polynomial shape functions which map a general quadrilateral or triangular surface element curved in three-dimensional space onto two-dimensional squares or triangles.

The choice of shape functions, i.e., order of interpolations, has a significant effect on the accuracy and efficiency of the numerical solution. With the simplest linear shape functions, flat elements are obtained and numerous elements are needed. Furthermore, discontinuities in strain and, hence, stress will be predicted at element boundaries. However, fewer nodes per element are required, and the resulting finite element model will be seen to be less stiff than those obtained using higher-order shape functions. With higher-order shape functions, more accurate results can be obtained because they provide better predictions of the equilibrium

TABLE 8.4.1   Quadrilateral Isoparametric Shape Functions

Element	Number of nodes	Shape functions
Linear	4	$H_1 = (1 + x_1)(1 + x_2)/4$ $H_2 = (1 - x_1)(1 + x_2)/4$ $H_3 = (1 - x_1)(1 - x_2)/4$ $H_4 = (1 + x_1)(1 - x_2)/4$
Quadratic	8	$H_1 = (1 + x_1)(1 + x_2)(x_1 + x_2 - 1)/4$ $H_2 = (1 - x_1)(1 + x_2)(-x_1 + x_2 - 1)/4$ $H_3 = (1 - x_1)(1 - x_2)(-x_1 - x_2 - 1)/4$ $H_4 = (1 + x_1)(1 - x_2)(x_1 - x_2 - 1)/4$ $H_5 = (1 - x_1)(1 - x_2)(1 + x_2)/2$ $H_6 = (1 - x_1)(1 + x_1)(1 + x_2)/2$ $H_7 = (1 - x_1)(1 + x_2)(1 - x_2)/2$ $H_8 = (1 - x_1)(1 + x_1)(1 - x_2)/2$
Cubic	12	$H_1 = (1 + x_1)(1 + x_2)(9x_1^2 + 9x_2^2 - 10)/32$ $H_2 = (1 - x_1)(1 + x_2)(9x_1^2 + 9x_2^2 - 10)/32$ $H_3 = (1 - x_1)(1 - x_2)(9x_1^2 + 9x_2^2 - 10)/32$ $H_4 = (1 + x_1)(1 - x_2)(9x_1^2 + 9x_2^2 - 10)/32$ $H_5 = (1 + x_1)(1 - x_2^2)(1 - 3x_2)9/32$ $H_6 = (1 + x_2)(1 - x_1^2)(1 + 3x_2)9/32$ $H_7 = (1 - x_1)(1 - x_2^2)(1 + 3x_2)9/32$ $H_8 = (1 - x_2)(1 - x_1^2)(1 - 3x_1)9/32$ $H_9 = (1 + x_1)(1 - x_2^2)(1 + 3x_2)9/32$ $H_{10} = (1 + x_2)(1 - x_1^2)(1 - 3x_1)9/32$ $H_{11} = (1 - x_1)(1 - x_2^2)(1 - 3x_2)9/32$ $H_{12} = (1 - x_2)(1 - x_1^2)(1 - 3x_1)9/32$

shape and discontinuities in stress are eliminated. However, they require intensive computations at the element level and lead to stiffer results as well as lengthier computation times.

In a subsequent subsection we will report results obtained by Lo [90] using linear, quadratic, and cubic shape functions presented in Ref. 204 for both quadrilateral and triangular elements. The family of shape functions for quadrilateral elements with various numbers of nodes per element is given in Table 8.4.1. For triangles, the natural coordinates are area coordinates which are deformed in terms of the two convected coordinates x_1, x_2 by

$$\zeta = x_1$$

$$\zeta_2 = x_2$$

$$\zeta_3 = 1 - x_1 - x_2$$

TABLE 8.4.2   Triangular Isoparametric Shape Functions

Element	Number of nodes	Shape functions
Linear	3	$H_1 = \zeta_1$ $H_2 = \zeta_2$ $H_3 = \zeta_3$
Quadratic	6	$H_1 = (2\zeta_1 - 1)\zeta_1$ $H_2 = (2\zeta_2 - 1)\zeta_2$ $H_3 = (2\zeta_3 - 1)\zeta_3$ $H_4 = 4\zeta_1\zeta_2$ $H_5 = 4\zeta_2\zeta_3$ $H_6 = 4\zeta_1\zeta_3$
Cubic	10	$H_1 = (3\zeta_1 - 1)(3\zeta_1 - 2)\zeta_1/2$ $H_2 = (3\zeta_2 - 1)(3\zeta_2 - 2)\zeta_2/2$ $H_3 = (3\zeta_3 - 1)(3\zeta_3 - 2)\zeta_3/2$ $H_4 = \zeta_1\zeta_2(3\zeta_1 - 1)9/2$ $H_5 = \zeta_1\zeta_2(3\zeta_2 - 1)9/2$ $H_6 = \zeta_2\zeta_3(3\zeta_2 - 1)9/2$ $H_7 = \zeta_2\zeta_3(3\zeta_3 - 1)9/2$ $H_8 = \zeta_1\zeta_3(3\zeta_3 - 1)9/2$ $H_9 = \zeta_1\zeta_3(3\zeta_1 - 1)9/2$ $H_{10} = 27\zeta_1\zeta_2\zeta_3$

The family of shape functions for triangular elements with various numbers of nodes per element is given in Table 8.4.2 in terms of ζ_1, ζ_2, ζ_3.

With Eqs. (8.4.1), the various geometric parameters of the reference surface $^@C$ may now be written as (a comma denotes partial derivatives with respect to index following)

$$^@\overline{A}_\alpha = H_{N,\alpha}\,{}^@Y_{Ni}\hat{e}_i \tag{8.4.3.a}$$

$$^@A_{\alpha\beta} = H_{N,\alpha}H_{M,\beta}\,{}^@Y_{Ni}\,{}^@Y_{Mi} \tag{8.4.3b}$$

$$^@A = \text{determinant of } {}^@A_{\alpha\beta} \tag{8.4.3c}$$

$$= (H_{M,1}H_{N,1}\,{}^@Y_{Mi}\,{}^@Y_{Ni})(H_{Q,2}H_{S,2}\,{}^@Y_{Qj}\,{}^@Y_{Sj})$$

$$- (H_{M,1}H_{N,2}\,{}^@Y_{Mi}\,{}^@Y_{Ni})(H_{Q,2}H_{S,1}\,{}^@Y_{Qj}\,{}^@Y_{Sj})$$

$$^@\hat{N} = \frac{e_{ijk}H_{M,1}H_{N,2}\,{}^@Y_{Mj}\,{}^@Y_{Nk}\hat{e}_i}{\sqrt{^@\overline{A}}} \tag{8.4.3d}$$

where e_{ijk} = permutation symbol

$$= \begin{cases} 0 \text{ when any two indices are repeated} \\ +1 \text{ when } ijk \text{ are even permuations of 1, 2, 3} \\ -1 \text{ when } ijk \text{ are odd permutations of 1, 2, 3} \end{cases}$$

The contravariant components of $^@A_{\alpha\beta}$ are

$$^@A^{11} = \frac{^@A_{22}}{^@A}$$

$$^@A^{22} = \frac{^@A_{11}}{^@A}$$

$$^@A^{12} = {^@A^{21}} - \frac{-^@A_{12}}{^@A}$$

8.4.2   Discretized equations of motion

As the surface C of the membrane deforms from the unstrained state ^0C to $'C$ at time t, we write the time-varying position vector of a point in the element as

$$'\overline{R}(x_1, x_2, t) = 'y_i(x_1, x_2, t)\grave{e}_i$$

$$= {^@\overline{R}}(x_1, x_2) + {_@^{\,\prime}\overline{V}}(x_1, x_2, t)$$

where $^@\overline{R}$ and $_@^{\,\prime}\overline{V}$ are given by Eqs. (8.4.2), and $_@^{\,\prime}y_i$ are the cartesian coordinates of the point $'C$ at time t. Thus

$$'y_i = {^@y_i} + {_@^{\,\prime}u_i}$$

Since the metric tensor of a differential arc length on $'C$ relative to $^@C$ is given by [using Eq. (8.4.1)],

$$_@^{\,\prime}A_{\alpha\beta} = 'y_{i,\alpha} \cdot 'y_{i,\beta} \tag{8.4.4}$$

$$= H_{N,\alpha}H_{M,\beta}(^@Y_{in} + {_@^{\,\prime}D_{iN}})(^@Y_{iM} + {_@^{\,\prime}D_{iM}})$$

and since the strain tensor of a differential arc length on $'C$ relative to $^@C$ is

$$_@^{\,\prime}\gamma_{\alpha\beta} = \tfrac{1}{2}({_@^{\,\prime}A_{\alpha\beta}} - {^@A_{\alpha\beta}})$$

we can rewrite the strain tensor previously given by Eq. (8.3.2) in Sec. 8.3 as

$$_@^{\,\prime}\gamma_{\alpha\beta} = \tfrac{1}{2}H_{N,\alpha}H_{M,\beta}(^@Y_{iN}{_@^{\,\prime}D_{iM}} + {^@Y_{iM}}{_@^{\,\prime}D_{iN}} + {_@^{\,\prime}D_{iN}}{_@^{\,\prime}D_{iM}}) \tag{8.4.5}$$

The stress resultants for a hookean material are

$$'_@n^{\alpha\beta} = {}^@C^{\alpha\beta\lambda\mu}\,'_@\gamma_{\alpha\beta} \tag{8.4.6}$$

where for an isotropic material

$$^@C^{\alpha\beta\lambda\mu} = \frac{^@hE\nu}{(1+\nu)(1-2\nu)}\,{}^@A^{\alpha\beta}\,{}^@A^{\lambda\mu} + \frac{^@hE\nu}{1+\nu}\left({}^@A^{\alpha\lambda}\,{}^@A^{\beta\mu} + {}^@A^{\alpha\mu}\,{}^@A^{\beta\lambda}\right) \tag{8.4.7}$$

Alternatively, for a rubberlike Mooney material which is isotropic and incompressible,

$$'_@n^{\alpha\beta} = 2'_@\lambda\left\{ {}^@A^{\alpha\beta}(C_1 + C_2{}'_@\lambda^2) + {}'_@A^{\alpha\beta}\left[\frac{C_2}{'_@\lambda^2}\right.\right.$$

$$\left.\left. - {}'_@\lambda^2(C_1 + C_2{}^@A^{\eta\rho}\,{}'_@A_{\eta\rho})\right]\right\} \tag{8.4.8a}$$

with

$$'_@A_{\eta\rho} = {}^@A_{\eta\rho} + {}'_@\lambda_{\eta\rho} \tag{8.4.8b}$$

$$'_@A = {}'_@A_{11}\,{}'_@A_{22} - {}'_@A_{12}\,{}'_@A_{21} \tag{8.4.8c}$$

$$'_@\lambda = \frac{\sqrt{^@A}}{'_@A} \tag{8.4.8d}$$

Most analyses of membrane structures assume isotropic or inextensible behavior, but membranes are commonly made up from strips of fabric materials. Considerable data are needed on material properties for fabrics, on constitutive relations, and on numerical models including orthotropy. Most finite element theories can readily accommodate orthotropic materials. However, in doing so, one needs to specify the principal directions. Thus, one should align the elements and the warp directions with the strips of fabric.

We will here develop the discretized equations of motion in terms of an arbitrary constitutive relation which will allow several extant material models to be used. The equations of motion can be developed from the principle of virtual work as written on the dynamic equilibrium state $'C$ but referred to $^@C$, as

$$0 = \int\int_{@C} ('_@n^{\alpha\beta}\,\delta\gamma_{\alpha\beta} - {}'_@p_i\,\delta u_i)\sqrt{^@A}\,dx_1\,dx_2$$

$$+ {}^@h\int\int_{@C} {}^@\rho\,'_@\ddot{u}_i\,\delta u_u\,\sqrt{^@A}\,dx_1\,dx_2 \tag{8.4.9}$$

where $\delta\gamma_{\alpha\beta}$ is the infinitesimal virtual strain corresponding to a small virtual displacement

$$\delta\overline{V} = \delta u_i \, \hat{e}_i = H_N \, \delta D_{Ni} \, \hat{e}_i \qquad (8.4.10)$$

superposed on $'C$, and the force vector

$$\overline{F}\sqrt{'A} \, dx_1 \, dx_2 = {}_{@}'p_i \sqrt{{}^{@}A} \, dx_1 \, dx_2 \, \hat{e}_i$$

has been expressed in terms of cartesian components ${}_{@}'p_i$ relative to ${}^{@}C$.

The second integral term in Eq. (8.4.9) represents the virtual work of inertial force (from kinetic energy) in which the principle of mass conservations has been invoked to replace $'h'\rho\sqrt{'A} \, dx_1 \, dx_2$ by ${}^{@}h{}^{@}\rho\sqrt{{}^{@}A} \, dx_1 \, dx_2$, where $'\rho$, ${}^{@}\rho$ = densities in $'C$ and ${}^{@}C$, and the acceleration of $'C$ is given by $'\ddot{y}_i = {}_{@}'\ddot{u}_i$.

If we now adopt a consistent mass approach, as described in Sec. 4.4.2 for cable elements, we assume

$$_{@}'\ddot{u}_i = H_{N@}'\ddot{D}_{Ni} \qquad (8.4.11)$$

Then substitute Eqs. (8.4.1), (8.4.5), (8.4.10), and (8.4.11) into Eq. (8.4.9) to obtain for an artibrary virtual displacement δD_{Mi} the following discretized equation of motion for the ith force component at the Nth mode:

$$^{@}M_{MNij@}'\ddot{D}_{Mi} + {}_{@}'R_{Ni} = {}_{@}'F_{Ni} \qquad (8.4.12a)$$

where the mass matrix is

$$^{@}M_{MNij} = \delta_{ij} \, {}^{@}\rho{}^{@}h \int_{@C} \int H_N H_M \sqrt{{}^{@}A} \, dx_1 \, dx_2 \qquad (8.4.12b)$$

the internal restoring force is

$$_{@}'R_{Ni} = \tfrac{1}{2} \int_{@C} \int {}_{@}'n^{\alpha\beta}(H_{M,\alpha}H_{N,\beta}$$

$$+ H_{N,\alpha}H_{M,\beta}) \sqrt{{}^{@}A} \, dx_1 \, dx_2 \cdot ({}^{@}Y_{Mi} + {}_{@}'D_{Mi}) \qquad (8.4.12c)$$

and the external force is

$$_{@}'F_{Ni} = \int_{@C} \int {}_{@}'p_i H_N \sqrt{{}^{@}A} \, dx_1 \, dx_2 \qquad (8.4.12d)$$

Note that $\sqrt{{}^{@}A}$ in Eqs. (8.4.12) is given by Eq. (8.4.3c) in terms of derivatives of the shape functions and the nodal coordinates, thus further complicating the integration required to form the element matrices.

Equations (8.4.12) represent a set of ordinary differential equations of dynamic equilibrium at time t of every degree of freedom on $'C$ and have been cast into the same form as Eqs. (5.1.1), solution techniques for which were discussed in Sec. 5.1.2. Equations (8.4.12) are nonlinear in that

1. Material nonlinearities are possible in the restoring force ${}_{@}^{\prime}R_{Ni}$ since ${}_{@}^{\prime}n^{\alpha\beta}$ can be replaced by any linear [hookean as in Eq. (8.4.6)] or nonlinear [Mooney as in Eqs. (8.4.8)] constitutive equations.

2. ${}_{@}^{\prime}R_{Ni}$ is nonlinear even if the constitutive relation is linear since ${}_{@}^{\prime}R_{Ni}$ includes finite displacement components ${}_{@}^{\prime}D_{Mi}$.

3. Nonconservative loads are possible in the external force ${}_{@}^{\prime}F_{Ni}$ if the magnitude or direction of ${}_{@}^{\prime}p_{i}$ is dependent on displacements, e.g., pressure.

In pressurized membranes, nonconservative loads occur during large displacements. The tractions ${}_{@}^{\prime}p_{i}$ due to net pressure p normal to the membrane surface $'C$ are

$$
{}_{@}^{\prime}p_{i} = \sqrt{\frac{'A}{@A}}\, p({}_{@}^{\prime}\hat{N} \cdot \hat{e}_{i}) \tag{8.4.13}
$$

where the dot between ${}_{@}^{\prime}\hat{N}$ and $\hat{e}_{i}$ represents their inner vector product. Since the unit normal vector ${}_{@}^{\prime}\hat{N}$ is defined in terms of the cross products of the deformed base vectors as

$$
{}_{@}^{\prime}\hat{N} = \frac{'\overline{A}_{1} \times '\overline{A}_{2}}{\sqrt{'A}}
$$

$$
= \frac{e_{kji}}{\sqrt{'A}} ('y_{k}\hat{e}_{k}) \times ('y_{j}\hat{e}_{j})\hat{e}_{k} \tag{8.4.14}
$$

we have, replacing $'y_{i} = {}^{@}y_{i} + {}_{@}u_{i}$ by Eqs. (8.4.1),

$$
{}_{@}^{\prime}p_{i} = \frac{p}{\sqrt{@A}} e_{kji}H_{Q,1}H_{M,2}({}^{@}Y_{Qk}{}^{@}y_{Mj} + {}^{@}Y_{Qi}{}_{@}^{\prime}D_{Mj}
$$

$$
+ {}^{@}Y_{Mk}{}_{@}^{\prime}D_{Qj} + {}_{@}^{\prime}D_{Qk}{}_{@}^{\prime}D_{Mj}) \tag{8.4.15}
$$

Therefore, for pressure loading, Eq. (8.4.12c) becomes

$$
{}_{@}^{\prime}F_{Ni} = pe_{kji}\left(\int \int_{@C} H_{N}H_{Q,1}H_{M,2}\, dx_{1}\, dx_{2}({}^{@}Y_{Qk}{}^{@}Y_{Mi} \right.
$$

$$
\left. + {}^{@}Y_{Qk}{}_{@}^{\prime}D_{Mj} + {}^{@}Y_{Qk}{}_{@}^{\prime}D_{Mj} + {}_{@}^{\prime}D_{Qk}{}_{@}^{\prime}D_{Mj}) \right. \tag{8.4.16}
$$

We see that ${}_{@}^{\prime}p_{i}$ and, hence ${}_{@}^{\prime}F_{Ni}$, are indeed nonlinear in the displacement components ${}_{@}^{\prime}D_{Mi}$.

To evaluate the matrices in Eqs. (8.4.12) we must integrate products of the shape function $H_M(x_1, x_2)$ and their derivatives over the area of each element. Because of the complexities of those products, the integrations must be done numerically. Guassian guadrature [201] can be used for the quadrilateral elements and Guass-Hammer formulas [203] for the triangular elements. Reduced integration, in which element matrices are evaluated by lower-order formulas, often leads to superior results over the higher-order formulas. This is because the stiffness of a discretized membrane is generally greater than the real membrane and reduced integration recovers some of the "lost" flexibility. Furthermore, reduced integration significantly reduces computational effort in each element. However, care must be exercised in selecting the order of reduced integration to avoid singularities. Thus, before adopting reduced integration formulas for a particular class of problems, numerical experimentation with various orders of integration is advised.

8.4.3   Free vibrations of membranes

In studying the linear response of prestressed membranes to dynamic in-service loads, it is important to determine their mode shapes and natural frequencies of small free oscillations about the equilibrium state. Information concerning such modes and frequencies is also valuable in selecting model discretizations and time steps for numerical integration in nonlinear simulations of finite element models.

In many instances of analyses of cable-reinforced membrane roofs, the membrane is sufficiently light in weight that the cables can be considered the main load-carrying members. Thus, it is not unreasonable to perform dynamic analyses of cable-reinforced membranes as cable nets [31–34, 91, 92] in which the membrane is treated solely as a means of load transfer to the nets.

In evaluating the vibrations of membrane roofs, it is important to consider their interactions with wind [4, 45, 144–153]. According to Morris [95] there are four classes of behavior of membrane roofs under wind: (1) vortex shedding—harmonic lift forces transverse to the wind flow are caused by periodic shedding of vortices at points of flow separation (this phenomenon can be avoided by proper shaping, detailing, and stiffening so as to modify the natural frequencies); (2) panel flutter—flow-induced wrinkling of membrane surfaces [154] as a function of wind velocity and membrane prestress (because of insufficient prestress capability in fabric membranes, this phenomenon may be a significant factor in membrane roofs); (3) galloping [155, 156]—aerodynamic motion due to negative damping in turbulence which cancels structural damping and leads to roof flutter [45, 46, 157]; and (4) wind gusting—dynamic pressures due to gusts in wind speed [158–162] may lead to dynamic overstressing of membrane components. Morris [95] also notes two other aspects of the dynamic analysis of membrane enclosures which should be considered. Vibrations induced by wind may set the enclosed air or fluid into motion [151, 162]. Neglect of the added mass term which should thereby be included in the mass matrix for the membrane is not conservative because any increase in mass may increase the possibility of large dynamic response under wind. Second, there is an increase in effective damping of the structure due to the aerodynamic damping from motion of enclosed air or fluid. [151].

To perform a free vibration analysis of a finite element model of a membrane about its static equilibrium state, we seek linear equations of motion in the form

$$[M]\{\ddot{u}\} + [K]\{u\} = \{0\}$$

solution techniques for which were described in Sec. 4.3.1. If we, therefore, adopt the static equilibrium surface $*C$ as the reference surface $^{@}C$, we can rewrite Eqs. (8.4.12) as

$$[M]\{{}'_*\dot{D}\} = \{{}'_*p\} \tag{8.4.17a}$$

where

$$M_{MNij} = \delta_{ij} \, {}^*\rho^*h \int \int_{*C} H_M H_N \sqrt{*A} \, dx_1 \, dx_2 \tag{8.4.17b}$$

$$\underset{*}{'}R_{Ni} = \frac{1}{2} \int \int_{*C} {}'_* n^{\alpha\beta} (H_{M,\alpha} H_{N,\beta}$$

$$+ \, H_{N,\alpha} H_{M,\beta}) \sqrt{*A} \, dx_1 \, dx_2 \cdot (Y_{Mi} + {}'_*D_{Mi}) \tag{8.4.17c}$$

$$'F_{Ni} = \int \int_{*C} {}'_* p_i H_N \sqrt{*A} \, dx_1 \, dx_2 \tag{8.4.17d}$$

If we now take the Taylor expansion of ${}'_*R_{Ni}$ about $*R_{Ni}$ and assume small motions such that ${}'_*p_i = {}^*p_i$ and only first-order terms in $\{{}'_*D\}$ are retained, then Eqs. (8.4.17) become

$$[M]\{{}'_*\dot{D}\} + \{{}^*R_N\} + [{}^*K]\{{}'_*D\} = \{{}^*p\} \tag{8.4.18a}$$

where $*R_{Ni} = {}^*p_{Ni}$ can be canceled and

$$*K_{NMij} {}'_*D_{Mj} = \frac{\partial {}'_*R_{Ni}}{\partial {}'_*D_{Mj}} \bigg|_{*C} {}'_*D_{Mj} + \frac{\partial {}'_*R_{Ni}}{\partial {}'_*n^{\alpha\beta}} \bigg|_{*C} \frac{\partial {}'_*n^{\alpha\beta}}{\partial \gamma_{\eta\xi}} \Delta {}'_*\gamma_{\eta\xi} \tag{8.4.18b}$$

in terms of small displacements ${}'_*D_{Mj}$ and small increments in strains $\Delta {}'_*\gamma_{\eta\xi}$. We approximate the constitutive equations

$$C^{\alpha\beta\eta\xi} = \frac{\partial'_* n_1^{\alpha\beta}}{\partial\gamma_{\eta\xi}} \tag{8.4.18c}$$

for a hookean material by Eq. (8.4.6), or for a Mooney material [from Eqs. (8.4.8)] by

$$C^{\alpha\beta\eta\xi} = 8*A^{\alpha\beta}*A^{\eta\xi}*\lambda^2 J + 2(e^{\alpha\eta}e^{\beta\xi} + e^{\alpha\xi}e^{\beta\eta}) \frac{*\lambda^2}{*A}\left(\frac{C_2}{*\lambda^4} - J\right)$$

$$- 4C_2*\lambda^2({}^0A^{\alpha\beta}*A^{\eta\lambda} + {}^0A^{\eta\xi}*A^{\alpha\beta}) \tag{8.4.18d}$$

in which $J = C_1 + {}^0A^{\alpha\beta}*A_{\alpha\beta}$. The small increments in strain $\Delta'_*\gamma_{\eta\xi}$ are obtained by retaining only the linear terms in Eq. (8.4.5):

$$\Delta'_*\gamma_{\eta\xi} = \tfrac{1}{2}(H_{N,\eta}H_{M,\xi} + H_{M,\eta}H_{N,\xi})* Y_{Nj}{}'_* D_{Mj} \tag{8.4.18e}$$

Thus, the tangent stiffness matrix at $*C$ is given by

$$*K_{MNij} = \tfrac{1}{2}\int\int [*n^{\alpha\beta}(H_{M,\alpha}H_{N,\beta} + H_{N,\alpha}H_{M,\beta})\,\delta_{ij} + \tfrac{1}{4}(H_{R,\alpha}H_{M,\beta}$$

$$+ H_{R,\beta}H_{M,\alpha})* Y_{Ri}C^{\alpha\beta\eta\xi}(H_{Q,N}H_{N,\xi} + H_{Q,\xi}H_{N,\eta})* Y_{Qj}]\sqrt{*A}\,dx_1\,dx_2 \tag{8.4.19}$$

The equation of free vibration is

$$[M]\{'_*\ddot{D}\} + [*K]\{D\} = \{0\} \tag{8.4.20}$$

with M_{MNij} given by Eq. (8.4.17b) and $*K_{MNij}$ by Eq. (8.4.19).

As an example, consider the problem of a square, flat, hookean, and uniformly prestressed membrane as shown in Fig. 8.4.1. Results for free vibration of several finite element models shown in Fig. 8.4.2 for this problem were compiled by Lo [90].

The natural frequencies of the first eight models are given in Table 8.4.3 along with the exact solution [90]. Note that since this is a flat membrane, the advantages of higher-order elements are somewhat lost in this problem. The model with 9 quadratic quadrilaterals provides the best overall estimates of the natural frequencies, followed by the model with 25 linear quadrilaterals. The model with 25 linear quadrilaterals has the same number of degrees of freedom as does the model with 9 quadratic quadrilaterals but only requires 70 percent of the computational time for the higher-order element. Cubic models [142] require the most computational time and do not provide sufficient accuracy to warrant their use in this problem, particularly in the higher modes.

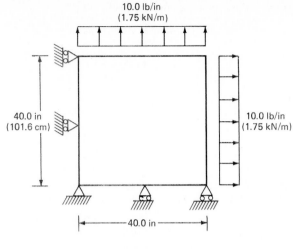

10.0 lb/in
(1.75 kN/m)

40.0 in
(101.6 cm)

10.0 lb/in
(1.75 kN/m)

40.0 in

h_0 = 0.1 in (2.54 mm)
E = 30 × 10⁶ psi (207 GPa)
ν = 0.3
ρ = 0.00075 lb·s²/in⁴ (8016 kg/m³)

Figure 8.4.1 Membrane geometry and properties.

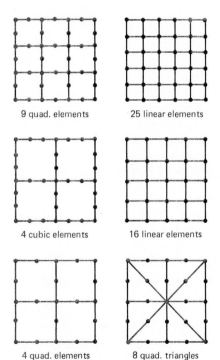

9 quad. elements

25 linear elements

4 cubic elements

16 linear elements

4 quad. elements

8 quad. triangles

Figure 8.4.2 Membrane finite element models.

TABLE 8.4.3 Comparison of Natural Frequencies of Membrane Models [90]

| Model | Number of interior nodes | Natural frequencies, rad/s (percent error in parentheses) | | | | |
| | | Mode | | | | |
		1	2, 3	4	5, 6	7, 8
Exact	—	40.55	64.13	81.12	90.69	103.40
9 quadratic quadrilaterials	15	40.59 (0.10)	64.76 (0.98)	82.98 (2.20)	91.65 (1.06)	109.54 (5.94)
4 cubic quadrilaterals	9	40.62 (0.17)	65.83 (2.65)	93.37 (15.00)	94.19 (3.80)	122.47 (18.38)
4 quadratic quadrilaterals	5	40.78 (0.56)	65.83 (2.65)	107.49 (32.50)	109.05 (20.24)	—
25 linear quadrilaterals	16	41.23 (1.60)	67.75 (5.60)	86.49 (6.62)	102.60 (13.13)	115.80 (11.99)
16 linear quadrilaterals	9	41.61 (2.61)	69.75 (8.76)	89.44 (10.26)	106.90 (17.89)	120.68 (16.71)
8 quadratic triangles	9	41.95	69.11	99.99	99.99	144.70

8.4.4 Nonlinear forced vibrations

In Sec. 8.4.2, we developed discretized equations of motion for a finite element model of the nonlinear dynamic motion of the membrane. In the present section we will develop a solution scheme for predicting those nonlinear dynamic motions. That scheme is a combined incremental and iterative scheme as described for general nonlinear dynamic equations in Sec. 5.1.2. We will adopt a total lagrangian approach in which we disregard nonlinear interactions with the embedding medium.

If we fix the reference surface $^{@}C$ to be the initial unstrained surface ^{0}C, we can rewrite the governing equations of motion, Eqs. (8.4.12), of surface $'C$ at time t as

$$[M]\{'_{@}\ddot{D}\} + \{'_0 R\} = \{'_0 F\} \tag{8.4.21a}$$

with

$$M_{MNij} = \delta_{ij}\,^0\rho^0 h \int_{^0C} \int H_M H_N \sqrt{^0 A}\ dx_1\ dx_2 \tag{8.4.21b}$$

$$'_0 R_{Ni} = \tfrac{1}{2}(^0 Y_{Mi} + '_0 D_{Mi}) \int_{^0C} \int '_0 n^{\alpha\beta}(H_{M,\alpha}H_{N,\beta}$$

$$+ H_{N,\alpha}H_{M,\beta})\sqrt{^0 A}\ dx_1\ dx_2 \tag{8.4.21c}$$

$$'_0F_{Ni} = \int_{^0C} \int {'_0p} H_N \sqrt{^0A}\ dx_1\ dx_2 \qquad (8.4.21d)$$

If we consider the surface $^\#C$ at a slightly later time $t + \Delta t$, where $\Delta t = $ small discrete time step, then we can rewite Eq. (8.4.21) at time $t + \Delta t$ as

$$[M]\{{^\#_0}\ddot{D}\} + \{{^\#_0}R\} = \{{^\#_0}F\} \qquad (8.4.22)$$

where $\{{^\#_0}R\}$ and $\{{^\#_0}F\}$ are obtainable for Eqs. (8.4.21) by replacing the superscript $'$ by $\#$.

We will use Eq. (8.4.22) as the basis for an implicit time integration scheme (Newmark's method [94]). If we assume, as in Sec. 5.1, that the exact value of $\{{^\#_0}D\}$ at time $t + \Delta t$ is given by an iteration $\{\Delta D\}$ on a prior guess $\{{^\#_0}\tilde{D}\}$ to the solution, i.e.,

$$\{{^\#_0}D\} = \{{^\#_0}\tilde{D}\} + \{{^\#_0}\Delta D\} \qquad (8.4.23a)$$

then, we can take Taylor expansions in Eq. (8.4.22) about the solution $\{{^\#_0}\tilde{D}\}$ and obtain

$$[M]\{{^\#_0}\ddot{D}\} + [{^\#_0}K]\{{^\#_0}\Delta D\} = (\{{^\#_0}\tilde{F}\} - \{{^\#_0}\tilde{R}\})\Big|_{\{{^\#_0}\tilde{D}\}} \qquad (8.4.23b)$$

in which $(\{{^\#_0}F\} - \{{^\#_0}\tilde{R}\})$ are evaluated at $\{{^\#_0}\tilde{D}\}$ and $[{^\#_0}K]$ is the tangent stiffness evaluated at $\{{^\#_0}\tilde{D}\}$:

$$[{^\#_0}K] = \frac{\partial(\{{^\#_0}\tilde{R}\} - \{{^\#_0}\tilde{F}\})}{\partial\{{^\#_0}\tilde{D}\}}\Bigg|_{\{{^\#_0}\tilde{D}\}} \qquad (8.4.23c)$$

If conservative loads are considered, $\partial\{{^\#_0}F\}/\partial\{{^\#_0}D\} = 0$. If nonconservative loads, such as pressure, are considered, $\{{^\#_0}F\}$ is obtained from Eq. (8.4.16) as

$$\{{^\#_0}F\} = pe_{kji}({^0}Y_{Qk}{^0}Y_{Mj} + {^0}Y_{Qk}{^\#_0}D_{Mj} + {^0}Y_{Mk}{^\#_0}D_{Qj}$$

$$+ {^\#_0}D_{Qk}{^\#_0}D_{Mj}) \int_{^0C}\int H_N H_{Q,1} H_{M,2}\ dx_1\ dx_2 \qquad (8.4.24a)$$

and therefore,

$$\frac{\partial\{{^\#_0}\tilde{F}\}}{\partial\{{^\#_0}\tilde{D}\}} = pe_{kji}[({^0}Y_{Qk} + {^\#_0}D_{Qk})\int_{^0C}\int H_N H_{Q,1} H_{M,2}\ dx_1\ dx_2$$

$$+ ({^0}Y_{Qk} - {^\#_0}D_{Qk})\int_{^0C}\int H_N H_{M,1} H_{Q,2}\ dx_1\ dx_2] \qquad (8.4.24b)$$

Because of the permutation symbol e_{kji} we see that Eq. (8.4.24b) will contribute nonsymmetric terms to the tangent stiffness matrix if nonconservative loads are included. In the results for the example [90] given subsequently, this contribution to the tangent stiffness matrix was not included because of computational convenience; however, the load $\{^{\#}_{0}\tilde{F}\}$ in Eq. (8.4.22) was treated iteratively to partially account for the nonlinear effect.

The principal term in the tangent stiffness matrix is denoted by

$$[\tilde{K}] = \frac{\partial\{^{\#}_{0}\tilde{R}\}}{\partial\{^{\#}_{0}\tilde{D}\}} \tag{8.4.25}$$

and is obtained by taking the indicated derivatives of Eq. (8.4.21c). Thus,

$$K_{MNij} = \frac{\partial^{\#}_{0}R_{Ni}}{\partial^{\#}_{0}D_{Mj}}\bigg|_{\tilde{D}} = \frac{\partial^{\#}_{0}\tilde{R}_{Ni}}{\partial^{\#}_{0}Y_{Rk}}\frac{\partial^{\#}_{0}Y_{Rk}}{\partial^{\#}_{0}D_{Mj}} + \frac{\partial^{\#}_{0}R_{Ni}}{\partial^{\#}_{0}n^{\alpha\beta}}\frac{\partial^{\#}_{0}n^{\alpha\beta}}{\partial\gamma_{\eta\xi}}\frac{\partial\gamma_{\eta\xi}}{\partial^{\#}_{0}D_{Mj}} \tag{8.4.25a}$$

with

$$\frac{\partial^{\#}_{0}\tilde{R}_{Ni}}{\partial^{\#}_{0}Y_{Rk}} = \tfrac{1}{2}\,\delta_{ik}\int_{0C}\int\, {}^{\#}_{0}\tilde{n}^{\alpha\beta}(H_{R,\alpha}H_{N,\beta}$$

$$+ H_{N,\alpha}H_{R,\beta})\sqrt{^{0}A}\,dx_1\,dx_2 \tag{8.4.25b}$$

$$\frac{\partial^{\#}_{0}Y_{Rk}}{\partial^{\#}_{0}D_{Nj}} = \delta_{RNkj} \tag{8.4.25c}$$

$$\frac{\partial^{\#}_{0}\tilde{R}_{Ni}}{\partial n^{\alpha\beta}} = -\tfrac{1}{2}\int_{0C}\int\, (H_{R,\alpha}H_{N,\beta}$$

$$+ H_{N,\alpha}H_{R,\beta})\sqrt{^{0}A}\,dx_1\,dx_2\cdot({}^{0}Y_{Ri} + {}^{\#}_{0}\tilde{D}_{Ri}) \tag{8.4.25d}$$

$$\frac{\partial^{\#}_{0}n^{\alpha\beta}}{\partial\gamma_{\eta\xi}} = {}^{0}C^{\alpha\beta\eta\xi} \tag{8.4.25e}$$

$$\frac{\partial\gamma_{\eta\xi}}{\partial^{\#}_{0}D_{Mi}} = \tfrac{1}{2}(H_{Q,\eta}H_{M,\xi} + H_{M,\eta}H_{Q,\xi})({}^{0}Y_{Qi} + {}^{\#}_{0}\tilde{D}_{Qi}) \tag{8.4.25f}$$

where ${}^{0}C^{\alpha\beta\eta\xi}$ = coefficients of the applicable constitutive equation; Eq. (8.4.7) for a hookean material or Eq. (8.4.18c) for a Mooney material. Substituting Eqs. (8.4.25b to f) into Eq. (8.4.25a), we obtain the tangent stiffness matrix as

$$\tilde{K}_{MNij} = \int_{0C}\int\, [\tfrac{1}{2}\,\delta_{ij}\,{}^{\#}_{0}n^{\alpha\beta}(H_{M,\alpha}H_{N,\beta} + H_{N,\alpha}H_{M,\beta})$$

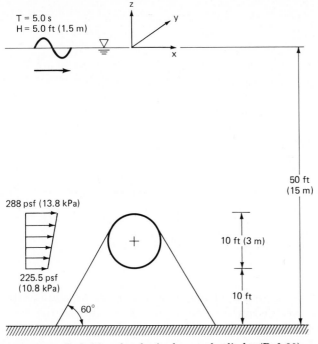

Figure 8.4.3 Definition sketch of submerged cylinder (Ref. 90).

$$+ \; \tfrac{1}{4} {}^0 C^{\alpha\beta\eta\xi} (H_{R,\alpha} H_{N,\beta} + H_{N,\alpha} H_{R,\beta})({}^0 Y_{Ri} + {}^{\#}_0 \tilde{D}_{Ri}) \; (H_{Q,\eta} H_{M,\xi}$$

$$+ \; H_{M,\eta} H_{Q,\xi})({}^0 Y_{Qj} + {}^{\#}_0 \tilde{D}_{Qj})\} \sqrt{{}^0 A} \; dx_1 \; dx_2 \qquad (8.4.26)$$

for use in Eq. (8.4.23).

As an example of the nonlinear dynamic solution process for cable-reinforced membranes, consider the problem solved by Lo [90] of an infinitely long, 10-ft (3 m) diameter cylinder filled with a fluid with specific gravity 0.9 and submerged in 50 ft (15 m) of water. The tube is pressurized and moored by cables so that the static position is as shown in Fig. 8.4.3. The membrane and cables have the following properties:

Property	Membrane	Cables
Thickness or area, in or in^2	0.2	0.785
E, psi	50,000	100,000
ν	0.3	0.3
${}^0\rho$, slug/ft^3	1.5528	4.6584

The finite element model of this problem is shown in Fig. 8.4.4 and consists of six cable elements and six cylindrical membranes in a typical sec-

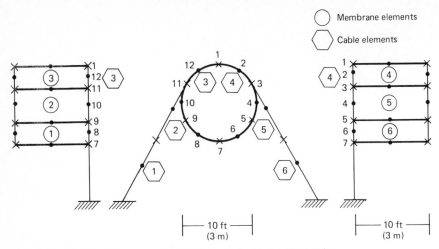

Figure 8.4.4 Finite element model of submerged cylinder (Ref. 90).

tion of the tube. The internal fluid was modeled as follows; the internal static pressure was included in the net pressure resultants on the membrane (in Fig. 8.4.3); the internal mass was lumped at membrane nodal points; and quiescent flow was assumed.

First, a static analysis was performed to determine the prestressed equilibrium shape. Second, a free vibration analysis, as described in Sec. 8.4.3, was performed to determine modal shapes and frequencies. The first six mode shapes and frequencies are given in Fig. 8.4.5. The first two modes represent rigid body sway and heave of the tube subject to mooring restraints. The second two modes represent a combination of rigid body roll and local deformation of the membrane. The higher modes are purely local deformation modes for the membrane with little cable deformation.

The third problem considered was that of subjecting the tube to a train of gravity waves [wave height = 5 ft (1.5 m), period = 5 s] traveling rightward on Fig. 8.4.3 and normally incident to the axis of the tube. The wave loading was treated approximately in two ways. In the first approach [90], no diffraction or radiation effects were considered, the Morison equation [95, 167] was used to calculate the fluid drag and added mass effects on the cables (see Sec. 5.3), and Froude-Kryloff [96] pressures due only to incident waves were applied to the membrane. The linear waves were assumed generated behind a wave front which at time $t = 0$ begins traveling rightward from a point 6 ft (2 m) to the left of the tube center at the wave group velocity [96]. The vertical and horizontal displacements of the top and bottom of the tube are displayed in Fig. 8.4.6 for the first 10 s (two wave periods) of the dynamic response of the system. Tension histories of the mooring cables are shown in Fig. 8.4.7.

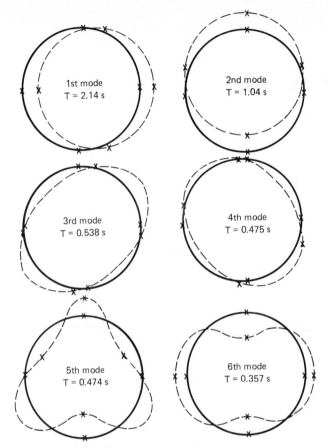

Figure 8.4.5 Mode shapes for submerged cylinder.

In the second approach, diffraction effects were included with the incident Froude-Kryloff pressure in an approximate sense. No radiation effects due to either rigid body motion or deformation of the tube were included. Frequency-dependent pressure coefficients for added mass and damping on an equivalent fixed and rigid cylinder were calculated from a finite element program [205] for predicting potential functions for wave diffraction effects. The vertical and horizontal displacements of the top of the tube obtained using this approach are displayed in Figs. 8.4.8 and 8.4.9. Results obtained using the first approach are also shown for comparison purposes. In addition to an anticipated phase shift due to the wave front, we see that inclusions of diffraction effects also modify the predicted amplitudes of motion.

The interaction of highly deformable membrane bodies with gravity waves is a nonlinear process not only involving nonlinearities in mem-

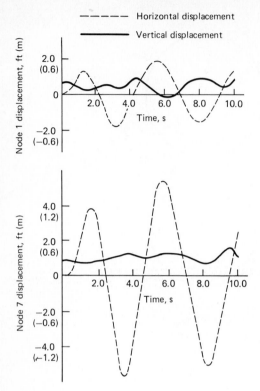

Figure 8.4.6 Displacement histories of top and bottom of cylinder (Ref. 90).

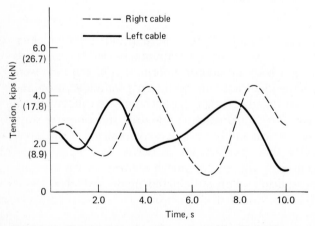

Figure 8.4.7 Tension histories of mooring cables (Ref. 90).

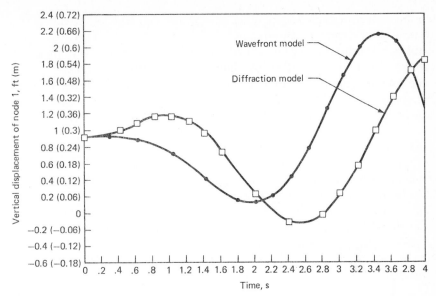

Figure 8.4.8 Comparison of vertical displacements of top of cylinder using different fluid loadings.

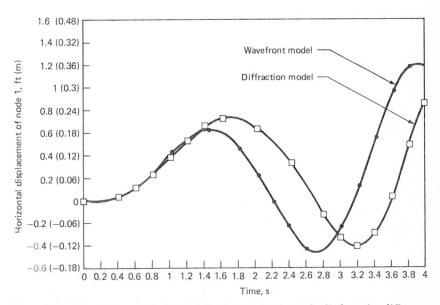

Figure 8.4.9 Comparisons of horizontal displacements of top of cylinder using different fluid loadings.

brane behavior but also coupling effects between the deformable membrane body shape and the embedding fluid. In addition to the fixed body diffraction effects described above, we should also include wave radiation effects for both rigid (floating) bodies and deformable bodies. If large motions are admitted, coupling effects from distortions of the overall body shape on predictions of wave potentials and, hence, of pressures should be included. An iterative solution process is indicated between the finite element model for the membrane and a time domain model for the fluid with nonlinear boundary motions [68].

Other problems of dynamic response which remain to be challenged include the dynamic wrinkling of membranes in cases where dynamic loads are such that pretension is reduced and the material buckles in one of the principal stress directions; and the interaction complications inherent in the heat and mass transfer problems of circulating fluids encapsulated in the membranes.

Bibliography

1. Clarke, G.: *The History of Airships,* Jenkins, London, 1961.
2. Dollfus, C.: *Balloons,* Prentice-Hall, London, 1962.
3. Kamrass, M.: "Pneumatic Buildings," *Scientific American,* vol. 194, no. 6, June 1956.
4. Allison, D.: "Those Ballooning Air Buildings," *Architectural Forum,* vol. 111, July 1959.
5. Otto, F.: *Tensile Structures,* vols. I and II, MIT Press, Cambridge, 1967, 1969.
6. Bird, W. W.: "The Development of Pneumatic Structures, Past, Present, and Future," *Proceedings, 1st International Colloquium on Pneumatic Structures,* IASS, Stuttgart, West Germany, 1967.
7. Bucholdt, H. A.: "Tension Structures," *Structural Engineer,* vol. 48, no. 2, February 1970.
8. Pugsley, A.: *The Theory of Suspension Bridges,* Edward Arnold Ltd., London, 1968.
9. Subcommittee on Air-Supported Structures: *Air-Supported Structures,* ASCE Spec. Publ., 1979.
10. Krishna, P.: *Cable-Suspended Roofs,* McGraw-Hill, New York, 1978.
11. Irvine, H. M.: *Cable Structures,* MIT Press, Cambridge, 1981.
12. Subcommittee on Cable-Suspended Structures: "Cable-Suspended Roof Construction State-of-the-Art," *Journal of Structural Division,* ASCE, vol. 97, no. ST6, June 1971.
13. Engelbrecht, R. M.: "Recent Developments and the Present State of the Art of Air Structures," 23d National Plant Eng. and Maintenance Tech. Conf., January 1972.
14. Leonard, J. W.: "State-of-the-Art in Inflatable Shell Research," *Journal of the Engineering Mechanics Division,* ASCE, vol. 100, no. EM7, February 1974.
15. Leonard, J. W.: "Inflatable Shells for Underwater Use," *Proceedings,* 3d Offshore Technical Conference, April 1971.
16. Tuah, H.: "Cable Dynamics in an Ocean Environment," Ph.D. dissertation, Oregon State University, Corvallis, Oregon, June 1983.
17. Crisp, J. D. C., and L. J. Hart-Smith: "Multi-Lobed Inflated Membranes: Their Stability Under Finite Deformation," *International Journal of Solids and Structures,* vol. 7, no. 7, July 1971.

18. Fung, Y. C.: *Foundations of Solid Mechanics,* Prentice-Hall, Englewood Cliffs, New Jersey, 1965.
19. Testa, R. B., N. Stubbs, and W. R. Spillers: "Bilinear Model for Coated Square Fabrics," *Journal of the Engineering Mechanics Division,* ASCE, vol. 104, no. EM5, October 1978.
20. Ma, D., J. W. Leonard, and K.-H. Chu: "Slack-Elasto-Plastic Dynamics of Cable Systems," *Journal of the Engineering Mechanics Division,* ASCE, vol. 105, no. EM2, April 1979.
21. Choo, Y., and M. J. Casarella: "Hydrodynamic Resistance of Towed Cables," *Journal of Hydronautics,* vol. 5, no. 4, October 1971.
22. Schuerch, H. W., and G. M. Schinder: "Analysis of Foldability in Expandable Structures," *AIAA Journal,* vol. 1, no. 4, 1963.
23. Shaviv, E., and D. P. Greenberg: "Funicular Surface Structures: A Computer Graphics Approach," *Bulletin of the IASS,* no. 37, March 1968.
24. Zuk, W., and R. H. Clark: *Kinetic Architecture,* van Nostrand Reinholdt, April 1970.
25. Lundy, V. A.: "Architectural and Sculptural Aspects of Pneumatic Structures," *Proceedings, 1st International Colloquium on Pneumatic Structures,* IASS, Stuttgart, West Germany, 1967.
26. Leonard, J. W.: "Inflatable Shells: Pressurization Phase," *Journal of the Engineering Mechanics Division,* ASCE, vol. 93, no. EM2, 1967.
27. McConnell, A. J.: *Applications of Tensor Analysis,* Dover Publications, New York, 1957.
28. Timoshenko, S. P., and Goodier, J. N.: *Theory of Elasticity,* 3d ed., McGraw-Hill, New York, 1970.
29. Bulson, P. E.: "Design Principles of Pneumatic Structures," *Structural Engineer,* vol. 51, no. 6, June 1973.
30. Rivlin, R. S.: "The Deformation of a Membrane Formed by Inextensible Cords," *Archive for Rational Mechanics and Analysis,* vol. 2, no. 5, 1959.
31. Cowan, H. J., and J. S. Gero: "Pneumatic Structures Constrained by Networks," *Proceedings, 1st International Colloquium on Pneumatic Structures,* IASS, Stuttgart, West Germany, 1967.
32. Malcomb, D. J., and P. G. Glockner: "Optimum Cable Configuration for Air-Supported Structures," *Journal of the Structural Division,* ASCE, vol. 105, no. ST2, February 1979.
33. Vinogradov, O. G., D. J. Malcomb, and P. G. Glockner: "Vibrations of Cable-Reinforced Inflatable Structures," *Journal of the Structural Division,* ASCE, vol. 107, no. ST10, October 1981.
34. Gero, J. S.: "Some Aspects of the Behavior of Cable Networks," *Civil Engineering Trans.,* Australia, vol. CE18, 1976.
35. Skop, R. A.: "Cable Spring Constants for Guy Tower Analysis," *Journal of the Structural Division,* ASCE, vol. 105, no. ST2, July 1979.
36. Fritzsche, E.: "Strain Measurements on Industrial Fabrics for Pneumatic Structures," *Proceedings, 1st International Colloquium on Pneumatic Structures,* IASS, Stuttgart, West Germany, 1967.
37. Migliore, H. J., and D. J. Meggitt: "Numerical Sensitivity Study of Ocean Cable Systems," *Journal of the WPOC Division,* ASCE, vol. 104, no. WW4, August 1978.
38. Leonard, J. W.: "Inflatable Shells: In-Service Behavior," *Journal of the Engineering Mechanics Division,* ASCE, vol. 93, no. EM6, 1967.
39. Leonard, J. W.: "Incremental Response of 3-D Cable Networks," *Journal of the Engineering Mechanics Division,* ASCE, vol. 99, no. EM3, June 1973.
40. Green, A. E., R. S. Rivlin, and R. T. Shield: "General Theory of Small Elastic Deformations Superposed on Finite Elastic Deformations," *Proceedings, Royal Society of London,* series A, vol. 211, 1952.
41. Li, C. T., and J. W. Leonard: "Nonlinear Response of a General Inflatable to In-Service Loads," *Proceedings, IASS Conference on Shell Structures and Climatic Influences,* Calgary, Canada, 1972.
42. Leonard, J. W.: "Dynamic Response of Initially Stressed Membrane Shells," *Journal of the Engineering Mechanics Division,* ASCE, vol. 95, no. EM5, October 1969.

43. Leonard, J. W., and W. W. Recker: "Nonlinear Dynamics of Cables with Low Initial Tension," *Journal of the Engineering Mechanics Division,* ASCE, vol. 98, no. EM2, April 1972.

44. Olofsson, B.: "Some Relationship Between Wind Pressures, Forces and Geometrical Characteristics of Air-Supported Tents," *Textile Research Journal,* vol. 44, no. 7, July 1974.

45. Kunieda, H. D., Y. Yokoyama, and M. Arakawa: "Cylindrical Pneumatic Membrane Structures Subject to Wind," *Journal of the Engineering Mechanics Division,* ASCE, vol. 107, no. EM5, October 1981.

46. Kunieda, H. D.: "Flutter of Hanging Roofs and Curved Membrane Roofs," *International Journal of Solids and Structures,* vol. 11, no. 4, 1975.

47. Christiano, P., G. R. Seeley, and H. Stefan: "Transient Wind Loads on Circularly Concave Cable Roofs, *Journal of the Structural Division,* ASCE, vol. 100, no. ST11, November 1974.

48. Knapp, R. H.: "Derivation of a New Stiffness Matrix for Helically Armored Cables Considering Tension and Torsion," *International Journal for Numerical Methods in Engineering,* vol. 14, no. 4, pp. 515–530, 1979.

49. Milburn, D. A., and M. Chi: "Research on Titanium Wire Rope for Marine Use," *Journal of the Structural Division,* ASCE, vol. 102, no. ST11, November 1976.

50. Treloar, L. R. G.: *The Physics of Rubber Elasticity,* Oxford University Press, 1958.

51. Murrell, V. W.: *Air-Supported Structures,* Chap. 2, ASCE Spec. Publ., 1979.

52. Schorn, G.: *Air-Supported Structures,* Chap. 5, ASCE Spec. Publ., 1979.

53. Skelton, J.: "Comparisons and Selection of Materials for Air-Supported Structures," *Journal of Coated Fibrous Material,* vol. 1, April 1972.

54. Harris, J. T.: "Silicon-Coated Airmat Cloth for Inflatable Structures," *Journal of Coated Fabrics,* vol. 4, July 1974.

55. Shrivastava, N. K., V. K. Handa, and S. Critchley: "Failure of Air-Supported Structures," Comité International de Belton IASS Symposium on Air-Supported Structures, Venice, 1977.

56. Custer, R. L. P.: "Test Burn and Failure Mode Analysis of an Air-Supported Structure," *Fire Technology,* vol. 8, no. 1, February 1972.

57. Morrison, A.: "The Fabric Roof," *Civil Engineer,* ASCE, August 1980.

58. Anon.: "Bringing Permanence to Membrane Buildings," *Modern Plastics,* vol. 54, no. 8, August 1977.

59. AISI Committee: *Tentative Criteria for Structural Applications of Steel Cables for Buildings,* American Iron and Steel Institute, New York, 1966.

60. Scalzi, J. B., W. Podolny, and W. C. Teng: *Design Fundamentals of Cable Roof Structures,* ADUSS 55-3580-01, United States Steel Corp., Pittsburgh, Pennsylvania, 1969.

61. Hornbeck, R. W.: *Numerical Methods,* Quantum Publications, New York, 1975.

62. Wilson, A. J., and R. J. Wheen: "Direct Designs of Taut Cables Under Uniform Loading," *Journal of the Structural Division,* ASCE, vol. 100, no. ST3, March 1974.

63. Wilson, A. J., and R. J. Wheen: "Inclined Cables Under Load-Design Expressions," *Journal of the Structural Division,* ASCE, vol. 103, no. ST5, May 1977.

64. Judd, B. J., and R. J. Wheen: "Nonlinear Cable Behavior," *Journal of the Structural Division,* ASCE, vol. 104, no. ST3, March 1978.

65. Michalos, J., and C. Birnstiel: "Movements of a Cable Due to Changes in Loading," *Journal of the Structural Division,* ASCE, vol. 86, no. ST12, December 1960.

66. O'Brien, W. T.: "Behavior of Loaded Cable Systems," *Journal of the Structural Division,* ASCE, vol. 94, no. ST10, October 1968.

67. Irvine, H. M.: "Statics of Suspended Cables," *Journal of the Engineering Mechanics Division,* ASCE, vol. 101, no. EM3, June 1975.

68. Lee, J. F.: "Finite Element Analysis of Wave-Structure Interactions in the Time Domain," Ph.D. dissertation, Oregon State University, Corvallis, Oregon, 1986.

69. Skop, R. A., and G. J. O'Hara: "The Method of Imaginary Reactions: A New Technique for Analyzing Structural Cable Systems," *Marine Technology Society Journal,* vol. 4, no. 1, January–February, 1970.

70. Skop, R. A., and G. J. O'Hara: "A Method for the Analysis of Internally Redundant

Structural Cable Arrays," *Marine Technology Society Journal,* vol. 6, no. 1, January–February, 1972.

71. Leonard, J. W.: "Convergence Acceleration Method for Oceanic Cables," *Preprint 80-131,* ASCE Convention, Portland, Oregon, April 1980.

72. Leonard, J. W.: "Newton-Raphson Iterative Method Applied to Circularly Towed Cable-Body System," *Engineering Structure,* vol. 1, no. 2, January 1979.

73. Chou, Y. I., and M. J. Casarella: "Configuration of a Towline Attached to a Vehicle Moving in a Circular Path," *Journal of Hydronautics,* vol. 6, 1972, pp. 51–57.

74. Watson, T. V., and J. E. Kineman: "Determination of the Static Configuration of Externally Redundant Submerged Cable Arrays," *Proceedings, 6th Offshore Technology Conference,* Paper OTC 2323, May 1975.

75. Subcommittee on Cable-Suspended Structures: "Tentative Recommendations for Cable-Stayed Bridge Structures," *Journal of the Structural Division,* ASCE, vol. 103, no. ST5, May 1977.

76. Subcommittee on Cable-Suspended Structures: "Commentary on the Tentative Recommendations for Cable-Stayed Bridge Structures," *Journal of the Structural Division,* ASCE, vol. 103, no. ST5, May 1977.

77. Shore, S., and G. N. Bathish: "Membrane Analysis of Cable Roofs," *Proceedings, International Conference on Space Structures,* London, 1966.

78. Siev, A., and J. Eidelman: "Shapes of Suspended Roofs," *Finite Difference Equations,* H. Levey and F. Lessman (eds.), Pitman Press, London, 1959.

79. Flügge, W.: *Stresses in Shells,* Springer-Verlag, Berlin, 1960.

80. Henghold, W. M., and J. J. Russell: "Equilibrium and Natural Frequencies of Cable Structures (a Nonlinear Finite Element Approach)," *Computers and Structures,* vol. 6, 1976, pp. 267–271.

81. Hood, C. G.: "A General Stiffness Method for the Solution of Nonlinear Cable Networks with Arbitrary Loadings," *Computers and Structures,* vol. 6, 1976, pp. 391–396.

82. Fellipa, C. A.: "Finite Element Analysis of Three-Dimensional Cable Structures," *Proceedings, International Conference on Computational Methods in Nonlinear Mechanics,* 1974, pp. 311–324.

83. Webster, R. L.: "Nonlinear Static and Dynamic Response of Underwater Cable Structures Using the Finite Element Method," *Proceedings, 7th Offshore Technology Conference,* Paper OTC 2322, Houston, May 1975.

84. Peyrot, A. H., and A. M. Goulois: "Analysis of Flexible Transmission Lines," *Journal of the Structural Division,* ASCE, vol. 104, no. ST5, May 1978.

85. Peyrot, A. H.: "Analysis of Marine Cable Structures," *Preprint 3640* ASCE Convention, Atlanta, Georgia, October 1979.

86. Bathe, K-J., H. Ozdemir, and E. L. Wilson: "Static and Dynamic Geometric and Material Nonlinear Analysis," *Report No. UCSESM 74-4,* Struct. Eng. Lab., Univ. of Calif., Berkeley, February 1974.

87. DeSalvo, G., and J. Swanson: *ANSYS Engineering Analysis System User's Manual,* Swanson Analysis Systems Corp., Houston, Pennsylvania, August 1978.

88. McCormick, C. W. (ed.): *MSC/NASTRAN USER'S MANUAL,* MacNeal-Schwendler Corp., Los Angeles, California, 1982.

89. Webster, R. L.: "On the Static Analysis of Structures with Strong Geometric Nonlinearities," *Computers and Structures,* vol. 11, 1980, pp. 137–145.

90. Lo, A.: "Nonlinear Dynamic Analysis of Cable and Membrane Structures," Ph.D. Dissertation, Oregon State University, December 1981.

91. Gero, J. S.: "Some Aspects of the Behavior of Cable Network Structures," *Bulletin of the IASS,* vol. XVII-3, no. 62, December 1976.

92. Gero, J. S.: "The Preliminary Designs of Cable Network Structures," *Bulletin of the IASS,* vol. XVII-3, no. 62, December 1976.

93. Clough, R. W., and J. Penzien: *Dynamics of Structures,* McGraw-Hill, New York, 1975.

94. Bathe, K-J.: *Finite Element Procedures in Engineering Analysis,* Prentice-Hall, Englewood Cliffs, New Jersey, 1982.

95. Morris, N. F.: "Chapter 4: Dynamic Structural Analysis," *Air-Supported Structures,* ASCE Spec. Publ., 1979.

96. Sarpkaya, T., and M.st.Q. Isaacson: *Mechanics of Wave Forces on Offshore Structures,* Van Nostrand Reinhold, New York, 1981.

97. Vickery, B. J.: "Wind Loads on Compliant Offshore Structures," *Proceedings, Ocean Structural Dynamics Symposium '82,* Oregon State University, Corvallis, Oregon, September 1982.

98. Jennings, A.: "The Free Cable," *The Engineer,* December 1962.

99. Henghold, W. H., J. J. Russell, and J. O. Morgan: "Free Vibrations of Cable in Three Dimensions," *Journal of the Structural Division,* vol. 103, no. ST5, May 1977.

100. Gambhir, M. L., and deV. B. Barrington: "Parametric Study of Free Vibration of Sagged Cables," *Computer and Structures,* vol. 8, 1978, pp. 651–648.

101. Wingert, J. M., and R. L. Huston: "Cable Dynamics—A Finite Segment Approach," *Computers and Structures,* vol. 6, 1976, pp. 475–480.

102. Ramberg, S. E., and O. M. Griffin: "Free Vibrations of Taut and Slack Marine Cables," *Journal of the Structural Division,* ASCE vol. 103, no. ST11, November 1977.

103. Sergev, S., and W. Q. Iwan: "The Natural Frequencies and Mode Shapes of Cables with Attached Masses," *TM No. M-44-79-3,* Naval Civil Engineering Laboratory, Port Hueneme, California April 1979.

104. Chandhari, B. S.: "Some Aspects of Dynamics of Cable Networks," Ph.D. dissertation, University of Pennslyvania, Philadelphia, 1969.

105. Soler, A. I., and H. Afshari: "On Analysis of Cable Network Vibrations Using Galerkin's Method," *Journal of Applied Mechanics,* vol. 37, September 1970.

106. Malcomb, D. J., and P. Glockner: "Collapse by Ponding of Air-Supported Membranes," *Journal of the Structural Div.,* vol. 104, no. ST9, September 1978.

107. McGuire, W., and R. H. Gallagher: *Matrix Structural Analysis,* John Wiley and Sons, New York, 1979.

108. Argyris, J. H., W. Aicher, and H. Fluh: "Experimental Analysis of Oscillations and Determination of Damping Values of Cables and Cable Net Structures," *Proceedings, International Conference on Tension Roof Structures,* Polytechnic of Central London, April 1974.

109. Geschwindner, L. F., and H. H. West: "Parametric Investigations of Vibrating Cable Networks," *Journal of the Structural Division,* ASCE, vol. 105, no. ST3, March 1979.

110. Morris, N. F.: "The Use of Modal Superposition in Nonlinear Dynamics," *Computers and Structures,* vol. 7, February 1977.

111. Dent, R. N.: *Principles of Pneumatic Architecture,* John Wiley and Sons, New York, 1977.

112. Herzog, T.: *Pneumatic Structures,* Crosby Lockwood Staples, London, 1977.

113. Dietz, A. E., R. G. Proffitt, R. S. Chabot, and E. L. Moak: "Design Manual for Ground-Mounted Air-Supported Structures," *Technical Report 67-35-ME,* U.S. Army Natick Lab, Natick, Massachusetts, 1969.

114. Brylka, R.: "Richtlinien fur den Bau und Betrieb von Traglufthauten in der Bundesrepublik Deutschland," *Proceedings, International Symposium on Pneumatic Structures,* IASS, Delft, 1972.

115. "Air-Supported Structures," *Proposed Standard 5367,* Canadian Standards Association, Toronto, 1978.

116. "Design and Standards Manual," Air Structures Institute, St. Paul, Minnesota, 1977.

117. "Draft for Development, Air Supported Structures," *(DD50:1976),* British Standards Institution, London, 1976.

118. Isono, Y.: "Abstract of Pneumatic Structure Design Standard in Japan," *Proceedings, International Symposium on Pneumatic Structures,* IASS, Delft, 1972.

119. Krauss, H.: *Thin Elastic Shell,* John Wiley and Sons, New York, 1967.

120. Pucher, A.: "Uber der Spannungszutand in gerkrümnten Flächen," *Beton und Eisen,* vol. 33, 1934, pp. 298–304.

121. Ambartsumyan, S. A.: "On the Calculation of Shallow Shells," *NACA TN 425,* December 1956.

122. Dean, D. L.: "Membrane Analysis of Shells," *Journal of the Engineering Mechanics Division,* ASCE, vol. 89, no. EM5, October 1963.

123. Fischer, L.: "Determination of Membrane Stresses in Elliptic Paraboloids Using Polynomials," *ACI Journal,* October 1960.
124. Nazarov, A. A.: "On the Theory of Thin Shallow Shells," *NACA TM 1426.*
125. Flügge, W., and D. A. Conrad: "A Note on the Calculation of Shallow Shells," *Journal of Applied Mechanics,* ASME, December 1959.
126. Chinn, J.: "Cylindrical Shell Analysis Simplified by Beam Method," *ACI Journal,* May 1959.
127. Chronowica, A., and D. Brohn: "Simplified Analysis of Cylindrical Shells," *Civil Engineering and Public Works Review,* London, August and September, 1964.
128. Fischer, L.: "Design of Cylindrical Shells with Edge Beams," *ACI Journal,* December 1955.
129. Tedesko, A.: "Multiple Ribless Shells," *Journal of the Structural Division,* ASCE, vol. 87, no. ST7, October 1961.
130. Modi, V. J., and A. K. Misra: "Response of an Inflatable Offshore Platform to Surface Wave Excitations," *Journal of Hydronautics,* vol. 14, 1979, pp. 10–18.
131. Topping, A. D.: "Shear Deflections and Buckling Characteristics of Inflated Members," *Journal of Aircraft,* vol. 1, no, 5, September 1964.
132. Comer, R. L., and S. Levy: "Deflections of an Inflated Circular Cylindrical Cantilever Beam," *AIAA Journal,* vol. 1, no. 7, July 1963.
133. Douglas, W. J.: "Bending Stiffness of an Inflated Cylindrical Cantilever Beam," *AIAA Journal,* vol. 7, no. 7, July 1969.
134. Li, C. T., and J. W. Leonard: "Finite Element Analysis of Inflatable Shells," *Journal of the Engineering Mechanics Division,* ASCE, vol. 99, no. EM3, June 1973.
135. Leonard, J. W., and V. K. Verma: "Double-Curved Element for Mooney-Rivlin Membranes," *Journal of the Engineering Mechanics Division,* ASCE, vol. 102, no. EM4, August 1976.
136. Verma, V. K., and J. W. Leonard: "Nonlinear Behavior of Cable-Reinforced Membranes," *Journal of the Engineering Mechanics Division,* ASCE, vol. 104, no. EM4, August 1978.
137. Haber, R. B., J. F. Abel, and D. P. Greenberg: "An Integrated Design System for Cable Reinforced Membranes Using Interactive Computer Graphics," *Computers and Structures,* vol. 14, no. 3–4, 1981, pp. 261–280.
138. Haber, R. B., and J. F. Abel: "Discrete Transfinite Mapping for the Description and Meshing of Three-Dimensional Surfaces Using Interactive Computer Graphics," *International Journal for Numerical Methods in Engineering,* vol. 18, 1982, pp. 41–66.
139. Haber, R. B., and J. F. Abel: "Initial Equilibrium Solution Methods for Cable Reinforced Membranes: Part I—Formulations," *Computer Methods in Applied Mechanics and Engineering,* vol. 30, 1982, pp. 263–284.
140. Haber, R. B., and J. F. Abel: "Initial Equilibrium Solution Methods for Cable Reinforced Membranes: Part II—Formulations," *Computer Methods in Applied Mechanics and Engineering,* vol. 30, 1982, pp. 285–306.
141. Haber, R. B., and J. F. Abel: "Contact-Slip Analysis Using Mixed Displacements," *Journal of the Engineering Mechanics Division,* ASCE, vol. 109, no. EM2, April 1983.
142. Benzley, S. E., and S. W. Key: "Dynamic Response of Membranes with Finite Elements," *Journal of the Engineering Mechanics Division,* ASCE, vol. 102, no. EM3, June 1976.
143. Huag, E.: "Finite Element Analysis of Pneumatic Structures," *Proceedings, International Symposium of Pneumatic Structures,* IASSD, Delft, 1972.
144. Rudolph, F.: "A Contribution to the Design of Air-Supported Structures," *Proceedings, 1st International Colloquium on Pneumatic Structures,* IASS, Stuttgart, West Germany, May 1967.
145. Davenport, A. G.: "Statistical Approaches to the Design of Shell Structures Against Wind and Other Natural Loads," *Proceedings, Conference on Shell Structures and Climatic Influences,* Calgary, Canada, July 1972.
146. Davenport, A. G.: "Structural Safety and Reliability Under Wind Action," *Proceedings, International Conference on Structural Safety and Reliability,* A. M. Freudenthal (ed.), Pergamon Press, New York, 1972.

147. Harris, R. I.: "The Nature of Wind," *Proceedings, Construction Industry Research and Information Association Seminar on the Modern Design of Wind-Sensitive Structures,* London, June 1970.

148. Howson, W. P., and L. R. Wooton: "Some Aspects of the Aerodynamics and Dynamics of Tension Roof Structures," *Proceedings, International Conference on Tension Roof Structures,* London, 1974.

149. Ostrowski, J. S., R. D. Marshal, and J. E. Cermak: "Vortex Formulation and Pressure Fluctuations on Buildings and Structures," *Proceedings, Ottawa Conference on Wind Effects on Buildings and Structures,* University of Toronto Press, 1967.

150. Walshe, D. E., and L. R. Wooton: "Preventing Wind-Induced Oscillations of Structures of Circular Sections," *Proceedings, Institute of Civil Engineers,* vol. 47, London, 1969.

151. Jensen, J. J.: "Dynamics of Tension Roof Structures," *Proceedings, International Conference on Tension Roof Structures,* London, 1974.

152. Sabzevari, A., and R. H. Scanlan: "Aerodynamic Instability of Suspension Bridges," *Journal of the Engineering Mechanics Division,* ASCE, vol. 94, no. 2, April 1968.

153. Fung, Y. C.: *An Introduction to the Theory of Aeroelasticity,* Dover Publications, New York, 1969.

154. Taylor, P. W., and D. J. Johns: "Effects of Turbulence on the Aeroelastic Behavior of Light Cladding Structures," *Proceedings, International Symposium on Wind Effects on Buildings and Structures,* Loughborough University of Technology, England, April 1968.

155. Novak, M.: "Galloping Oscillations of Prismatic Structures," *Journal of the Engineering Mechanics Division,* ASCE, vol. 98, no. EM1, February 1972.

156. Parkinson, G. V., and S. D. Smith: "The Square Prism as an Aeroelastic Nonlinear Oscillator," *Quarterly Journal of Mechanics and Applied Mathematics,* vol. 17, 1964.

157. Kunieda, H.: "Parametric Resonance of Suspension Roofs in the Wind," *Journal of the Engineering Mechanics Division,* ASCE, vol. 102, no. EM1, February 1976.

158. Davenport, A. G.: "Gust Loading Factors," *Journal of the Structural Division,* ASCE, vol. 93, no. ST3, June 1967.

159. Simiu, E.: "Gust Factors and Alongwind Pressure Correlations," *Journal of the Structural Division,* ASCE, vol. 99, no. ST4, April 1973.

160. Vellozzi, J., and E. Cohen: "Gust Response Factors," *Journal of the Structural Division,* ASCE, vol. 94, no. ST6, June 1968.

161. Berger, G., and E. Macher: "Results of Wind Tunnel Tests on Some Pneumatic Structures," *Proceedings, First International Symposium on Pneumatic Structures,* IASS, Stuttgart, West Germany, May 1967.

162. Newman, B. G., and D. Goland: "Two-Dimensional Inflated Buildings in a Cross Wind," *Journal of Fluid Mechanics,* vol. 117, April 1982, pp. 507–530.

163. Baron, F., and M. S. Vendatesan: "Nonlinear Analysis of Cable and Truss Structures," *Journal of the Structural Division,* ASCE, vol. 97, no. ST2, February 1971, pp. 679–710.

164. Pugsley, A. E.: "On the Natural Frequency of Suspension Cables," *Quarterly Journal of Applied Mathematics,* no. 2, 1949.

165. Cook, R. D.: *Concepts and Applications of Finite Element Analysis,* John Wiley and Sons, New York, 1981.

166. Timoshenko, S. J., and D. H. Young: *Vibration Problems in Engineering,* D. Van Nostrand, New York, 1955.

167. Leonard, J. W., C. J. Garrison, and R. T. Hudspeth: "Review of Deterministic Fluid Forces on Structures," *Journal of the Structural Division,* ASCE, vol. 107, no. ST6, June 1981, pp. 1041–1050.

168. Leonard, J. W., and J. H. Nath: "Comparison of Finite Element and Lumped Parameter Methods for Oceanic Cables," *Engineering Structures,* vol. 3, no. 3, July 1981, pp. 153–167.

169. Migliore, H. J.: "Comparison of SEADYN to Analytical and Experimental Results," *TM No. M-44-77-5,* Civil Engineering Laboratory, Naval Construction Battalion, Port Hueneme, California, March, 1977.

170. Leonard, J. W.: "Inflatable Shells: Non-Symmetric In-Service Behavior," *Journal of the Engineering Mechanics Division,* ASCE, vol. 94, no. EM5, October 1968.
171. Leonard, J. W.: "Pressure-Stabilized Drop Shaped Tanks," *Journal of the Engineering Mechanics Division,* ASCE, vol. 99, no. EM9, October 1973.
172. Green, A. E., and W. Zerna: *Theoretical Elasticity,* Oxford University Press, 1954.
173. Wu, C. H.: "Nonlinear Wrinkling of Nonlinear Membrane of Revolution," *Journal of Applied Mechanics,* vol. 45, 1978, pp. 533–538.
174. Mansfield, E. H.: "Load Transfer Via a Wrinkled Membrane," *Proceedings, Royal Society,* series A, vol. 316, 1970, pp. 269–289.
175. Zak, M. A.: "Wrinkling Phenomenon in Structures," *Solid Mechanics Archives,* vol. 8, 1983, pp. 181–216, 279–311.
176. Ramaswarmy, G.: *Design and Construction of Concrete Shell Roofs,* Krieger Publ. Co., Melbourne, Florida, 1984.
177. Parme, A. L.: "Hyperbolic Paraboloids and Other Shells of Double Curvature," *Journal of the Structural Division,* ASCE, vol. 82, no. ST5, September 1956.
178. Firt, V.: *Static Formfinding and Dynamics of Air-Supported Membrane Structures,* Martinus Nijhoff, the Hague, 1983.
179. Niemann, H. J.: "Wind Tunnel Experiments on Aeroelastic Models of Air-Supported Structures," *Proceedings, International Symposium on Pneumatic Structures,* Delft, 1972.
180. Ross, E. W.: "Large Deflections of an Inflated Cylindrical Tent," *Journal of Applied Mechanics,* vol. 36, no. 4, 1969, p. 845.
181. Blevins, R. D.: *Formulas for Natural Frequencies and Mode Shapes,* Van Nostrand Reinhold, New York, 1979.
182. Costello, G. A., and J. W. Phillips: "Effective Modulus of Twisted Wire Cables," *Journal of the Engineering Mechanics Division,* ASCE, vol. 102, no. EM1, January 1976, pp 171–181.
183. Costello, G. A.: "Analytical Investigations of Wire Rope," *Applied Mechanics Review,* vol. 31, no. 7, 1978, pp. 897–900.
184. Timoshenko, S. J., and S. Woinowsky-Krieger: *Theory of Plates and Shells,* McGraw-Hill, New York, 1959.
185. Magrab, E. B.: *Vibration of Elastic Structural Members,* Sijhoff and Noordhoff Co., The Netherlands, 1979.
186. Afshari, H., and A. I. Soler: "Vibration of Cable Gridworks with Small Initial Deformation," *Journal of Applied Mechanics,* vol. 41, no. 1, March 1974.
187. Mooney, M.: "A Theory of Large Deformation," *Journal of Applied Physics,* vol. 11, 1940.
188. Poincare, N.: *Les Methodes Nouvelles de la Mecanique Celeste,* vols. I, II, III (1892–1899), reprinted by Dover, New York, 1957.
189. Leonard, J. W.: "Behavior of Pressure-Stabilized Inflatable Shells of Revolution," Ph.D. dissertation, University of Illinois, Champaign, 1966.
190. Ince, E. L.: *Ordinary Differential Equations,* Dover, New York, 1956.
191. Moulton, F. R.: *Differential Equations,* Dover, New York, 1958.
192. Green, A. E., and J. E. Adkins: *Large Elastic Deformations,* Oxford University Press, London, 1960.
193. Oden, J. T., and W. K. Kubitza: "Numerical Analysis of Nonlinear Pneumatic Structures," *Proceedings, 1st International Colloquium on Pneumatic Structures,* IASS, Stuttgart, West Germany, 1967.
194. Oden, J. T., and T. Sato: "Finite Strains and Displacements of Elastic Membranes by the Finite Element Method," *International Journal of Solids and Structures,* vol. 3, 1967.
195. Li, C. T., and J. W. Leonard: "Strongly-Curved Finite Elements for Shell Analysis," *Journal of the Engineering Mechanics Division,* ASCE, vol. 99, no. EM3, June 1973.
196. Coons, S. A.: "Surfaces for Computer Aided Design of Space Forms," *Project MAC-TR-41 Report,* Massachusetts Institute of Technology, June 1967.
197. Leonard, J. W.: "Use of Pressurized Membranes as a Low-Cost Erection Scheme for Concrete Structures," *Proceedings, 2d International Symposium on Lower Cost Hous-*

ing Problems Related to Urban Renewal and Development, University of Missouri, Rolla, April 1972.

198. Verma, V. K.: "Finite Element Analysis of Nonlinear Cable Reinforced Membranes," *Ph.D. dissertation,* Illinois Institute of Technology, Chicago, 1974.

199. Oden, J. T.: "Finite Deformation of Elastic Plates, Shells, and Membranes by the Finite Element Method," *Proceedings, IASS Conference on Shell Structures and Climatic Influences,* Calgary, Canada, 1972.

200. Glockner, P. G., and T. Vishwanath: "On the Analysis of Nonlinear Membranes," *Dept. of Civil Engineering Report,* University of Calgary, Alberta, Canada, 1971.

201. Atkinson, K. E.: *An Introduction to Numerical Analysis,* John Wiley and Sons, New York, 1978.

202. Oden, J. T., J. E. Key, and R. B. Fost: "A Note on the Analysis of Nonlinear Dynamics of Elastic Membranes by the Finite Element Method," *Computer Structures,* vol. 4, 1974, pp. 445–452.

203. Cowper, G. R.: "Gaussian Quadrature Formulas for Triangles," *International Journal for Numerical Methods in Engineering,* vol. 7, 1973, pp. 405–408.

204. Zienkiewicz, O. C.: *The Finite Element Method,* McGraw-Hill, London, 1977.

205. Huang, M.-C., R. T. Hudspeth, and J. W. Leonard: "FEM Solution of 3-D Wave Interference Problems," *Journal of Waterways, Ports, Coastal and Ocean Engineering,* ASCE, vol. 111, no. 4, July 1985.

B

Tensor Calculus and Differential Geometry

In the text, vector and tensor calculus and the differential geometry of the surfaces are used to determine the field equations of membranes. In this appendix some of the useful relations of tensor calculus and differential geometry are given. This is only a brief introduction and summary. For more complete treatments of the subjects see Refs. 27 and 172. The nomenclature used here is essentially the same as is used in Ref. 172. First, the fundamentals of tensor analysis are considered.

B.1 Tensor Conventions

In tensor analysis a compact notation is used, based on the following: (1) a small Latin suffix is understood to have the range 1, 2, 3 unless specified otherwise (Greek indices denote a range of 1, 2 only); (2) when a suffix is repeated—once as a superscript and once as a subscript—summation with respect to that suffix is implied, the range of summation being 1, 2, 3.

Definition. A set of quantities T^r are the components of a contravariant vector if they transform, on change of coordinates, according to [27]

$$T^{r} = T^2 \frac{\partial x'^r}{\partial x^s}$$

The above definition can be extended to include tensors of higher order, e.g., T^{rsnm}. It is immediately obvious that a tensor of zero order—an invariant—transforms according to $T' = T$.

Definition. A set of quantities T^r are the components of a covariant vector if

$$T'_r = T_s \frac{\partial x^s}{\partial x'^r}$$

This definition can also be extended to tensors of higher order. Using these definitions, we can define mixed tensors, e.g., T^{rs}_{mn}.

A symbol (not a tensor) useful in handling cross products of vectors is the *permutation symbol* of the third order. It is defined as

$$e_{rst} = e^{rst} = \begin{cases} 0\text{—when any two indices are the same} \\ +1\text{—when } r, s, t \text{ is an even permutation of 1, 2, 3} \\ -1\text{—when } r, s, t \text{ is an odd permutation of 1, 2, 3} \end{cases}$$

where the permutation of the numbers 1, 2, 3 into the required sequence is by either an odd or an even number of interchanges of any pair of numbers.

Another useful symbol is the *Kronecker delta*. It is a tensor defined by

$$\delta^r_s = \begin{cases} 1 & \text{if } r = s \\ 0 & \text{if } r \neq s \end{cases}$$

The Kronecker delta may be used to change a suffix of another symbol, e.g.,

$$T^r_{ij} = \delta^r_s T^s_{ij}$$

B.2   Tensor Algebra

Two tensors of the same order may be added together to give another tensor, e.g., $T^{pq}_r = U^{pq}_r + V^{pq}_r$. A tensor is symmetrical if $T_{rs} = T_{sr}$, and is skew-symmetric if $T_{rs} = -T_{sr}$. To multiply, we may take two tensors of different order and simply write the suffixes in juxtaposition, e.g., $T^{mn}_{rsp} = U_{rs}V^{mn}_p$. This is called the *outer product*. The inner product of two tensors is obtained from the outer product by contraction, i.e., the same letter is written as a superscript and as a subscript. Therefore, $T^m_{rp} = U_{rs}V^{ms}_p$.

The *quotient law* is the means by which we are able to identify a tensor without satisfying the tensor transformation law directly. *Theorem:* "If the inner product $A_r X^r$ is an invariant, and X^r are the components of an arbitrary contravariant tensor, then A_r are the components of an arbitrary covariant tensor of the first order" [27]. The test is not restricted to tensors of the first order.

B.3   Tensor Calculus

Let x^1, x^2, x^3 be any system of coordinates. Then the element of length is

$$(ds)^2 = g_{mn} \, dx^m \, dx^n \tag{B.1}$$

where g_{mn} is a function of x^r. By the quotient law, g_{mn} is a symmetric tensor. It is called the metric tensor. It is possible to obtain the *conjugate metric tensor* g^{mn} from the metric tensor by means of the definition

$$g_{sn} \, g^{mn} = \delta_s^m$$

The existence of g^{mn} and g_{mn} permits the raising and lowering of indices by inner multiplication.

Equations involving derivations become simpler with the use of the following two symbols (not tensors):

$$\Gamma_{mn,r} = \frac{1}{2} \left(\frac{\partial g_{mr}}{\partial x^n} + \frac{\partial g_{nr}}{\partial x^m} - \frac{\partial g_{mn}}{\partial x^r} \right) \tag{B.2}$$

$$\Gamma_{mn}^r = g^{rs} \Gamma_{mn,s} \tag{B.2}$$

These are called the Christoffel symbols of the first and second kind, respectively. By virtue of the above symbols, three identities used in the text can be written:

$$\Gamma_{mn,r} = g_{rs} \Gamma_{mn}^s \tag{B.3}$$

$$\frac{\partial g_{mn}}{\partial x^r} = \Gamma_{rm,n} + \Gamma_{rn,m} \tag{B.4}$$

$$\frac{\partial g^{mn}}{\partial x^r} = - g^{np} \Gamma_{pr}^m - g^{pm} \Gamma_{pr}^n \tag{B.5}$$

Let ϕ be an invariant function. Then

$$\frac{\partial \phi}{\partial x^{\prime r}} = \frac{\partial \phi}{\partial x^s} \frac{\partial x^s}{\partial x^{\prime r}} \tag{B.6}$$

Therefore, the derivative of a tensor of zero order is a covariant tensor of the first order. However, this is the only case in which the partial derivative of a tensor is another tensor. By examination of the partial derivatives of tensors and of the Christoffel symbols, the equivalents of partial derivatives in tensor calculus can be written. The "covariant derivative" is defined as

$$T^r \Big|_n = \frac{\partial T^r}{\partial x^n} + \Gamma^r_{mn} \, T^m \tag{B.7}$$

The above definition can be extended, e.g.,

$$T^{rs}_p \Big|_n = \frac{\partial T^{rs}_p}{\partial x^n} + \Gamma^r_{mn} \, T^{ms}_p + \Gamma^s_{mn} T^{rm}_p - \Gamma^m_{pn} T^{rs}_m \tag{B.8}$$

It can be proved that the covariant derivatives of the metric tensors and the Kronecker delta are zero.

Covariant differentiation satisfies the two fundamental rules of calculus concerning the derivatives of sums and the derivatives of products. In cartesian coordinates all the Christoffel symbols vanish. Therefore, in a rectangular coordinate system covariant derivatives are identical to partial derivatives. It is to be noted that it is possible to define the tensor equivalents of total derivatives. Since these are not needed in the text, they are not considered here.

B.4   Physical Significance of Tensors

The results of tensor analysis can be related to operations in vector analysis. The magnitude T of a contravariant vector T^r is the positive real quantity satisfying

$$T = \sqrt{g_{mn} T^m T^n} \tag{B.9}$$

for a covariant vector T_r,

$$T = \sqrt{g^{mn} T_m T_n} \tag{B.10}$$

Let y^i be cartesian coordinates and x^i any curvilinear coordinates. The position vector to a point $\overline{P}$ is $\overline{r} = \overline{r}\,(x^i)$ (see Fig. B.1). The total differential of the position vector is

$$d\overline{r} = \frac{\partial \overline{r}}{\partial x^m} \, dx^m$$

The line element is then given by

$$(ds)^2 = d\overline{r} \cdot d\overline{r} = \frac{\partial \overline{r}}{\partial x^p} \cdot \frac{\partial x \overline{r}}{\partial x^q} \, dx^p \, dx^q$$

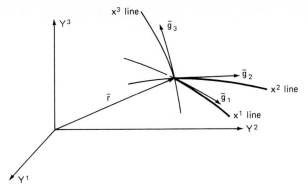

Figure B.1 Coordinate vectors.

Therefore,

$$g_{pq} = \frac{\partial \bar{r}}{\partial x^p} \cdot \frac{\partial \bar{r}}{\partial x^q}$$

The vector $\partial \bar{r}/\partial x^p$ is directed tangentially to the x^p coordinate line and is called the base vector $\bar{g}^p$. Hence,

$$g_{pq} = \bar{g}_p \cdot \bar{g}_q \tag{B.11}$$

Now, we define the contravariant vase vectors $\bar{g}^p$ by

$$\bar{g}_p \cdot \bar{g}^q = \delta_p^q$$

Also, the following relations hold:

$$\bar{g}^1 = \frac{\bar{g}_2 \times \bar{g}_3}{\sqrt{g}}$$

$$\bar{g}^1 = \frac{\bar{g}^2 \times \bar{g}^3}{\sqrt{g}} \tag{B.12}$$

(similar equations for $\bar{g}^2$ and $\bar{g}^3$ can be obtained by permutation) where g is the determinant of the components of the metric tensor,

$$g = \left| g_{pq} \right|$$

In terms of the contravariant base vectors, the conjugate metric tensor is

$$g^{pq} = \bar{g}^p \cdot \bar{g}^q \tag{B.13}$$

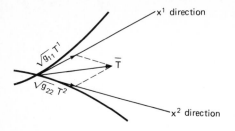

Figure B.2 Contravariant components in a 2-space.

The geometric significance of the contravariant components T^r of a first-order tensor is that these, when multiplied by $\sqrt{g_{rr}}$ (no summation of r) are the lengths of the vectors along the tangents to the curved directions for which the parallelpiped constructed on those lengths has $\overline{T}$ for its diagonal (see Fig. B.2). Therefore, the "physical contravariant components" of a contravariant vector are $\sqrt{g_{rr}}\,T^r$ (no sum on r).

The geometric significance of the covariant components T_r is that they, when divided by $\sqrt{g_{rr}}$ (no sum), are the orthogonal projections of $\overline{T}$ to the tangents to the coordinate curves (see Fig. B.3). The physical covariant components of a covariant vector are $T_r/\sqrt{g_{rr}}$ (no sum). Only if the coordinate system is orthogonal will the covariant and contravariant components be the same.

B.5   Differential Geometry of Surfaces

In the following sections the differential geometry of surfaces is summarized. As stated previously, Greek indices take the values 1, 2 and the two-dimensional surface is embedded in a three-dimensional euclidean space.

B.5.1   Metric tensor of a surface

The parametric equation of a surface is

$$\bar{r}_0 = \bar{r}_0\,(x^1, x^2) \tag{B.14}$$

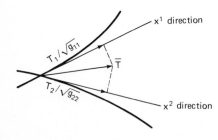

Figure B.3 Covariant components in a 2-space.

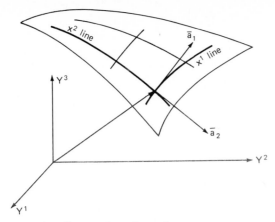

Figure B.4   Base vectors of a surface.

where $\bar{r}_0$ is the position vector of the surface, and x^1 and x^2 are two independent parameters which serve as coordinates on the surface.

Through each point on the surface pass two coordinate curves with their associated tangent vectors (see Fig. B.4). These base vectors of the surface are

$$\bar{a}_\alpha = \frac{\partial \bar{r}_0}{\partial x^\alpha} \tag{B.15}$$

The metric tensor of the surface is $a_{\alpha\gamma}$ and is expressible in terms of the base vectors as

$$a_{\alpha\gamma} = \bar{a}_\alpha \cdot \bar{a}_\gamma \tag{B.16}$$

The conjugate metric tensor of a surface is

$$a^{\alpha\gamma} a_{\gamma\beta} = \delta^\alpha_\gamma \tag{B.17}$$

Also,

$$\bar{a}^\alpha \cdot \bar{a}_\gamma = \delta^\alpha_\gamma \tag{B.18}$$

$$a^{\alpha\gamma} = \bar{a}^\alpha \cdot \bar{a}^\gamma \tag{B.19}$$

B.5.2   Surface permutation tensors

The surface permutation tensors can be obtained from the surface permutation symbols as

$$\varepsilon_{\alpha\gamma} = \sqrt{a}\, e_{\alpha\gamma}$$

$$\varepsilon_{\alpha\gamma} = \frac{e^{\alpha\gamma}}{\sqrt{a}} \tag{B.20}$$

where a is the determinant of the surface metric tensor,

$$a = |a_{\alpha\gamma}| = a_{11}a_{22} - a_{12}a_{21} \tag{B.21}$$

It can be shown that

$$\varepsilon^{\gamma\beta}\varepsilon_{\alpha\beta} = \delta^{\gamma}_{\alpha} \tag{B.22}$$

B.5.3   Surface area and the unit normal

The area dA of an infinitesimal element of the surface is the area of the parallelogram formed by $\bar{a}_1\, dx^1$ and $\bar{a}_2\, dx^2$. Therefore,

$$dA = \sqrt{\bar{a}}\, dx^1\, dx^2 \tag{B.23}$$

The unit vector normal to the surface at a point is determined by the cross product of the base vectors,

$$\hat{n} = \frac{\bar{a}_1 \times \bar{a}_2}{|\bar{a}_1 \times \bar{a}_2|} \tag{B.24}$$

The relations between the surface normal and the base vectors are

$$\bar{a}_\alpha \times \bar{a}_\gamma = \varepsilon_{\alpha\gamma}\hat{n} \tag{B.25a}$$

$$\bar{a}^\alpha \times \bar{a}^\gamma = \varepsilon^{\alpha\gamma}\hat{n} \tag{B.25b}$$

$$\hat{n} \times \bar{a}_\alpha = \varepsilon_{\alpha\gamma}\bar{a} \tag{B.25c}$$

$$\hat{n} \times \bar{a}^\alpha = \varepsilon^{\alpha\gamma}\bar{a} \tag{B.25d}$$

B.6   Derivatives on a Surface

The Christoffel symbols of a surface are formed in the same manner as before:

$$\Gamma_{\alpha\beta,\gamma} = \frac{1}{2}\left(\frac{\partial a_{\beta\gamma}}{\partial x^\alpha} + \frac{\partial a_{\alpha\gamma}}{\partial x^\beta} - \frac{\partial a_{\alpha\beta}}{\partial x^{\gamma\gamma}}\right)$$

$$\Gamma^\alpha_{\beta\gamma} = a^{\alpha\rho}\Gamma_{\beta\gamma,\rho} \tag{B.26}$$

Using the Christoffel symbols, we can form the covariant derivatives of surface tensors in the same way as they were formed for a general space.

B.7   Curvature of Surfaces

So far, only the metric tensor of a surface (first fundamental form), which need not be related to the embedding space, has been considered. Now the second fundamental form, $b_{\alpha\gamma}$, which is associated with the curvature of the surface, will be introduced. The second fundamental form is meaningful only when considered from a three-dimensional viewpoint.

B.7.1   Frenet formulas

First, the parametric equation of a curve in space, $\bar{r} = \bar{r}(u)$, must be considered along with the vector derivatives associated with it. The three principal vector derivatives are $\bar{t}_{(1)}$ = the unit tangent vector, $\bar{t}_{(2)}$ = the principal unit normal, and $\bar{t}_{(3)}$ = the unit binormal vector. If the arc length s of the curve is used as the parameter,

$$\bar{t}_{(1)} = \frac{d\bar{r}}{ds}$$

$$\frac{d\bar{t}_{(1)}}{ds} = k\bar{t}_{(2)} \tag{B.27}$$

$$\bar{t}_{(3)} = \bar{t}_{(1)} \times \bar{t}_{(2)} \tag{B.28}$$

$$\frac{d\bar{t}_{(3)}}{ds} = -\tau\bar{t}_{(2)} \tag{B.29}$$

$$\frac{d\bar{t}_{(2)}}{ds} = \tau\bar{t}_{(3)} - k\bar{t}_{(1)} \tag{B.30}$$

where k is the magnitude of the curvature vector $d\bar{t}_{(1)}/ds$ and is called the *first curvature*. Also, τ is the torsion of the curve. Equations (B.28), (B.29), and (B.30) are known as the *Frenet formulas* for a curve in space.

B.7.2   Curvature tensor

The Frenet formulas can be applied to a plane curve c on a surface. For a plane curve, $\tau = 0$. Therefore, from Eq. (B.30)

$$k_n = \frac{d\bar{r}_0}{ds} \frac{d\hat{n}}{ds}$$

where the curvature of a normal plane section of a surface is denoted by k_n. Therefore,

$$k_n = -\frac{1}{2}\left(\bar{a}_\alpha \cdot \frac{\partial n}{\partial x^\gamma} + \bar{a}_\gamma \cdot \frac{\partial \hat{n}}{\partial x^\alpha}\right)\frac{dx^\alpha\,dx^\gamma}{(ds)^2}$$

The second fundamental form of a surface is defined as

$$b_{\alpha\gamma} = \frac{1}{2}\left(\bar{a}_\alpha \cdot \frac{\partial \hat{n}}{\partial x^\gamma} + \bar{a}_\gamma \cdot \frac{\partial \hat{n}}{\partial x^\alpha}\right) \tag{B.31}$$

Therefore,

$$k_n = \frac{b_{\alpha\gamma}\,dx^\alpha\,dx^\gamma}{a_{\rho\beta}\,dx^\rho\,dx^\beta} \tag{B.32}$$

In addition

$$b_{\alpha\gamma} = \hat{n} \cdot \frac{\partial \bar{a}}{\partial x^\gamma} \tag{B.33}$$

$$b_\alpha^\gamma = a^{\gamma\rho}b_{\rho\alpha} \tag{B.34}$$

$$b = |b_{\alpha\gamma}| = b_{11}b_{22} - b_{12}b_{21} \tag{B.35}$$

B.7.3   Gauss-Weingarten equations

The partial derivatives of the surface base vectors and the unit normal can be expressed in terms of their components in the $\bar{a}_\alpha$ and $\hat{n}$ directions. The Gauss-Weingarten equations are

$$\frac{\partial \bar{a}_\alpha}{\partial x^\gamma} = b_{\alpha\gamma}\hat{n} + \Gamma^\rho_{\alpha\gamma}\bar{a}_\rho \tag{B.36}$$

$$\frac{\partial \bar{a}^\alpha}{\partial x^\gamma} = b^\alpha_\gamma\hat{n} - \Gamma^\alpha_{\gamma\rho}\bar{a}^\rho \tag{B.37}$$

$$\frac{\partial \hat{n}}{\partial x^\alpha} = -b^\gamma_\alpha\bar{a}_\gamma \tag{B.38}$$

B.7.4   Gauss-Codazzi relations

In the study of membranes, the deformed membrane is related to the undeformed membrane by the strain tensor. This is shown in the text to be a function of the first and second fundamental tensors. There must be restrictions on $a_{\alpha\gamma}$ and $b_{\alpha\gamma}$ in order to guarantee integrability. The Gauss-Codazzi equations are the necessary restrictions. They are

$$b_{\alpha\rho}|_\gamma - b_{\alpha\gamma}|_\rho = 0 \tag{B.39}$$

$$b_{\alpha\rho}b_{\nabla\gamma} - b_{\alpha\gamma}b_{\nabla\rho} = R_{\nabla\alpha\rho\gamma} \tag{B.40}$$

Note that there are only three independent equations. It can also be shown that the surface is uniquely determined if $a_{\alpha\gamma}$ and $b_{\alpha\gamma}$ are defined such that the Gauss-Codazzi equations are identically satisfied and if the element of length is positive definite.

B.7.5 Gaussian and mean curvatures

Let us at this point introduce an invariant of the surface, the gaussian curvature, defined as

$$K = \tfrac{1}{4}\varepsilon^{\alpha\rho}\varepsilon^{\nabla\gamma}R_{\alpha\rho\nabla\gamma} \tag{B.41}$$

Note that K is a function only of the metric tensor $a_{\alpha\gamma}$. By means of Eqs. (B.20) and (B.40),

$$K = \frac{R_{1212}}{a} = \frac{b}{a} \tag{B.42}$$

Another useful invariant of the surface can be formed by the contraction of the conjugate metric tensor and the second fundamental tensor. Therefore, the "mean curvature" is

$$M = \tfrac{1}{2}\, a^{\alpha\gamma}b_{\alpha\gamma} \tag{B.43}$$

Index